# STUDENT SOLUTIONS MANUAL

TO ACCOMPANY

# ELEMENTARY ALGEBRA: DISCOVERY AND VISUALIZATION
## THIRD EDITION

Hubbard/Robinson

## Frank C. Wilson
### Green River Community College

Houghton Mifflin Company    Boston  New York

Vice President and Publisher: Jack Shira
Assistant Editor: Marika Hoe
Senior Manufacturing Coordinator: Marie Barnes
Marketing Manager: Ben Rivera

Copyright © 2003 by Houghton Mifflin Company. All rights reserved.

No part of this work may be reproduced or transmitted in any form or by any means, electronic or mechanical, including photocopying and recording, or by any information storage or retrieval system without the prior written permission of Houghton Mifflin Company unless such copying is expressly permitted by federal copyright law. Address inquiries to College Permissions, Houghton Mifflin Company, 222 Berkeley Street, Boston, MA 02116-3764.

Printed in the U.S.A.

ISBN: 0-618-22389-4

23456789 – MA – 06 05 04 03 02

# Table of Contents

**CHAPTER**

**1 THE REAL NUMBER SYSTEM**
1.1  The Real Numbers ................................................................................................1
1.2  Operations with Real Numbers ............................................................................3
1.3  Properties of the Real Numbers ...........................................................................5
1.4  Addition .................................................................................................................7
1.5  Addition with Rational Numbers ........................................................................10
1.6  Subtraction ..........................................................................................................15
1.7  Multiplication ......................................................................................................18
1.8  Division ...............................................................................................................23
     Chapter 1 Review Exercises ...............................................................................28
     Chapter 1 Test .....................................................................................................31

**2 ALGEBRA BASICS, EQUATIONS, AND INEQUALITIES**
2.1  Algebraic Expressions ........................................................................................33
2.2  The Coordinate Plane .........................................................................................36
2.3  The Graph of an Expression ...............................................................................39
2.4  Equations and Estimated Solutions ....................................................................43
2.5  Properties of Equations ......................................................................................48
2.6  Solving Linear Equations ...................................................................................51
2.7  Inequalities: Graphing Methods .........................................................................57
2.8  Inequalities: Algebraic Methods ........................................................................61
     Chapter 2 Review Exercises ...............................................................................65
     Chapter 2 Test .....................................................................................................67

**3 MODELING AND APPLICATIONS**
3.1  Ratio and Proportion ..........................................................................................69
3.2  Percents ...............................................................................................................74
3.3  Formulas .............................................................................................................76
3.4  Translation ..........................................................................................................80
3.5  Modeling and Problem Solving .........................................................................81
3.6  Applications ........................................................................................................87
     Chapter 3 Review Exercises ...............................................................................94
     Chapter 3 Test .....................................................................................................97
     Cumulative Test, Chapters 1 – 3 ........................................................................98

**4 PROPERTIES OF LINES**
4.1  Linear Equations in Two Variables ..................................................................101
4.2  Intercepts and Special Cases ............................................................................107
4.3  Slope of a Line ..................................................................................................111
4.4  Slope and Graphing ..........................................................................................114
4.5  Parallel and Perpendicular Lines ......................................................................121
4.6  Equations of Lines ............................................................................................124
4.7  Graphs of Linear Inequalities ...........................................................................130
     Chapter 4 Review ..............................................................................................136
     Chapter 4 Test ...................................................................................................140

# 5 SYSTEMS OF LINEAR EQUATIONS

- 5.1 The Graphing Method ... 143
- 5.2 The Addition Method ... 149
- 5.3 The Substitution Method ... 155
- 5.4 Applications ... 161
- 5.5 Systems of Linear Inequalities ... 170
- Chapter 5 Review ... 173
- Chapter 5 Test ... 177
- Cumulative Test, Chapters 4 – 5 ... 179

# 6 EXPONENTS AND POLYNOMIALS

- 6.1 Properties of Exponents ... 181
- 6.2 Introduction to Polynomials ... 185
- 6.3 Addition and Subtraction ... 189
- 6.4 Multiplication ... 192
- 6.5 Special Products ... 198
- 6.6 Division ... 202
- 6.7 Negative Exponents ... 208
- 6.8 Scientific Notation ... 212
- Chapter 6 Review ... 214
- Chapter 6 Test ... 217

# 7 FACTORING

- 7.1 Common Factors and Grouping ... 219
- 7.2 Special Factoring ... 221
- 7.3 Factoring Trinomials of the Form $x^2 + bx + c$ ... 224
- 7.4 Factoring Trinomials of the Form $ax^2 + bx + c$ ... 226
- 7.5 General Strategy ... 232
- 7.6 Solving Equations by Factoring ... 234
- 7.7 Applications ... 240
- Chapter 7 Review ... 248
- Chapter 7 Test ... 251

# 8 RATIONAL EXPRESSIONS

- 8.1 Introduction to Rational Expressions ... 253
- 8.2 Multiplication and Division ... 256
- 8.3 Addition and Subtraction (Like Denominators) ... 260
- 8.4 Least Common Denominators ... 264
- 8.5 Addition and Subtraction (Unlike Denominators) ... 267
- 8.6 Complex Fractions ... 272
- 8.7 Equations with Rational Expressions ... 276
- 8.8 Applications ... 282
- Chapter 8 Review ... 288
- Chapter 8 Test ... 291
- Cumulative Test, Chapters 6 – 8 ... 293

# 9 RADICAL EXPRESSIONS
- 9.1 Radicals ................................................................................................. 295
- 9.2 Product Rule for Radicals ..................................................................... 297
- 9.3 Quotient Rule for Radicals .................................................................... 300
- 9.4 Operations with Radicals ...................................................................... 303
- 9.5 Equations with Radicals ........................................................................ 307
- 9.6 Applications with Right Triangles ........................................................ 313
- 9.7 Rational Exponents ............................................................................... 316
- Chapter 9 Review ................................................................................. 320
- Chapter 9 Test ....................................................................................... 323

# 10 QUADRATIC EQUATIONS AND FUNCTIONS
- 10.1 Special Methods ................................................................................... 325
- 10.2 Completing the Square ......................................................................... 330
- 10.3 The Quadratic Formula ........................................................................ 337
- 10.4 Complex Numbers ............................................................................... 346
- 10.5 Graphs of Quadratic Equations ............................................................ 351
- 10.6 Functions .............................................................................................. 356
- Chapter 10 Review ............................................................................... 358
- Chapter 10 Test ..................................................................................... 361
- Cumulative Test, Chapters 9 – 10 ........................................................ 362

# Acknowledgements

I would like to thank my typists, Selwyn and Sharon Brown, for their assistance with this project. I would also like to thank my family for their support. Any comments or suggestions concerning this *Student Solutions Manual* may be referred to the publisher.

Frank C. Wilson
Green River Community College

# Chapter 1

# The Real Number System

## Section 1.1 The Real Numbers

1. The whole numbers contain all of the natural numbers plus the number zero.

3. $\{0,1,2,3\}$

5. $\{...,-2,-1,0,1,2,3\}$

7. $\{8,10,12,14\}$

9. $\{0,1,2,3....\}$

11. $\{-2,-3,-4,-5\}$

13. You may write every integer as a rational number by writing the number over 1. For example, 3 may be written as $\frac{3}{1}$, a rational number.

15. True. The whole numbers, $\{0,1,2,....\}$ may be written as $\left\{\frac{0}{1},\frac{1}{2},\frac{2}{1},...\right\}$.

17. True. The integers are $\{...-2,-1,0,1,2,...\}$ and the whole numbers are $\{0,1,2,...\}$. The number zero is in both sets.

19. False. The natural numbers include $\{1,2,3,...\}$ but not $\left\{\frac{1}{2},\frac{1}{3},\frac{1}{4},...\right\}$ and other positive real numbers.

21. False. The negative integers $\{...,-2,-1\}$ may be written as rational numbers $\left\{...,\frac{-2}{1},\frac{-1}{1}\right\}$.

23. True. By definition, the real numbers consist of the rational and irrational numbers.

25. The number zero is the only whole number that is not a natural number.

27. The whole numbers make up the integers that are not negative.

29. The number $0.35 = \frac{35}{100}$ and $0.\overline{35} = \frac{35}{99}$.

31. $\frac{7}{5} = 1.4$, terminating.

33. $\frac{5}{6} = 0.8\overline{3}$, repeating.

35. $-\frac{5}{11} = -0.4\overline{5}$, repeating.

37. (a) $5, \frac{35}{5}$

   (b) $0, 5, \frac{35}{5}$

   (c) $-3, 0, 5, \frac{35}{5}$

(d) $-3, -\frac{2}{3}, 0, 5, \frac{35}{5}$

(e) none

39. (a) $1, \sqrt{9}$

(b) $1, \sqrt{9}$

(c) $1, \sqrt{9}$

(d) $1, \sqrt{9}, \frac{4}{3}$

(e) $-\sqrt{7}$

41. (a) $532, \frac{18}{6}$

(b) $532, \frac{18}{6}$

(c) $532, \frac{18}{6}$

(d) $-2\frac{1}{3}, 532, -0.35, \frac{18}{6}, 1.1\overline{4}$

(e) $\sqrt{5}$

43.

45.

47.

49.

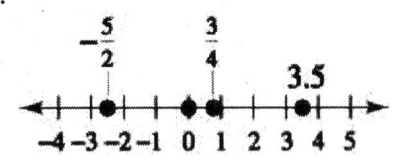

51. $5 > 0, 0 < 5$

53. $-12 \geq -20, -20 \leq -12$

55. $24 > -17, -17 < 24$

57. $n \geq -5, -5 \leq n$

59. $y \leq 4, 4 \geq y$

61. False since $-2$ is not bigger than 1.

63. True since $-8 = -8$.

65. True since $0 = 0$.

67. False since $\frac{5}{6} = 0.8\overline{33}$ and $\frac{6}{7} \approx 0.8571$.

69. $<$

71. $=$

73. $>$

75. $<$

77. $>$

79. $\{\ldots, -9, -8, -7\}$

81. $\{0, 1\}$

83. $\{0\}$

85. $=$ since $7 \cdot 45 = 315$ and $9 \cdot 35 = 315$

87. $9.1\% < 13.1\%$
$13.1\% < 13.3\%$
$13.3\% > 10.8\%$
$10.8\% > 9.6\%$

89. Since $\frac{1}{4} = 25\%$, we are looking for a two-month period that accounts for 25% or more of the fatal fires. The two months are December and January, since $13.1\% + 13.3\% = 26.4\%$.

91. (a) False since 0 is not a natural number.

(b) True since 3 is both a whole and a natural number.

(c) True since 7 is the only element in $\{7\}$.

(d) False since $-0.5$ is not an integer.

93. $1.8$
$= \dfrac{18}{10}$
$= \dfrac{9}{5}$

95. $0.\overline{6}$
$= \dfrac{6}{9}$
$= \dfrac{2}{3}$

97. (a) $\dfrac{1}{9} = 0.\overline{1}$

$\dfrac{2}{9} = 0.\overline{2}$

(b) $\dfrac{3}{9} = 0.\overline{3}$

$\dfrac{4}{9} = 0.\overline{4}$

(c) $0.\overline{7} = \dfrac{7}{9}$

## Section 1.2 Operations with Real Numbers

1. Addends are the numbers in a sum; factors are the numbers in a product.

3. (c)

5. (b)

7. (g)

9. (f)

11. $2(5)$ or $2 \cdot 5$ or $(2)(5)$

13. A natural number exponent indicates repeated multiplication.

15. $6^4 = 6 \cdot 6 \cdot 6 \cdot 6$

17. $t^5 = t \cdot t \cdot t \cdot t \cdot t$

19. $5^4$

21. $y^4$

23. $x^{14}$

25. $64 = 8 \cdot 8$

27. $32 = 2 \cdot 2 \cdot 2 \cdot 2 \cdot 2$

29. 1024 (use calculator)

31. 759,375 (use calculator)

33. $\sqrt{81} = 9$ since $9 \cdot 9 = 81$

35. $\sqrt{49} = 7$ since $7 \cdot 7 = 49$

37. $\sqrt{676} = 26$ since $26 \cdot 26 = 676$

39. The $-$ symbol is used for subtraction, as in $6 - 2$; for negative number, as in $-4$; and for opposites, as in $-t$.

41. 13

43. $-(8) = -8$

45. The opposite of 7 is $-7$; $|7| = 7$

47. The opposite of $-2$ is 2;
    $|-2| = 2$

49. The opposite of $-\frac{5}{8}$ is $\frac{5}{8}$;
    $\left|-\frac{5}{8}\right| = \frac{5}{8}$

51.

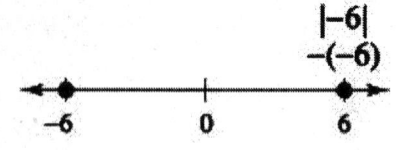

```
         |-6|
         -(-6)
<---•----|----•--->
   -6    0    6
```

53.

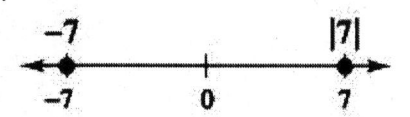

```
   -7         |7|
<---•----|----•--->
   -7    0    7
```

55. $|-8| = 8$ and $|8| = 8$

57. 3 and $-3$ are three units from 0.

59. $-6$ and 2 are four units from $-2$ since
    $-2 - 4 = -6$ and $-2 + 4 = 2$.

61. $|-3| > -3$

63. $-|7| = -7$
    since $-(7) = 7$

65. $-(-4) = 4$
    since $4 = 4$

67. $-|-6| < |-10|$
    since $-(6) < 10$

69. $7 + 3(2)$
    $= 7 + 6$
    $= 13$

71. $2 + 3^2$
    $= 2 + 9$
    $= 11$

73. $8(5) - 3 \cdot 7$
    $= 40 - 21$
    $= 19$

75. $4(3-1)^2$
    $= 4(2)^2$
    $= 4(4)$
    $= 16$

77. $3^2 - \sqrt{3^2}$
    $= 9 - \sqrt{9}$
    $= 9 - 3$
    $= 6$

79. $\sqrt{3^2 - 4(1)(2)}$
    $= \sqrt{9 - 8}$
    $= \sqrt{1}$
    $= 1$

81. $17 - 3(2^2 + 1)$
    $= 17 - 3(4 + 1)$
    $= 17 - 3(5)$
    $= 17 - 15$
    $= 2$

83. $7 - 2[5 - (6 - 4)]$
    $= 7 - 2[5 - 2]$
    $= 7 - 2(3)$
    $= 7 - 6$
    $= 1$

85. $\dfrac{8^2 - 5 \cdot 3}{5(10) - 1}$
    $= \dfrac{64 - 15}{50 - 1}$
    $= \dfrac{49}{49}$
    $= 1$

87. $59 + 370$
    $= 429$

The Real Number System

89. $87 \cdot 236$
$= 20{,}532$

91. $|-4| - |4|$
$= 4 - 4$
$= 0$

93. $|-12| - |-10|$
$= 12 - 10$
$= 2$

95. $\dfrac{|-8|}{|-2|}$
$= \dfrac{8}{2}$
$= 4$

97. Griffey:
$\dfrac{1}{12.625} \cdot 606$
$= 48$ homeruns

McGwire:
$\dfrac{1}{8.015} \cdot 521$
$= 65$ homeruns

99. $\dfrac{1}{4} = 25\%$, so we select categories with a percentage of 25% or higher. They are presidential debates, newspaper stories, and TV news.

101. No, because some people, getting information from the newspaper, read both the stories and the editorials.

103. $\sqrt{x} = a$, where $x$ is a whole number.

$a^2 = \left(\sqrt{x}\right)^2$
$= x$

So $a^2 = x$

105. $7 \cdot (2+3) \cdot 1 = 35$

107. $(4^2 - 1) \cdot 3 = 45$

## Section 1.3 Properties of the Real Numbers

1. In a sum, the addends may be added in any order. In a product, the factors may be multiplied in any order.

3. $73 + (92 + 8)$
$= 73 + 100$
$= 173$
Associative Property of Addition

5. $(5 \cdot 4) \cdot 7$
$= 20 \cdot 7$
$= 140$
Associative Property of Multiplication

7. Associative Property of Addition, since the addends were regrouped.

9. Property of Additive Inverses.

11. Additive Identity Property

13. Distributive Property

15. Commutative Property of Multiplication, since the order of the factors was changed.

17. For addition, the number 0 is the additive identity, since the sum of any number and zero is equal to the first number. This is, $a + 0 = a$ for all $a$.

19. $c = 1 \cdot c$

21. $-7x + 7x = 0$

23. $5x + 5y = 5(x+y)$

25. $x \cdot \dfrac{1}{x} = 1, x \neq 0$

27. $4(6+x) = 4(x+6)$

29. (a) Associative Property of Addition

    (b) Commutative Property of Addition

    (c) Associative Property of Addition

    (d) Distributive Property

    (e) Commutative Property of Multiplication

31. (a) Associative Property of Multiplication

    (b) Commutative Property of Multiplication

    (c) Associative Property of Multiplication

33. (d) since $(5 \cdot 0)^{17}$
$= 0^{17}$
$= 0$

35. (e) since $(5-4)^{12}$
$= 1^{12}$
$= 1$

37. (b) since $\dfrac{x+12}{x+12} = 1$

39. $n \cdot 11 = n(10+1)$
$= n \cdot 10 + n \cdot 1$
$= 10n + n$

41. $2(x+3)$
$= 2x + 2 \cdot 3$
$= 2x + 6$

43. $4(y-3)$
$= 4 \cdot y - 4 \cdot 3$
$= 4y - 12$

45. $x(2+y)$
$= x \cdot 2 + x \cdot y$
$= 2x + xy$

47. $a(x-5)$
$= a \cdot x - a \cdot 5$
$= ax - 5a$

49. $4x + 4y$
$= 4(x+y)$

51. $7y - 7z$
$= 7(y-z)$

53. $ax + ay$
$= a(x+y)$

55. $3x + 6$
$= 3x + 3 \cdot 2$
$= 3(x+2)$

57. $8a - 16$
$= 8a - 8 \cdot 2$
$= 8(a-2)$

59. $15 \cdot 102$
$= 15 \cdot (100+2)$
$= 15 \cdot 100 + 15 \cdot 2$

61. $98 \cdot 7$
$= (100-2) \cdot 7$
$= 100 \cdot 7 - 2 \cdot 7$
$= 7 \cdot 100 - 7 \cdot 2$

63. $98 \cdot 17 + 2 \cdot 17$
$= (98+2)17$
$= 17(98+2)$

# The Real Number System

65. $31 \cdot 13 - 31 \cdot 3$
    $= 31(13 - 3)$

67. $8 \cdot 140 + 8 \cdot 132$ and
    $8(140 + 132)$

69. (a) The sum of the table entries must equal 100%.

    $49\% + 12\% + 12\% + 11\% + 1\% = 85\%$, so "3 to 4 weekdays" must equal $100\% - 85\% = 15\%$.

    (b) $12\% + 15\% = 27\%$

71. $73 \cdot 5 + 73 \cdot 3 + 73 \cdot 2$
    $= 73(5 + 3 + 2)$

73. $15 \cdot 98$
    $= 15(100 - 2)$

75. $\frac{3}{4}(8x) = \left(\frac{3}{4} \cdot 8\right)x;$
    Associative Property of Multiplication

77. $-2 \cdot 5 + (-2)y = (-2)(5 + y);$
    Distributive Property

## Section 1.4 Addition

1. Add the absolute values and use the sign of the addends.

3. 3

5. $-4$

7. $-10$

9. $-8$

11. 0

13. 2

15. 8

17. 8

19. $-7$

21. 20

23. $-55$

25. 8

27. 0

29. $-22$

31. $-1$

33. $-11$

35. $-14 + (-2) + (-1)$
    $= -14 + (-3)$
    $= -17$

37. $-3 + [6 + (-9)]$
    $= -3 + (-3)$
    $= -6$

39. $6 + (-7) + 2$
    $= 6 + (-5)$
    $= 1$

41. $-12 + (-2) + (-5) + 10$
    $= -14 + 5$
    $= -9$

43. $-6 + 8 + [(-5) + 3]$
    $= 2 + (-2)$
    $= 0$

45. negative; $-6.8$

47. negative; −51.57

49. negative; −10.31

51. 9

53. $6+(-8)+(-2)$
$= 6+(-10)$
$= -4$

55. $-6+|-8|$
$= -6+8$
$= 2$

57. 0

59. −45

61. $-10+(-2)+15$
$= -10+13$
$= 3$

63. −18.3

65. $-12+(3)+10+(-2)$
$= -15+8$
$= -7$

67. $|-7|+(-7)$
$= 7+(-7)$
$= 0$

69. −0.222

71. The Commutative Property of Addition allows us to reorder, and the Associative Property of Addition allows us to regroup.

73. $-3+11+(-4)$
$= -3+(-4)+11$    (a) Commutative Property of Addition
$= [-3+(-4)]+11$  (b) Associative Property of Addition

75. $3x + y + (-3x)$
 $= 3x + (-3x) + y$    (a) Commutative Property of Addition
 $= \left[3x + (-3x)\right] + y$    (b) Associative Property of Addition
 $= 0 + y$    (c) Property of Additive Inverses
 $= y$    (d) Additive Identity Property

77. $-8 + 3^2$
 $= -8 + 9$
 $= 1$

79. $-4 + 5(-4 + 5)$
 $= -4 + 5(1)$
 $= -4 + 5$
 $= 1$

81. $\dfrac{-7 + 11}{-6 + 2(5)}$
 $= \dfrac{4}{-6 + 10}$
 $= \dfrac{4}{4}$
 $= 1$

83. $-14 + 30 - 2^3$
 $= -14 + 30 - 8$
 $= -14 + 22$
 $= 8$

85. $-6 + \sqrt{4^2 + (-12)}$
 $= -6 + \sqrt{16 - 12}$
 $= -6 + \sqrt{4}$
 $= -6 + 2$
 $= -4$

87. $-964 + 351 = -613$

89. $-9 + 5 = -4$

91. $(-5 + 2) + 12$
 $= -3 + 12$
 $= 9$

93. $-7 + 2 = -5$

95. $-1+4+(-12)$
    $= 3+(-12)$
    $= -9$

97. $-9+4+7 = 2$
    There was a 2-yard gain.

99. $216+(-5)+1+(-2)+(-1)+4+(-3)+2$
    $= 216+(-4)+(-3)+1+2$
    $= 212+(-2)+2$
    $= 212+0$
    $= 212$ pounds

101. $|a|+b$
     $= -a+b$
     The sum is positive since $-a$ is positive and $b$ is positive

103. $a+(-b)$
     The sum is negative since $a$ is negative and $-b$ is negative.

105. Each set of one step forward and two steps backwards results in a total gain of one step backward. After 26 sets, the person is 26 steps behind their starting point. They take one step forward and two steps backwards and end up 27 steps behind the starting point. Each set was made up of 3 steps: two forward and one backward. $27(3) = 81$ steps.

## Section 1.5 Addition with Rational Numbers

1. Equivalent fractions are fractions with the same decimal value.

3. $\dfrac{1}{6} \cdot \dfrac{3}{3} = \dfrac{3}{18}$
   $\dfrac{1}{6} \cdot \dfrac{5}{5} = \dfrac{5}{30}$
   $\dfrac{1}{6} \cdot \dfrac{7}{7} = \dfrac{7}{42}$

5. $-\dfrac{2}{7} \cdot \dfrac{2}{2} = -\dfrac{4}{14}$
   $-\dfrac{2}{7} \cdot \dfrac{5}{5} = -\dfrac{10}{35}$
   $-\dfrac{2}{7} \cdot \dfrac{8}{8} = -\dfrac{16}{56}$

7. $\dfrac{3}{5} \cdot \dfrac{3}{3} = \dfrac{9}{15}$
   The numerator is 9.

9. $\dfrac{7}{4} \cdot \dfrac{3}{3} = \dfrac{21}{12}$
   The numerator is 21.

11. $-\dfrac{5}{6} \cdot \dfrac{5}{5} = -\dfrac{25}{30}$
    The numerator is 25.

13. $-\dfrac{6}{1} \cdot \dfrac{2}{2} = -\dfrac{12}{2}$
    The numerator is 12.

The Real Number System

15. $\dfrac{7}{36} \cdot \dfrac{4}{4} = \dfrac{28}{144}$
The numerator is 28.

17. A fraction can be reduced, if there is a common factor in the numerator and denominator.

19. $\dfrac{4}{6} = \dfrac{2 \cdot 2}{2 \cdot 3} = \dfrac{2}{3}$

21. $\dfrac{6}{48} = \dfrac{6 \cdot 1}{6 \cdot 8} = \dfrac{1}{8}$

23. $-\dfrac{30}{12} = -\dfrac{6 \cdot 5}{6 \cdot 2} = -\dfrac{5}{2}$

25. $\dfrac{12}{35} = \dfrac{3 \cdot 2 \cdot 2}{5 \cdot 7} = \dfrac{12}{35}$

There are no common factors in the numerator and denominator, so the fraction is already reduced.

27. $-\dfrac{72}{16} = -\dfrac{8 \cdot 9}{8 \cdot 2} = -\dfrac{9}{2}$

29. Although 8 is a common denominator, 4 is the least common denominator.

$\dfrac{1}{2} + \dfrac{1}{4} = \dfrac{2}{4} + \dfrac{1}{4}$

31. $\dfrac{2}{5} + \dfrac{7}{5} = \dfrac{9}{5}$

33. $\dfrac{5}{8} + \dfrac{3}{8} = \dfrac{8}{8} = 1$

35. $-\dfrac{3}{7} + \dfrac{5}{7} = \dfrac{2}{7}$

37. $\dfrac{5}{4} + \dfrac{-3}{4} = \dfrac{2}{4} = \dfrac{1}{2}$

39. $\dfrac{1}{2} + \dfrac{1}{3}$

$= \dfrac{1}{2} \cdot \dfrac{3}{3} + \dfrac{1}{3} \cdot \dfrac{2}{2}$

$= \dfrac{3}{6} + \dfrac{2}{6}$

$= \dfrac{5}{6}$

41. $\dfrac{5}{7} + \dfrac{-3}{14}$

$= \dfrac{10}{14} + \dfrac{-3}{14}$

$= \dfrac{7}{14}$

$= \dfrac{1}{2}$

43. $3 + \dfrac{-2}{7}$

$= \dfrac{3}{1} \cdot \dfrac{7}{7} + \dfrac{-2}{7}$

$= \dfrac{21}{7} + \dfrac{-2}{7}$

$= \dfrac{19}{7}$

45. $\dfrac{5}{9} + \dfrac{7}{12}$

$= \dfrac{5}{9} \cdot \dfrac{4}{4} + \dfrac{7}{12} \cdot \dfrac{3}{3}$

$= \dfrac{20}{36} + \dfrac{21}{36}$

$= \dfrac{41}{36}$

47. $\dfrac{4}{15} + \dfrac{5}{18}$

$= \dfrac{4}{15} \cdot \dfrac{6}{6} + \dfrac{5}{18} \cdot \dfrac{5}{5}$

$= \dfrac{24}{90} + \dfrac{25}{90}$

$= \dfrac{49}{90}$

49. $\dfrac{5}{7}+\dfrac{-3}{14}$

$=\dfrac{5}{7}\cdot\dfrac{2}{2}+\dfrac{-3}{14}$

$=\dfrac{10}{14}+\dfrac{-3}{14}$

$=\dfrac{7}{14}$

$=\dfrac{1}{2}$

51. $\dfrac{5}{3}+\dfrac{-2}{3}+\dfrac{-1}{3}$

$=\dfrac{5+(-2)+(-1)}{3}$

$=\dfrac{2}{3}$

53. $\dfrac{-2}{5}+\dfrac{3}{5}+\dfrac{1}{2}$

$=\dfrac{1}{5}+\dfrac{1}{2}$

$=\dfrac{1}{5}\cdot\dfrac{2}{2}+\dfrac{1}{2}\cdot\dfrac{5}{5}$

$=\dfrac{2}{10}+\dfrac{5}{10}$

$=\dfrac{7}{10}$

55. $\dfrac{1}{8}+\dfrac{1}{4}+\dfrac{1}{2}$

$=\dfrac{1}{8}+\dfrac{1}{4}\cdot\dfrac{2}{2}+\dfrac{1}{2}\cdot\dfrac{4}{4}$

$=\dfrac{1}{8}+\dfrac{2}{8}+\dfrac{4}{8}$

$=\dfrac{1+2+4}{8}$

$=\dfrac{7}{8}$

57. $2\dfrac{1}{4}+3\dfrac{1}{3}$

$=2+\dfrac{1}{4}+3+\dfrac{1}{3}$

$=2+3+\dfrac{1}{4}+\dfrac{1}{3}$

$=5+\dfrac{1}{4}\cdot\dfrac{3}{3}+\dfrac{1}{3}\cdot\dfrac{4}{4}$

$=5+\dfrac{3}{12}+\dfrac{4}{12}$

$=5+\dfrac{7}{12}$

$=5\dfrac{7}{12}$

59. $1\dfrac{5}{6}+4\dfrac{1}{2}$

$=1+\dfrac{5}{6}+4+\dfrac{1}{2}$

$=1+4+\dfrac{5}{6}+\dfrac{1}{2}$

$=5+\dfrac{5}{6}+\dfrac{1}{2}\cdot\dfrac{3}{3}$

$=5+\dfrac{5}{6}+\dfrac{3}{6}$

$=5+\dfrac{8}{6}$

$=5+\dfrac{4}{3}$

$=5+1\dfrac{1}{3}$

$=6\dfrac{1}{3}$

61. $\dfrac{-1}{3}+\dfrac{-4}{3}$

$=\dfrac{-5}{3}$

The Real Number System

63. $\dfrac{-7}{8}+\dfrac{3}{2}$

$=\dfrac{-7}{8}+\dfrac{3}{2}\cdot\dfrac{4}{4}$

$=\dfrac{-7}{8}+\dfrac{12}{8}$

$=\dfrac{5}{8}$

65. $4+\dfrac{3}{7}$

$=\dfrac{4}{1}\cdot\dfrac{7}{7}+\dfrac{3}{7}$

$=\dfrac{28}{7}+\dfrac{3}{7}$

$=\dfrac{31}{7}$

67. $\dfrac{2}{5}+\dfrac{1}{3}$

$=\dfrac{2}{5}\cdot\dfrac{3}{3}+\dfrac{1}{3}\cdot\dfrac{5}{5}$

$=\dfrac{6}{15}+\dfrac{5}{15}$

$=\dfrac{11}{15}$

69. $\dfrac{1}{2}+\dfrac{1}{4}+\dfrac{1}{8}+\dfrac{1}{16}$

$=\dfrac{1}{2}\cdot\dfrac{8}{8}+\dfrac{1}{4}\cdot\dfrac{4}{4}+\dfrac{1}{8}\cdot\dfrac{2}{2}+\dfrac{1}{16}$

$=\dfrac{8}{16}+\dfrac{4}{16}+\dfrac{2}{16}+\dfrac{1}{16}$

$=\dfrac{8+4+2+1}{16}$

$=\dfrac{15}{16}$

71. $1\dfrac{1}{8}+\dfrac{3}{8}$

$=1+\dfrac{1}{8}+\dfrac{3}{8}$

$=1+\dfrac{4}{8}$

$=1+\dfrac{1}{2}$

$=1\dfrac{1}{2}$

73. $2\dfrac{2}{3}+\dfrac{-5}{6}$

$=2+\dfrac{2}{3}\cdot\dfrac{2}{2}+\dfrac{-5}{6}$

$=2+\dfrac{4}{6}+\dfrac{-5}{6}$

$=2+\dfrac{-1}{6}$

$=\dfrac{12}{6}+\dfrac{-1}{6}$

$=\dfrac{11}{6}$

$=\dfrac{6}{6}+\dfrac{5}{6}$

$=1\dfrac{5}{6}$

75. $25+\dfrac{3}{8}+\dfrac{3}{4}$

$=25+\dfrac{3}{8}+\dfrac{3}{4}\cdot\dfrac{2}{2}$

$=25+\dfrac{3}{8}+\dfrac{6}{8}$

$=25+\dfrac{9}{8}$

$=25+1\dfrac{1}{8}$

$=26\dfrac{1}{8}$

The stock selling price was $\$26\dfrac{1}{8}$.

77. (a) $\dfrac{41}{100} + \dfrac{7}{50}$

$= \dfrac{41}{100} + \dfrac{7}{50} \cdot \dfrac{2}{2}$

$= \dfrac{41}{100} + \dfrac{14}{100}$

$= \dfrac{55}{100}$

$= \dfrac{5 \cdot 11}{5 \cdot 20}$

$= \dfrac{11}{20}$ of the people consume the giblets.

(b) $1 - \left( \dfrac{41}{100} + \dfrac{1}{5} + \dfrac{7}{50} + \dfrac{2}{25} + \dfrac{1}{25} \right)$

$= 1 - \left( \dfrac{41}{100} + \dfrac{1}{5} \cdot \dfrac{20}{20} + \dfrac{7}{50} \cdot \dfrac{2}{2} + \dfrac{2}{25} \cdot \dfrac{4}{4} + \dfrac{1}{25} \cdot \dfrac{4}{4} \right)$

$= 1 - \left( \dfrac{41}{100} + \dfrac{20}{100} + \dfrac{14}{100} + \dfrac{8}{100} + \dfrac{4}{100} \right)$

$= 1 - \left( \dfrac{61 + 26}{100} \right)$

$= \dfrac{100}{100} - \left( \dfrac{87}{100} \right)$

$= \dfrac{13}{100}$ of the people did not know what giblets were.

79. $\dfrac{142}{100{,}000} = \dfrac{2 \cdot 71}{2 \cdot 50{,}000} = \dfrac{71}{50{,}000}$

$\dfrac{50}{100{,}000} = \dfrac{50 \cdot 1}{50 \cdot 2{,}000} = \dfrac{1}{2{,}000}$

$\dfrac{34}{100{,}000} = \dfrac{2 \cdot 17}{2 \cdot 50{,}000} = \dfrac{17}{50{,}000}$

$\dfrac{25}{100{,}000} = \dfrac{25 \cdot 1}{25 \cdot 4{,}000} = \dfrac{1}{4{,}000}$

81. $\dfrac{142}{100{,}000} + \dfrac{50}{100{,}000} + \dfrac{34}{100{,}000} + \dfrac{25}{100{,}000}$

$= \dfrac{251}{100{,}000}$

In these four occupations, there are 251 fatalities per 100,000 workers.

The Real Number System

83. $\dfrac{3}{a} + \dfrac{7}{a}$
$= \dfrac{10}{a}$

85. $\dfrac{a}{b} \cdot \dfrac{d}{d} = \dfrac{ad}{bd}$

$\dfrac{c}{d} \cdot \dfrac{b}{b} = \dfrac{bc}{bd}$

So, (ii) $bd$ is a common denominator, but not necessarily the least common denominator.

## Section 1.6 Subtraction

1. Change the minus sing to a plus sign and change 8 to –8.

3. $4 - 15 = -9$

5. $-7 - 3 = -10$

7. $9 - (-4) = 9 + 4 = 13$

9. $-2 - (-8) = -2 + 8 = 6$

11. $-5 - 8 = -13$

13. $0 - (-52) = 0 + 52 = 52$

15. $13 - (-5) = 13 + 5 = 18$

17. $-12 - 12 = -24$

19. $-6 - (-3) = -6 + 3 = -3$

21. $24 - (-24) = 24 + 24 = 48$

23. $-4 - 16 = -20$

25. $-8 - (-5) = -8 + 5 = -3$

27. Take the opposite of a difference changes the order in which the numbers are subtracted.

29. $-(-5 - 7) = -(-12) = 12$

31. $-[4 - (-6)]$
$= -[4 + 6]$
$= -(10)$
$= -10$

33. $-(a - 9)$
$= 9 - a$

35. $15 - 20 - 5$
$= 15 - 25$
$= -10$

37. $10 - (6 - 9)$
$= 10 - (-3)$
$= 10 + 3$
$= 13$

39. $-8 - 10 - (-3) + 7$
$= -18 + 3 + 7$
$= -18 + 10$
$= -8$

41. $-3 + (-2) - (-1 + 10)$
$= -3 - 2 - (9)$
$= -5 - 9$
$= -14$

43. $16-[(-3)+10-(-9)]$
$=16-[-3+10+9]$
$=16-(-3+19)$
$=16-(16)$
$=0$

45. $-17-(-18)$
$=-17+18$
$=1$

47. $-1-9=-10$

49. $-\dfrac{3}{5}-\left(-\dfrac{1}{2}\right)$
$=-\dfrac{3}{5}+\dfrac{1}{2}$
$=-\dfrac{3}{5}\cdot\dfrac{2}{2}+\dfrac{1}{2}\cdot\dfrac{5}{5}$
$=-\dfrac{6}{10}+\dfrac{5}{10}$
$=-\dfrac{1}{10}$

51. $-5-\left(-\dfrac{1}{3}\right)$
$=-5+\dfrac{1}{3}$
$=\dfrac{-5}{1}\cdot\dfrac{3}{3}+\dfrac{1}{3}$
$=\dfrac{-15}{3}+\dfrac{1}{3}$
$=\dfrac{-14}{3}$

53. $\dfrac{7}{16}-\left(-\dfrac{5}{8}\right)$
$=\dfrac{7}{16}+\dfrac{5}{8}$
$=\dfrac{7}{16}+\dfrac{5}{8}\cdot\dfrac{2}{2}$
$=\dfrac{7}{16}+\dfrac{10}{16}$
$=\dfrac{17}{16}$

55. $-34$

57. $74.23$

59. $3-5=-2$
So the unknown number is 5.

61. $6-7=-1$
So the unknown number is 6.

63. $-7-(-2)$
$=-7+2$
$=-5$

65. $7-(-36)$
$=7+36$
$=43$

67. $1-8 1-|-3|$
$=8-3$
$=5$

69. $-\dfrac{2}{3}-\dfrac{4}{5}$
$=-\dfrac{2}{3}\cdot\dfrac{5}{5}-\dfrac{4}{5}\cdot\dfrac{3}{3}$
$=-\dfrac{10}{15}-\dfrac{12}{15}$
$=-\dfrac{22}{15}$

71. $-6-10+(-5)$
   $=-21$

73. $-16-16$
   $=-32$

75. $0-(-32)$

   $=0+32$
   $=32$

77. $23.89-(-35.87)$
   $=23.89+35.87$
   $=59.76$

79. $-24-56$
   $=-80$

81. $9-27+2-(-1)$
   $=-18+2+1$
   $=-18+3$
   $=-15$

83. $-4-(-3)+7$
   $=-4+3+7$     (a) Definition of Subtraction
   $=-4+(3+7)$   (b) Associative Property of Addition
   $=-4+10$
   $=6$

85. $-(a-b)+a$
   $=(b-a)+a$     (a) Property of the Opposite of a Difference
   $=[b+(-a)]+a$  (b) Definition of Subtraction
   $=b+(-a+a)$   (c) Associative Property of Addition
   $=b+0$         (d) Property of Additive Inverses
   $=b$            (e) Additive Identity Property

87. $7-4(3)$
   $=7-12$
   $=-5$

89. $-3-3^2-3(4)$
   $=-3-9-12$
   $=-12-12$
   $=-24$

91. $-2-\sqrt{3^2-(-16)}$
$=-2-\sqrt{9+16}$
$=-2-\sqrt{25}$
$=-2-5$
$=-7$

93. $\dfrac{23-2^3}{0-(-15)}$
$=\dfrac{23-2\cdot 2\cdot 2}{0+15}$
$=\dfrac{23-8}{15}$
$=\dfrac{15}{15}$
$=1$

95. $29{,}108-(-1290)$
$=29{,}108+1290$
$=30{,}398$ feet

97. $2.3-3.1-1.7+0.6$
$=-1.9$
$=1.9$ inches below the yearly average

99. Colorado: $14{,}431-2500=11{,}931$ ft.
Florida: $325-2500=-2175$ ft.
Illinois: $1241-2500=-1259$ ft.
Rhode Island: $805-2500=1695$ ft.
Tennessee: $6642-2500=4142$ ft.
Washington: $14{,}408-2500=11{,}908$ ft.

101. $\dfrac{14{,}431}{5280}$
$=2.73$ miles

103. (a) $a-b$; negative

   (b) $b-a$; positive

105. $3-(7-2)=-2$
$3-(5)=-2$
$-2=-2$

107. $2(3-2)-5=-3$
$2(1)-5=-3$
$2-5=-3$
$-3=-3$

## Section 1.7 Multiplication

1. In both cases, multiply the absolute values of the numbers. For like signs, the product is positive; for unlike signs, the product is negative.

3. $7(-4)=-28$

5. $-5(2)=-10$

7. $(-3)(-4)=12$

9. $-6(-2)=12$

11. $-10\cdot 10=-100$

13. $11(-4)=-44$

15. $-7\cdot 1=-7$

17. $(0)(-45)=0$

19. positive; 6.21

21. negative; –4.83

23. $-4\cdot 7=-28$

25. $-8\cdot(-1)=8$

27. Because there is an even number of negative factors, the product is positive.

The Real Number System

29. $(-2)(-3)(-4)$
$= 6(-4)$
$= -24$

31. $(-2)(-1)(3)(2)$
$= 2 \cdot 6$
$= 12$

33. $(-1)(-2)(3)(1)(-6)$
$2 \cdot 3 \cdot (-6)$
$= 6 \cdot (-6)$
$= -36$

35. $(-3)^5$
$= (-3)(-3)(-3)(-3)(-3)$
$= 9 \cdot 9 \cdot (-3)$
$= 81 \cdot (-3)$
$= -243$

37. $(-2)^8$
$= (-2)(-2)(-2)(-2)(-2)(-2)(-2)(-2)$
$= 4 \cdot 4 \cdot 4 \cdot 4$
$= 16 \cdot 16$
$= 256$

39. $(-1)^{33}$
$= -1$

41. The Multiplication Property of $-1$ states that multiplying a number by $-1$ results in the opposite of the number.

43. $-\dfrac{1}{3} \cdot \dfrac{3}{5}$
$= -\dfrac{3}{15}$
$= -\dfrac{1}{5}$

45. $-5 \cdot \left(-\dfrac{2}{3}\right)$
$= \dfrac{(-5)(-2)}{1 \cdot (3)}$
$= \dfrac{10}{3}$

47. $-\dfrac{34}{5} \cdot \left(-\dfrac{25}{4}\right)$
$= \dfrac{(-34)(-25)}{5 \cdot 4}$
$= -\dfrac{17 \cdot \cancel{2} \cdot (-5)(\cancel{5})}{\cancel{5} \cdot \cancel{2} \cdot 2}$
$= -\dfrac{17(-5)}{2}$
$= \dfrac{85}{2}$

49. $(-5.76)(-4.95)$
$= 28.512$

51. $-5|-4|$
$= -5 \cdot 4$
$= -20$

53. $5(-2)(4)$
$= -10(4)$
$= -40$

55. $-\dfrac{3}{7} \cdot \left(-\dfrac{4}{9}\right)$
$= \dfrac{12}{63}$
$= \dfrac{3 \cdot 4}{3 \cdot 21}$
$= \dfrac{4}{21}$

57. $|9| \cdot |-8|$
$= 9 \cdot 8$
$= 72$

59. $-4(-24)$
   $= 96$

61. $(-7)(-2)(0)(2)(-3)$
   $= 0$

63. $-9-(-3)\cdot 2$
   $= -9-(-6)$
   $= -9+6$
   $= -3$

65. $(-4)^2 - 5(-4)$
   $= 16 + 20$
   $= 36$

67. $\dfrac{-3+\sqrt{(-3)^2 - 4(2)(-2)}}{2}$

   $= \dfrac{-3+\sqrt{9-8(-2)}}{2}$

   $= \dfrac{-3+\sqrt{9+16}}{2}$

   $= \dfrac{-3+\sqrt{25}}{2}$

   $= \dfrac{-3+5}{2}$

   $= \dfrac{2}{2}$

   $= 1$

69. $-3(4-6)^3 \div (2^2 \cdot 3)$

   $= \dfrac{-3(-2)^3}{2^2 \cdot 3}$

   $= \dfrac{-3(-8)}{4 \cdot 3}$

   $= \dfrac{24}{12}$

   $= 2$

71. $29 \cdot (0.63) = 18.27$

73. $\dfrac{-3}{4} \cdot 32$

$= \dfrac{-3}{4} \cdot \dfrac{4 \cdot 8}{1}$

$= \dfrac{-3 \cdot \cancel{4} \cdot 8}{\cancel{4}}$

$= -24$

75. $-6 \cdot 2 \cdot (-1) \cdot 3$
$= 2 \cdot 3 \cdot (-6) \cdot (-1)$  (a) Commutative Property of Multiplication
$= (2 \cdot 3) \cdot [(-6) \cdot (-1)]$  (b) Associative Property of Multiplication
$= 6 \cdot 6$
$= 36$

77. $-\dfrac{1}{6}(6x) = \left(-\dfrac{1}{6} \cdot 6\right)x$  (a) Associative Property of Multiplication
$= -1 \cdot x$
$= -x$  (b) Multiplication Property of $-1$

79. $2 + 3 = 5$
$2 \cdot 3 = 6$

81. $4 + (-3) = 1$
$4 \cdot (-3) = -12$

83. $-6 + (-2) = -8$
$-6 \cdot (-2) = 12$

85. $-8 + 1 = -7$
$-8 \cdot 1 = -8$

87. $-3 \cdot (-8) = 24$

89. $6(-3) = -18$

91. Total payments $= 5(225.39) = \$1126.95$
Interest $=$ Total payments $-$ loan amount
$= 1126.95 - 1075.00$
$= \$51.75$

93. $1(-2)+3(-1)+9(0)+2(1)+3(2)$
    $= -2 + (-3) + 0 + 2 + 6$
    $= -5 + 8$
    $= 3$ over par

    Score $= 72 + 3$
    $\phantom{Score} = 75$

95. (a)

| Country | Total Annual Consumption (billions of gallons) |
|---|---|
| Brazil | 1.235 |
| China | 0.305 |
| Germany | 1.030 |
| Mexico | 1.892 |
| Russia | 0.055 |
| United States | 5.638 |
| Zimbabwe | 0.041 |

To calculate each value, we multiply population by per–person consumption to get the number of 8–ounce servings consumed in millions. To convert to millions of gallons, we divide by 16, since there are 16 8–ounce servings in a gallon. To convert to billions of gallons from millions of gallons, we divide by 1000, since there are 1000 million gallons in a billion gallons. For example,

Brazil:

$162 \text{ million} \cdot 122 \text{ servings} \cdot \dfrac{1 \text{ gallon}}{16 \text{ servings}} \cdot \dfrac{1 \text{ billion}}{1000 \text{ million}}$

$= \dfrac{19{,}764 \text{ billion gallons million servings}}{16{,}000 \text{ million servings}}$

$= 1.235$ billion gallons

(b) In the United States, the average per person consumption is 343 servings per year. This is nearly one serving a day, since there are 365 days in a year.

97. $a \cdot |b| \cdot c$
    $= a \cdot (-b) \cdot c$
    The product is negative, since $a$ is negative and $-b$ and $c$ are positive.

99. $a \cdot a$
    The product is positive, since a negative times a negative is a positive.

101. $-5 \cdot (6-8) = 10$
$-5 \cdot (-2) = 10$
$10 = 10$

## Section 1.8 Division

1. We call $-12$ the dividend, 2 the divisor, and $-6$ the quotient.

3. $\dfrac{8}{-4} = \dfrac{-4 \cdot (-2)}{-4} = -2$

5. $\dfrac{15}{-5} = \dfrac{-5 \cdot (-3)}{-5} = -3$

7. $\dfrac{-40}{-5} = \dfrac{-5 \cdot (8)}{-5} = 8$

9. $\dfrac{-16}{-2} = \dfrac{-2 \cdot 8}{-2} = 8$

11. $\dfrac{-80}{-5} = \dfrac{-5 \cdot 6}{-5} = 16$

13. $\dfrac{-21}{7} = \dfrac{7 \cdot (-3)}{7} = -3$

15. $\dfrac{-20}{5} = \dfrac{5 \cdot (-4)}{5} = -4$

17. $\dfrac{-4}{0}$ = undefined

19. $\dfrac{0}{-1} = 0$

21. $\dfrac{-6}{3} \div (-4)$
$= \dfrac{-2 \cdot 3}{3} \div (-4)$
$= -2 \div (-4)$
$= \dfrac{-2}{-4}$
$= \dfrac{-2 \cdot 1}{-2 \cdot 2}$
$= \dfrac{1}{2}$

23. $24 \div [-6 \div (-2)]$
$= 24 \div \left[\dfrac{-6}{-2}\right]$
$= 24 \div \left[\dfrac{-2 \cdot 3}{-2}\right]$
$= 24 \div 3$
$= \dfrac{3 \cdot 8}{3}$
$= 8$

25. $-16 \cdot 2 \div 8$
$= \dfrac{-16 \cdot 2}{8}$
$= \dfrac{-2 \cdot 8 \cdot 2}{8}$
$= \dfrac{-2 \cdot 2 \cdot 8}{8}$
$= -2 \cdot 2$
$= -4$

27. $18 \div (-6 \cdot 3)$
$= \dfrac{18}{-6 \cdot 3}$
$= \dfrac{18}{-18}$
$= -1$

29. positive; 2.19

31. negative; −0.51

33. $-\dfrac{x}{y}, \dfrac{-x}{y}$; Property of Signs in Quotients

35. $\dfrac{-5}{-8} = \dfrac{5}{8}$

37. $-\dfrac{3}{-5} = \dfrac{-3}{-5} = \dfrac{3}{5}$

39. $-\dfrac{-y}{-5} = -\dfrac{y}{5}$

41. Invert the divisor and multiply.

43. (a) −5

   (b) $\dfrac{1}{5}$

45. (a) $\dfrac{3}{4}$

   (b) $-\dfrac{4}{3}$

47. (a) 0

   (b) none

49. $\dfrac{5}{16} \div \left(\dfrac{-3}{8}\right)$
$= \dfrac{5}{16} \cdot \left(\dfrac{-8}{3}\right)$
$= \dfrac{5 \cdot (-8)}{16 \cdot 3}$
$= \dfrac{5 \cdot 8 \cdot (-1)}{2 \cdot 8 \cdot 3}$
$= \dfrac{5(-1)}{2 \cdot 3}$
$= \dfrac{-5}{6}$

51. $-5 \div \left(-\dfrac{2}{3}\right)$
$= -5 \cdot \left(-\dfrac{3}{2}\right)$
$= \dfrac{(-5)(-3)}{2}$
$= \dfrac{15}{2}$

53. $\left(-\dfrac{3}{4}\right) \div 8$
$= -\dfrac{3}{4} \cdot \dfrac{1}{8}$
$= -\dfrac{3}{32}$

55. $-\dfrac{2}{3} \cdot \dfrac{27}{8}$
$= -\dfrac{2 \cdot 27}{3 \cdot 8}$
$= -\dfrac{1 \cdot 2 \cdot 3 \cdot 9}{3 \cdot 2 \cdot 4}$
$= -\dfrac{1 \cdot 9 \cdot 2 \cdot 3}{4 \cdot 2 \cdot 3}$
$= -\dfrac{9}{4}$

The Real Number System

57. $-56 \div 4 \cdot 2$
$= -\dfrac{56}{4} \cdot 2$
$= -\dfrac{14 \cdot 4}{4} \cdot 2$
$= -14 \cdot 2$
$= -28$

59. $-7 \div \left(-\dfrac{2}{5}\right)$
$= -7 \cdot \left(-\dfrac{5}{2}\right)$
$= -\dfrac{7 \cdot (-5)}{2}$
$= \dfrac{35}{2}$

61. $-\dfrac{9}{3}$
$= -\dfrac{3 \cdot (-3)}{-3}$
$= -3$

63. $\dfrac{86.47}{-4.47}$
$= -19.34$

65. $\dfrac{-20}{-10} \div (-6)$
$= 2 \div -6$
$= -\dfrac{2}{6}$
$= \dfrac{2 \cdot 1}{2 \cdot (-3)}$
$= -\dfrac{1}{3}$

67. $\dfrac{|-48|}{|-12|}$
$= \dfrac{48}{12}$
$= \dfrac{12 \cdot 4}{12}$
$= 4$

69. $27 \div (-3)$
$= -\dfrac{27}{3}$
$= \dfrac{-3 \cdot (-9)}{-3}$
$= -9$

71. $\dfrac{0}{0} =$ undefined

73. $\dfrac{3-(-9)}{-5-(-1)}$
$= \dfrac{3+9}{-5+1}$
$= \dfrac{12}{-4}$
$= \dfrac{-4 \cdot (-3)}{-4}$
$= -3$

75. $-18 \div (-3)^2$
$= \dfrac{-18}{(-3)^2}$
$= \dfrac{-18}{9}$
$= \dfrac{-2 \cdot (9)}{9}$
$= -2$

77. $\dfrac{5(3-4)-5^2}{3(9)-2(1+5)}$

$=\dfrac{5(-1)-25}{27-2(6)}$

$=\dfrac{-5-25}{27-12}$

$=\dfrac{-30}{15}$

$=\dfrac{-2 \cdot 15}{15}$

$=-2$

79. $\dfrac{-16.4}{-4.1}=4$

81. $\dfrac{-10}{-20}=\dfrac{-10 \cdot 1}{-10 \cdot 2}=\dfrac{1}{2}$

83. $\dfrac{3}{7}+\dfrac{8}{7}=3 \cdot \dfrac{1}{7}+8 \cdot \dfrac{1}{7}$   (a) Multiplication Rule for Fractions

$=\dfrac{1}{7}(3+8)$   (b) Distributive Property

$=\dfrac{3+8}{7}$   (c) Multiplication Rule for Fractions

$=\dfrac{11}{7}$

85. For $x \neq 0, \dfrac{x^2}{x}=x^2 \cdot \dfrac{1}{x}$   (a) Multiplication Rule for Fractions

$= x \cdot x \cdot \dfrac{1}{x}$

$= x \cdot \left(x \cdot \dfrac{1}{x}\right)$   (b) Associative Property of Multiplication

$= x \cdot 1$   (c) Property of Multiplication Inverses

$= x$   (d) Multiplicative Identity Property

87. $\dfrac{14}{-7}=\dfrac{-7 \cdot (-2)}{-7}=-2$

The dividend is 14.

89. $\dfrac{-40}{-4}=\dfrac{-4 \cdot 10}{-4}=10$

The dividend is $-40$.

The Real Number System

91. $\dfrac{-39 \text{ points}}{8 \text{ days}} = -4\dfrac{7}{8}$ points per day

93. (a) 
```
   10,400
   12,000
      450
      400
      260
  +   150
   _____
   23,660
```

(b)

United States $\quad \dfrac{10,400}{23,660} = \dfrac{40}{91}$

Russia $\quad \dfrac{12,000}{23,660} = \dfrac{600}{1183}$

France $\quad \dfrac{450}{23,660} = \dfrac{45}{2366}$

China $\quad \dfrac{400}{23,660} = \dfrac{20}{1183}$

United Kingdom $\quad \dfrac{260}{23,660} = \dfrac{1}{90}$

Israel $\quad \dfrac{150}{23,660} = \dfrac{15}{2366}$

95. Army: $10,367 + 60,787 = 71,154$
Navy: $7,777 + 42,261 = 50,038$
Marines: $854 + 8,928 = 9,782$
Air Force: $11,971 + 53,542 = 65,513$

97. Army: $\dfrac{61,000}{10,000 + 61,000} = \dfrac{61,000}{71,000} = \dfrac{61}{71}$

Navy: $\dfrac{42,000}{8,000 + 42,000} = \dfrac{42,000}{50,000} = \dfrac{21}{25}$

Marines: $\dfrac{9,000}{1,000 + 9,000} = \dfrac{9,000}{10,000} = \dfrac{9}{10}$

Air Force: $\dfrac{54,000}{12,000 + 54,000} = \dfrac{54,000}{66,000} = \dfrac{9}{11}$

99. $\dfrac{a}{b}$; negative

101. $\dfrac{-a}{-b}$; positive since $-a$ and $b$ are positive.

103. $\dfrac{-\dfrac{5}{8}}{-\dfrac{3}{4}}$

$= -\dfrac{5}{8} \cdot \left(-\dfrac{4}{3}\right)$

$= \dfrac{-5 \cdot (-4)}{8 \cdot 3}$

$= \dfrac{-5 \cdot 4 \cdot (-1)}{2 \cdot 4 \cdot 3}$

$= \dfrac{-5 \cdot (-1)}{2 \cdot 3}$

$= \dfrac{5}{6}$

## Chapter 1 Review Exercises

1. $\{-3, -2, -1, 0\}$

3. (a) $-0.625$, terminating

   (b) $0.1\overline{6}$, repeating

5. (a) $\{0, 1, 2, 3\}$

   (b) $\{..., -1, 0, 1, 2\}$

7. (a) True. The irrational numbers are a subset of the real numbers.

   (b) True. $0.47 = \dfrac{47}{100}$ while $0.\overline{47} = \dfrac{47}{99}$

   (c) False. $\dfrac{1}{3}$ is a rational number, but its decimal form, $0.\overline{3}$, is nonterminating.

   (d) True. The natural numbers are a subset of the whole numbers, the integers, the rational numbers, and the real number.

9. (a) Multiplication
   (b) Division
   (c) Subtraction
   (d) Addition

11. $b \cdot b \cdot b \cdot b = b^4$

13. $\sqrt{2116} = 46$ since $46 \cdot 46 = 2116$

15. $-(-(-10))$

    $= -(10)$

    $= -10$

17. (a) $|-2| + |5| = 2 + 5$ since $|-2| = 2$ and $|5| = 5$.

The Real Number System

(b) $\dfrac{4}{|-2|} > -\dfrac{4}{2}$ since

$\dfrac{4}{2} > -\dfrac{4}{2}$

$2 > -2$

(c) $|0| < |-1|$ since

$0 < 1$

19. Property of Multiplicative Inverses

21. Commutative Property of Addition

23. Property of Additive Inverses

25. $6(x+2)$
$= 6x + 6 \cdot 2$
$= 6x + 12$

27. Additive Identity, 0

29. $-8 + (-11)$
$= -19$

31. $17.4 + (-8.1)$
$= 9.3$

33. $-23 + (-1) + 23$
$= -23 + 23 + (-1)$
$= 0 + (-1)$
$= -1$

35. $-16 + 18 + (-3)$
$= 2 + (-3)$
$= -1$

37. $(-9+7)^2 + (-11+4)$
$= (-2)^2 + (-7)$
$= 4 + (-7)$
$= -3$

39. (ii) since $\dfrac{14}{21} = 0.66 \neq 0.7 = \dfrac{7}{10}$

41. $-\dfrac{5}{8} \cdot \dfrac{5}{5} = -\dfrac{25}{40}$

43. $8 = 2 \cdot 2 \cdot 2$ and $12 = 2 \cdot 2 \cdot 3$
so the LCD is $2 \cdot 2 \cdot 2 \cdot 3 = 24$

45. $-\dfrac{2}{9} + \dfrac{4}{15}$
$= -\dfrac{2}{9} \cdot \dfrac{5}{5} + \dfrac{4}{15} \cdot \dfrac{3}{3}$
$= -\dfrac{10}{45} + \dfrac{12}{45}$
$= \dfrac{2}{45}$

47. The number $n$ must contain a factor of 2 or 3.

49. Change minus to plus and change $-7$ to 7.

51. $18 - (-17)$
$= 18 + 17$
$= 35$

53. $-12 - (-3)$
$= -12 + 3$
$= -9$

55. $-2 - (-9)$
$= -2 + 9$
$= 7$

57. $\dfrac{16 - (-8)}{|2 - 20| - (7 - 13)}$
$= \dfrac{16 + 8}{|-18| - (-6)}$
$= \dfrac{24}{18 + 6}$
$= \dfrac{24}{24}$
$= 1$

59. $7(-4) = -28$

61. $(-2)(-1)(-7)(-6)$
$= 2(42)$
$= 84$

63. $-\dfrac{14}{3} \cdot \dfrac{9}{7}$
$= -\dfrac{14 \cdot 9}{3 \cdot 7}$
$= -\dfrac{2 \cdot 7 \cdot 3 \cdot 3}{3 \cdot 7}$
$= -\dfrac{2 \cdot 3 \cdot 3 \cdot 7}{3 \cdot 7}$
$= -2(3)$
$= -6$

65. $2(-1)^5 (-4)^2$
$= 2(-1)(16)$
$= -2(16)$
$= -32$

67. $\sqrt{(6)^2 - 4(1)(-16)}$
$= \sqrt{36 - 4(16)}$
$= \sqrt{36 + 64}$
$= \sqrt{100}$
$= 10$

69. $-\dfrac{28}{4} = -\dfrac{7 \cdot 4}{4} = -7$

71. $\dfrac{48}{-4} \div (-3)$
$= \dfrac{-4 \cdot (-12)}{-4} \div (-3)$
$= -12 \div (-3)$
$= \dfrac{-12}{-3}$
$= \dfrac{-3 \cdot 4}{-3}$
$= 4$

73. $\dfrac{3}{16} \div \left(-\dfrac{9}{4}\right)$
$= \dfrac{3}{16} \cdot \left(-\dfrac{4}{9}\right)$
$= \dfrac{3 \cdot (-4)}{16 \cdot 9}$
$= \dfrac{3 \cdot (-4)}{(-4)(-4) \cdot 3 \cdot 3}$
$= \dfrac{3 \cdot (-4) \cdot 1}{3 \cdot (-4) \cdot 3(-4)}$
$= \dfrac{1}{3(-4)}$
$= -\dfrac{1}{12}$

75. $-6 \div \left(-\dfrac{2}{3}\right)$
$= -\dfrac{6}{1} \cdot \left(-\dfrac{3}{2}\right)$
$= \dfrac{(-6)(-3)}{2}$
$= \dfrac{2(-3)(-3)}{2}$
$= (-3)^2$
$= 9$

77. $\dfrac{(-3)^2 - |2 - (-12)|}{4(-1) - 1}$
$= \dfrac{9 - |2 + 12|}{-4 - 1}$
$= \dfrac{9 - |14|}{-5}$
$= \dfrac{9 - 14}{-5}$
$= \dfrac{-5}{-5}$
$= 1$

The Real Number System

## Chapter 1 Test

1. (a) True. By definition, rational numbers can be written as repeating or terminating decimals, whereas irrationals cannot.

   (b) True. Repeating decimals do not terminate.

   (c) False. Since $\pi$ is irrational, $\dfrac{\pi}{3}$ is irrational.

   (d) True. All whole numbers are nonnegative.

3. $n \leq 7$ or $7 \geq n$

5. (a) $\sqrt{492.84} = 22.2$ since $22.2 \cdot 22.2 = 492.84$

   (b) $-|-4| = -(4) = -4$

   (c) $-(-4) = 4$

7. (a) Associative Property of Addition

   (b) Multiplication Property of $-1$

   (c) Property of Additive Inverses

   (d) Commutative Property of Multiplication.

9. $3(a+2)$
   $= 3 \cdot a + 3 \cdot 2$
   $= 3a + 6$

11. $-23 + 22 = -1$

13. $-\dfrac{-30}{-6}$
    $= -\dfrac{30}{6}$
    $= -\dfrac{6 \cdot 5}{6}$
    $= -5$

15. $\dfrac{-12}{5} \div \dfrac{3}{20}$
    $= \dfrac{-12}{5} \cdot \dfrac{20}{3}$
    $= \dfrac{-12 \cdot 20}{5 \cdot 3}$
    $= \dfrac{3 \cdot (-4)(4) \cdot 5}{5 \cdot 3}$
    $= \dfrac{5 \cdot 3 \cdot (-4)(4)}{5 \cdot 3}$
    $= -4(4)$
    $= -16$

17. $\dfrac{27}{6}$
    $= \dfrac{3 \cdot 9}{3 \cdot 2}$
    $= \dfrac{9}{2}$

19. $-8 + 3 = -5$
    $-8 \cdot 3 = -24$
    So the two numbers are $-8$ and $3$.

20. $292 - 15 - 35 + 27 - 16 + 5$
    $= 292 - 50 + 11 + 5$
    $= 242 + 16$
    $= 258$ Democrats

# Chapter 2

# Algebra Basics, Equations, and Inequalities

## Section 2.1 Algebraic Expressions

1. A numerical expression is an algebraic expression that does not contain a variable.

3. $2(3) - 7$
   $= 6 - 7$
   $= -1$

5. $-3[1 - (1)]$
   $= -3(1 + 1)$
   $= -3(2)$
   $= -6$

7. $-2 - [1 - 3(-2)]$
   $= -2 - (1 + 6)$
   $= -2 - (7)$
   $= -2 - 7$
   $= -9$

9. $2(3)^2 + 3(3) + 2$
   $= 2(9) + 9 + 2$
   $= 18 + 11$
   $= 29$

11. $3(2) - (-4)$
    $= 6 + 4$
    $= 10$

13. $(-3)^2 + (-3)(1) + 3(1)^2$
    $= 9 + (-3) + 3(1)$
    $= 6 + 3$
    $= 9$

15. $\dfrac{5 - (-1)}{3 - 2}$
    $= \dfrac{5 + 1}{1}$
    $= \dfrac{6}{1}$
    $= 6$

17. $\dfrac{3 + (-2)}{4}$
    $= \dfrac{1}{4}$

19. $[3 + (-2)]^2 - 3(4)$
    $= (1)^2 - 12$
    $= 1 - 12$
    $= -11$

21. $-3.92 - 2(-3.92 - 1) = 5.92$

23. $\dfrac{-1.2 + 5}{2(-1.2) - 1} = -1.12$

Copyright © Houghton Mifflin Company. All rights reserved.

33

25. There are two terms in $2x + 1$, but only one term in $2(x + 1)$. In $2(x + 1)$, both 2 and $x + 1$ are factors.

27. $3y$, $x$, and $-8$ are the terms. The coefficients are 3, 1, and $-8$.

29. $x^2$, $3x$, and $-6$ are the terms. The coefficients are 1, 3, and $-6$.

31. $7b$, $-c$, and $\dfrac{3a}{7}$ are the terms. The coefficients are 7, $-1$, and $\dfrac{3}{7}$.

33. $3x + 8x$
$= (3+8)x$
$= 11x$

35. $-y + 3y$
$= (-1+3)y$
$= 2y$

37. $-3x - 5x = -8x$

39. $4t^2 - t^2 = 3t^2$

41. $6x - 7x + 3$
$= -x + 3$

43. $9x - 12 - 3x + 5$
$= 9x - 3x - 12 + 5$
$= 6x - 7$

45. $2m - 7n - 3m + 6n$
$= 2m - 3m - 7n + 6n$
$= -m - n$

47. $10 - 2b + 3b - 9$
$= -2b + 3b + 10 - 9$
$= b + 1$

49. $2a^2 + a^3 - a^2 - 5a^3$
$= a^3 - 5a^3 + 2a^2 - a^2$
$= -4a^3 + a^2$

51. $ab + 3ac - 2ab + 5ac$
$= ab - 2ab + 3ac + 5ac$
$= -ab + 8ac$

53. $\dfrac{1}{3}a + \dfrac{1}{2}b - a + \dfrac{2}{3}b$
$= \dfrac{1}{3}a - a + \dfrac{1}{2}b + \dfrac{2}{3}b$
$= \left(\dfrac{1}{3} - 1\right)a + \left(\dfrac{1}{2} + \dfrac{2}{3}\right)b$
$= \left(\dfrac{1}{3} - \dfrac{3}{3}\right)a + \left(\dfrac{3}{6} + \dfrac{4}{6}\right)b$
$= -\dfrac{2}{3}a + \dfrac{7}{6}b$

55. $2.1x - 4.2y + 0.3y + 1.2x$
$= 2.1x + 1.2x - 4.2y + 0.3y$
$= 3.3x - 3.9y$

57. Remove parentheses and combine like terms.

59. $-5(3x)$
$= -15x$

61. $-\dfrac{1}{4}(4y)$
$= \left(-\dfrac{1}{4} \cdot 4\right)y$
$= -y$

63. $-\dfrac{5}{6} \cdot \left(-\dfrac{9}{10}x\right)$
$= \left[-\dfrac{5}{6} \cdot \left(-\dfrac{9}{10}\right)\right]x$
$= \dfrac{45}{60}x$
$= \dfrac{15 \cdot 3}{15 \cdot 4}x$
$= \dfrac{3}{4}x$

## Algebra Basics, Equations, and Inequalities

65. $5(4+3b)$
$= 5 \cdot 4 + 5 \cdot 3b$
$= 20 + 15b$
$= 15b + 20$

67. $-3(5-2x)$
$= -3 \cdot 5 - (3)(2x)$
$= -15 - (-6x)$
$= -15 + 6x$
$= 6x - 15$

69. $2(3x+y)$
$= 2 \cdot 3x + 2 \cdot y$
$= 6x + 2y$

71. $-3(-x+y-5)$
$= -3 \cdot (-x) + (-3)y - (-3)(5)$
$= 3x - 3y - (-15)$
$= 3x - 3y + 15$

73. $-\dfrac{3}{5}(5x+20)$
$= -\dfrac{3}{5} \cdot 5x + \left(-\dfrac{3}{5}\right) \cdot 20$
$= \left(-\dfrac{3}{5} \cdot 5\right)x + \left(-\dfrac{60}{5}\right)$
$= -3x - 12$

75. $-(x-4)$
$= -x - (-4)$
$= -x + 4$

77. $-(3+n)$
$= -3 - n$

79. $-(2x+5y-3)$
$= -2x - 5y - (-3)$
$= --2x - 5y + 3$

81. $2(x-3)+4$
$= 2x - 2 \cdot 3 + 4$
$= 2x - 6 + 4$
$= 2x - 2$

83. $5 - 3(x-1)$
$= 5 - 3x + 3$
$= -3x + 8$

85. $-2(x-3y) + y + 2x$
$= -2x + 6y + y + 2x$
$= 7y + 0x$
$= 7y$

87. $(2a+7) + (5-a)$
$= 2a - a + 7 + 5$
$= a + 12$

89. $2 + (2a+b) + a - b$
$= 2 + 2a + a + b - b$
$= 2 + 3a + 0b$
$= 3a + 2$

91. $4(1-3x) - (3x+4)$
$= 4 - 12x - 3x - 4$
$= -12x - 3x + 4 - 4$
$= -15x$

93. $2[5-(2x-3)] + 3x$
$= 2(5 - 2x + 3) + 3x$
$= 2(-2x + 8) + 3x$
$= -4x + 16 + 3x$
$= -x + 16$

95. $4 + 5[x - 3(x+2)]$
$= 4 + 5(x - 3x - 6)$
$= 4 + 5(-2x - 6)$
$= 4 - 10x - 30$
$= -10x - 26$

97. There are 8 hours between 8 a.m. and 4 p.m., so $t = 8$. $8(8) + 20 = 64 + 20 = 84$. The predicted grade is 84.

99. $55(2.8) + 75$
    $= 154 + 75$
    $= \$229$

101. $2(3n+5) - 7 - [3(n+1) - 2(n-4)]$
    $= 6n + 10 - 7 - (3n + 3 - 2n + 8)$
    $= 6n + 3 - (n + 11)$
    $= 6n + 3 - n - 11$
    $= 5n - 8$

103. $\dfrac{x+y}{2x+3y}$
    $= \dfrac{1.2 + (-3.1)}{2(1.2) + 3(-3.1)}$
    $= 0.28$

105. $2a^2 + (c-4)^2$
    $= 2(-5.3)^2 + (2.15 - 4)^2$
    $= 59.60$

## Section 2.2 The Coordinate Plane

1. The number lines are called the $x$-axis and $y$-axis. Their point of intersection is called the origin.

3. Quadrant I since $x > 0$ and $y > 0$.

5. Quadrant III since $x < 0$ and $y < 0$.

7. Quadrant II since $x < 0$ and $y > 0$.

9. Quadrant IV since $x > 0$ and $y < 0$.

11. $x$-axis since $y = 0$.

13. $y$-axis since $x = 0$.

15. $A(5, 0)$, $B(-7, 2)$, $C(2, 6)$, $D(0, -4)$, $E(-3, -5)$, $F(4, -4)$

17. $A(0, 6)$, $B(-5, 0)$, $C(7, 3)$, $D(-6, -3)$, $E(-2, 8)$, $F(3, -3)$

19. The order of the coordinates is different.

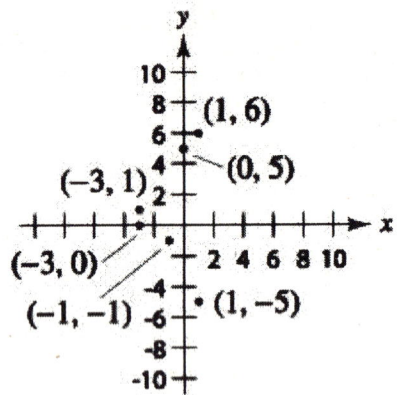

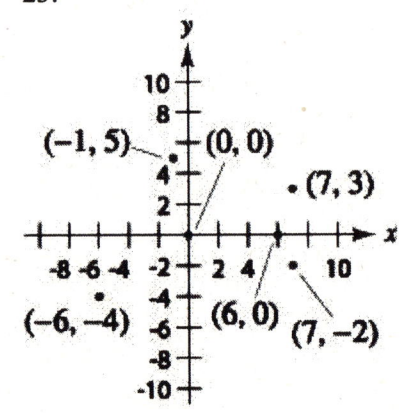

21.

23.

# Algebra Basics, Equations, and Inequalities

25. (2, 4)

29. (7, 10)

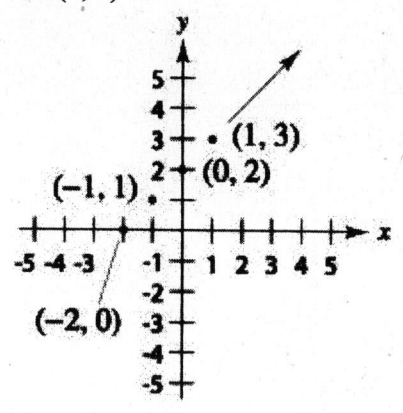

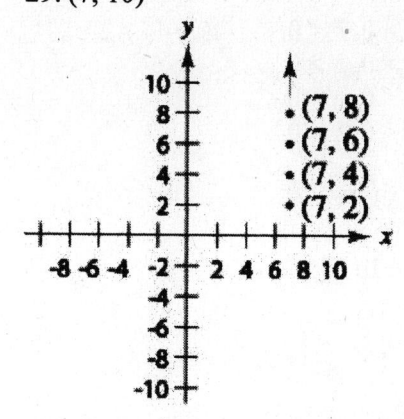

27. (6, −12)

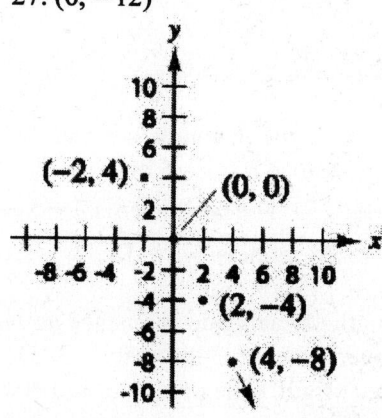

31. (10, −7)

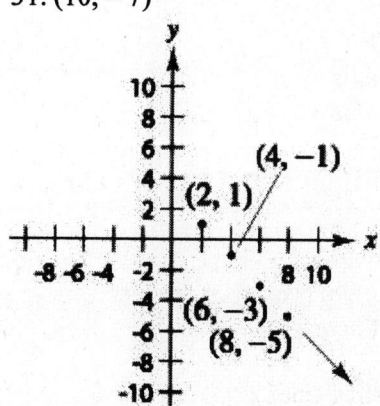

33. (a) $a > 0, b < 0$
    (b) $a > 0, b > 0$
    (c) $a < 0, b > 0$
    (d) $a < 0, b < 0$

35. $x$–axis, I, IV

37. $y$–axis, III, IV

39. I, III

41. $y$–axis

43. II, IV

45. (a) $x$–axis
    (b) $y$–axis
    (c) IV
    (d) I, III

47. (0, 13)

49. (9, 9)

51. (– 12, 6)

53.

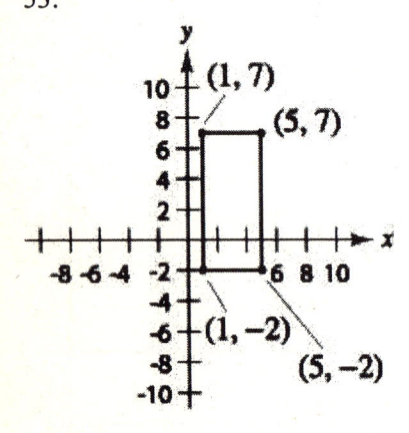

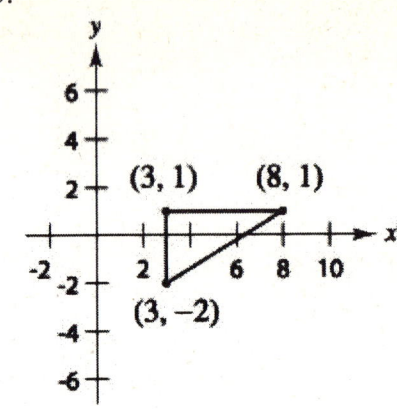

55.

57. (– 4, 6)

59. (17, 0)

61. The length of the rectangle is 5 and the width is 3, since $6 - 1 = 5$ and $-1 - (-4) = 3$. The perimeter is $3 + 5 + 3 + 5 = 16$. The area is $3 \cdot 3 = 15$.

63. The length of each side is 3, since you had to go over 3 and up 3 to get to (– 3, 3) from the origin. The perimeter is $3 + 3 + 3 + 3 = 12$. The area is $3 \cdot 3 = 9$.

65. (– 3, – 1), (– 1, 1), (1, 3), (3, 5), (5, 7); $y = x + 2$

67. (– 3, 3), (– 1, 1), (0, 0), (1, – 1), (3, – 3); $y = -x$

69. (a) (– 40, – 10)
    (b) The length is 20 units and the width is 50 units, since $|-30 - (-10)| = 20$ and $|-40 - 10| = 50$. The area is $50 \cdot 20 = 1000$ square feet.

71. (a) Quadrant I, since $x > 0$ and $y > 0$.
    (b) (0, 25), (30, 17), (70, 11)

73. Sales/marketing and written communication.

Algebra Basics, Equations, and Inequalities

75. The points are associated with the areas in which more than 50% of the companies offer training.

77. $\dfrac{-2+6}{2} = \dfrac{4}{2} = 2$
The midpoint is (2, 3).

79. $\dfrac{1+3}{2} = \dfrac{4}{2} = 2$
The midpoint is (– 4, 2).

81. The coordinates of point P are (7, 5).

## Section 2.3 The Graph of an Expression

1. It is not necessary to enter the expression each time it is to be evaluated.

3.

| X | Y1 |
|---|----|
| -1 | -7 |
| 0 | -6 |
| 4 | -2 |
| 11 | 5 |

X=

5.

| X | Y1 |
|---|----|
| -8 | 23.8 |
| 3 | .7 |
| 6 | -5.6 |
| 12 | -18.2 |

X=

7.

| X | Y1 |
|---|----|
| -9 | 24 |
| -4 | -21 |
| 3 | 0 |
| 5 | 24 |

X=

9.

| X | Y1 |
|---|----|
| -8 | 0 |
| -1 | 14 |
| 2 | 20 |
| 10 | 36 |

X=

11.

| X | Y1 |
|---|----|
| -5 | 9 |
| -1 | 1 |
| -.5 | 0 |
| 8 | 17 |

X=

13.

| X | Y1 |
|---|----|
| -5 | .4 |
| 1 | 1 |
| 25 | .52 |
| 50 | .51 |

X=

15.

```
-2→X
         -2
Y₁
         -5
■
```

| $x-3$ | $-5$ | $-2$ | 1 | 4 | 7 |
|---|---|---|---|---|---|
|  | $(-2,-5)$ | $(1,-2)$ | $(4,1)$ | $(7,4)$ | $(10,7)$ |

17.

```
3→X
         3
Y₁
        11
■
```

| $2x+5$ | 11 | 19 | 5 | $-3$ | $-5$ |
|---|---|---|---|---|---|
|  | $(3,11)$ | $(7,19)$ | $(0,5)$ | $(-4,-3)$ | $(-5,-5)$ |

19.

```
1→X
         1
Y₁
        -6
■
```

| $x^2-3x-4$ | $-6$ | 0 | 14 | 14 | 36 |
|---|---|---|---|---|---|
|  | $(1,-6)$ | $(4,0)$ | $(6,14)$ | $(-3,14)$ | $(-5,36)$ |

21. You obtain an error message for $x-1$, because division by 0 is not defined.

23. When $x=3$, $y=4$.

25. When $x=-1$, $y=1$.

Algebra Basics, Equations, and Inequalities

27.

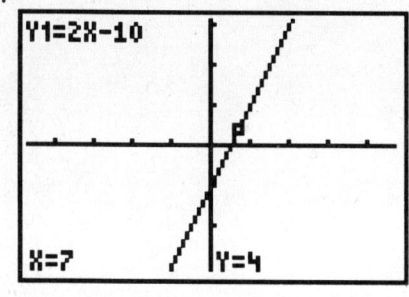

(2, −6), (5, 0), (7, 4)

29.

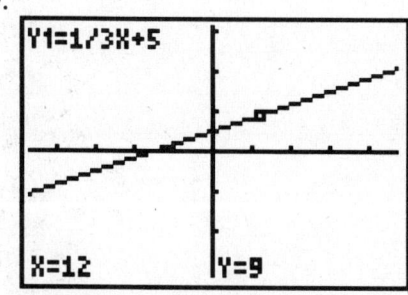

(−30, −5), (−9, 2), (12, 9)

31.

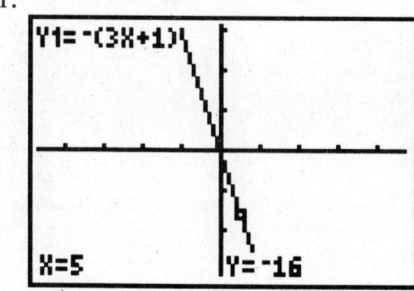

(−6, 17), (0, −1), (5, −16)

33. The first coordinate, 3, is the value of the variable; the second coordinate, 7, is the value of the expression for $x = 3$.

35.

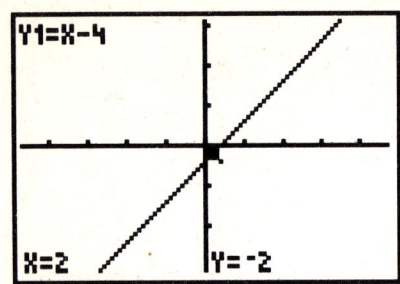

| $x$     | 2  | 4 | 0  | 5 | −1 | −2 |
|---------|----|---|----|---|----|----|
| $x-4$   | −2 | 0 | −4 | 1 | −5 | −6 |

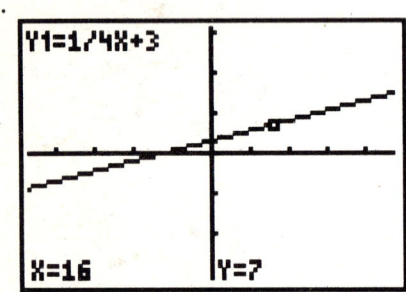

37.

| $x$           | −32 | −8 | −12 | 16 | 6   | 28 |
|---------------|-----|----|-----|----|-----|----|
| $\frac{1}{4}x+3$ | −5  | 1  | 0   | 7  | 4.5 | 10 |

39. (a) 3
    (b) 9
    (c) −5
    (d) −10

41. (a) −1
    (b) 2
    (c) 5
    (d) 8

43. (a) −23
    (b) 6
    (c) 20
    (d) 33

45. (a) −2, 11
    (b) −1, 10
    (c) 1, 8
    (d) 2, 7

Algebra Basics, Equations, and Inequalities

47. (a) −12, 22
(b) −3, 13
(c) 5
(d) 1, 9

49. $2(1)+1 = 4-1$
$2+1 = 3$
$3 = 3$
So, $x = 1$

51. $3(2) = 8-2$
$6 = 6$
So, $x = 2$

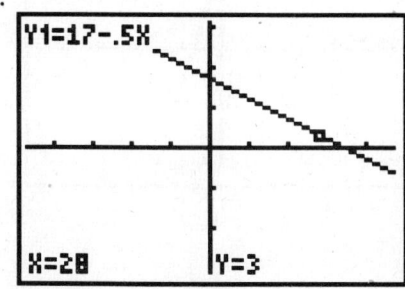

53.

| X | Y1 |
|---|---|
| 0 | 675.3 |
| 5 | 1943.3 |
| 10 | 3211.3 |
| 15 | 4479.3 |
| 20 | 5747.3 |
| 25 | 7015.3 |
| 30 | 8283.3 |

X=0

3 pizzas

55.

| X | Y1 |
|---|---|
| 3 | 955.5 |
| 4 | 950 |
| 5 | 944.5 |

X=

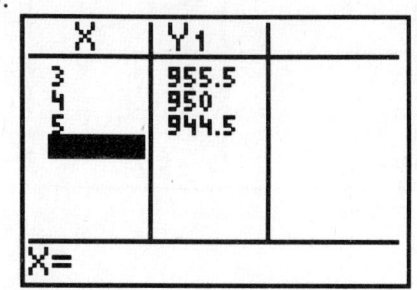

The model is least accurate for 1999.

57.

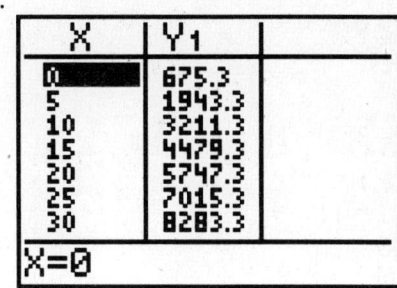

59. 1985

61. 1 more than twice $x$.

63. Twice the sum of $x$ and 1.

65. (a) 30.75

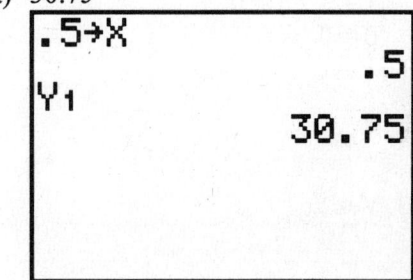

.5→X         .5
Y1
            30.75

(b) 30.75

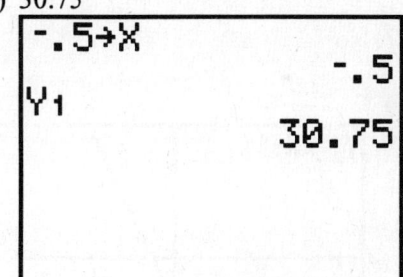

-.5→X        -.5
Y1
            30.75

## Section 2.4 Equations and Estimated Solutions

1. An equation has an = symbol but an expression does not.

3. expression

5. equation

7. equation

9. expression

11. A solution of an equation is a value of the variable that makes the equation true.

13. $14 - 5 = 9$
    $9 = 9$
    Yes

15. $5 - 3(2) = 11$
    $5 - 6 = 11$
    $-1 = 11$
    No

17. $10(1) - 3 = 5(1) + 1$
    $10 - 3 = 5 + 1$
    $7 = 6$
    No

19. $(-1)^2 + 3(-1) + 2 = 0$
    $1 - 3 + 2 = 0$
    $0 = 0$
    Yes

21. Yes

23. Yes

25. Yes

27. No

29.

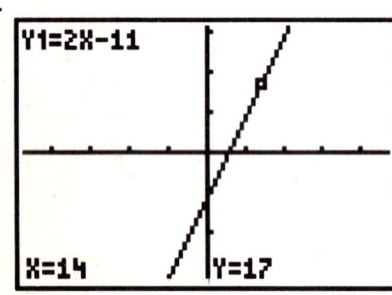

$x = 14$

31.

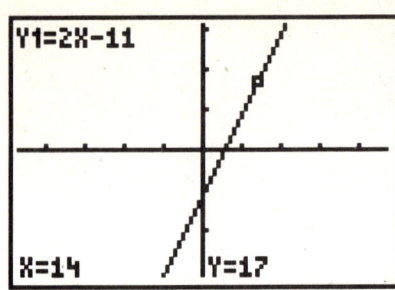

$x = 12$

33.

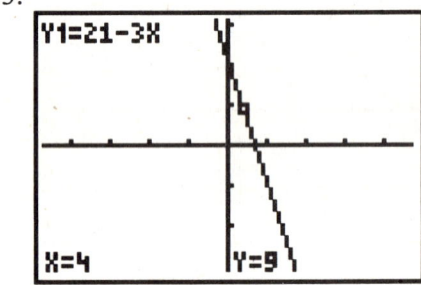

$x = 4$

35.

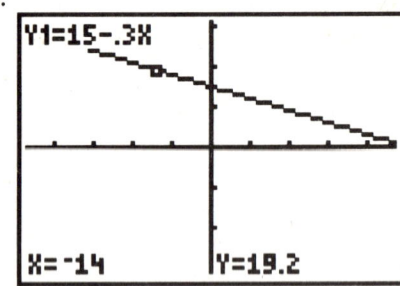

$x = -14$

37.

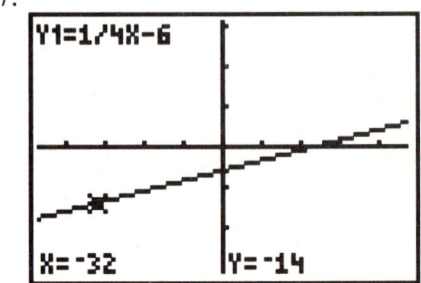

$x = -32$

Algebra Basics, Equations, and Inequalities    45

39.

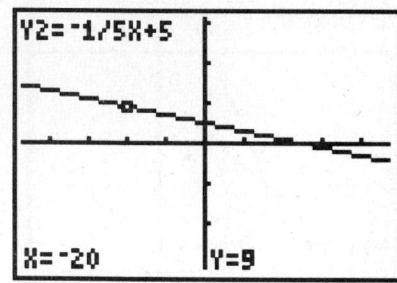

$x = -20$

41. Graph each side of the equation and trace to the point of intersection. The x–coordinate of that point is the estimated solution.

43.

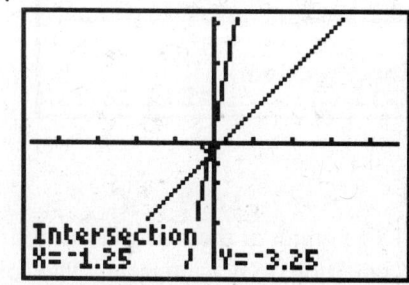

$x = -1.25$

45.

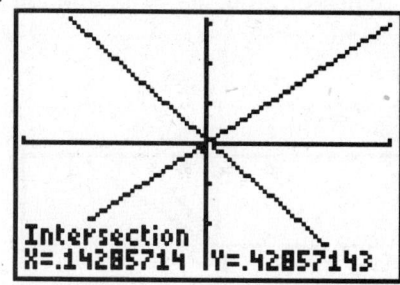

$x = 0.14$

47.

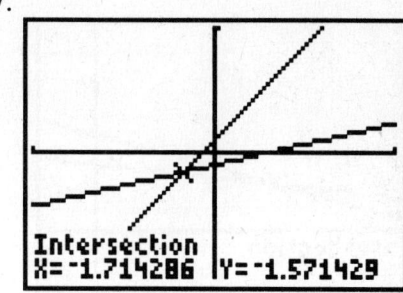

$x = -1.71$

49.

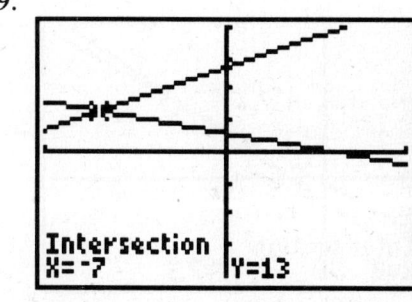

$x = -7$

51.

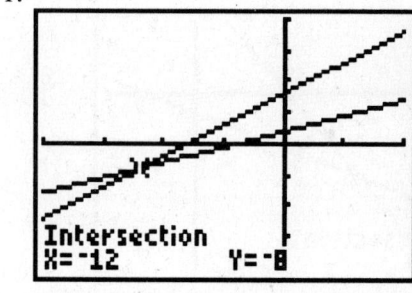

$x = -12$

53.

55. $x = -20$

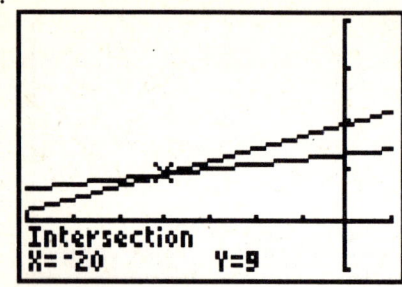

57. $x = 8$

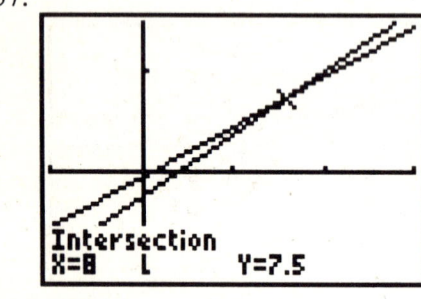

59. $x = -9$

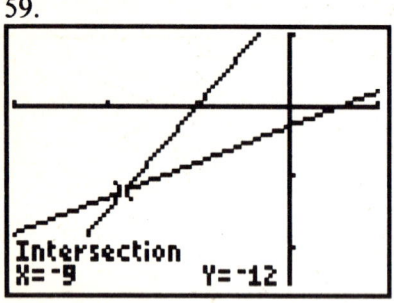

61. $x = 15$

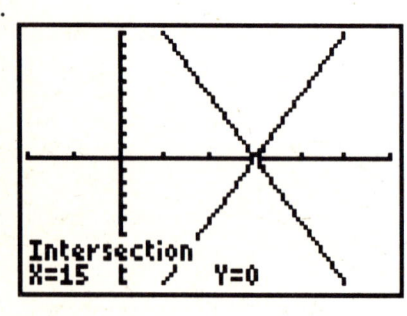

63. $x = -9$

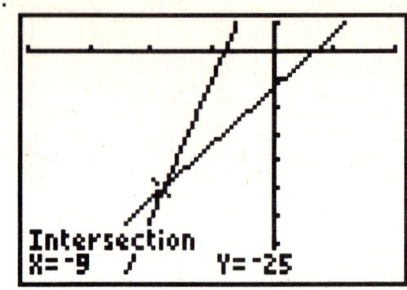

65. $x = 15$

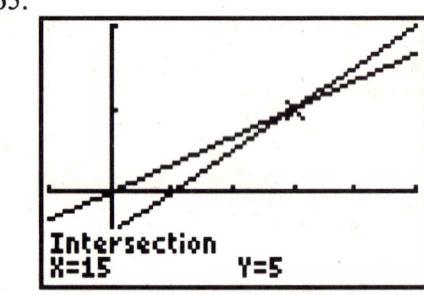

67. (a) The graph of the two sides of the equation do not intersect. The solution set is $\varnothing$.

(b) The graphs of the two sides of the equation coincide. The solution set is $\mathbb{R}$.

69. $x = 0$

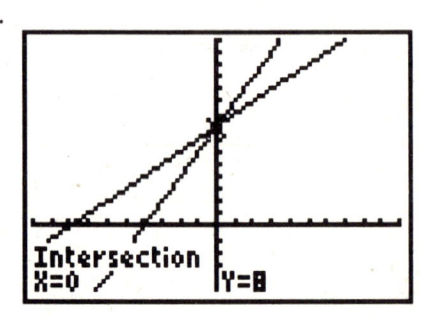

Algebra Basics, Equations, and Inequalities

47

71.

∅

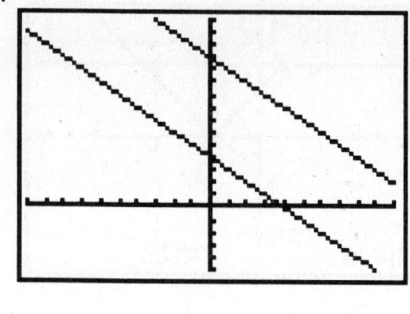

73.

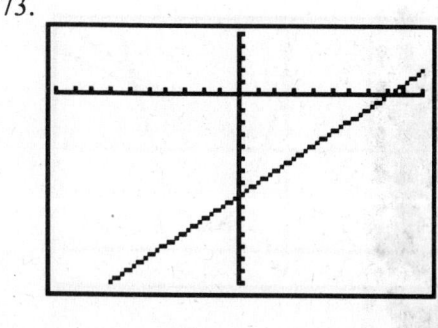

ℝ

75.

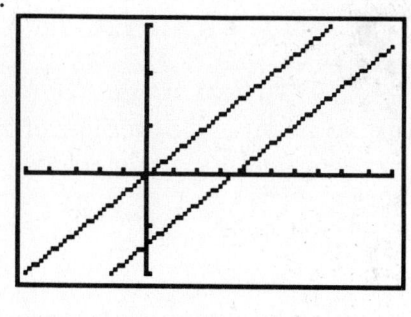

∅

77.

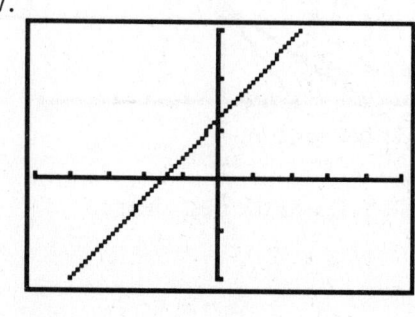

ℝ

79.

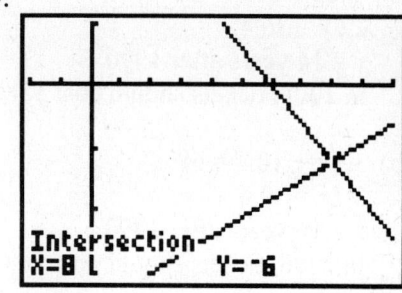

$x = 8$

81.

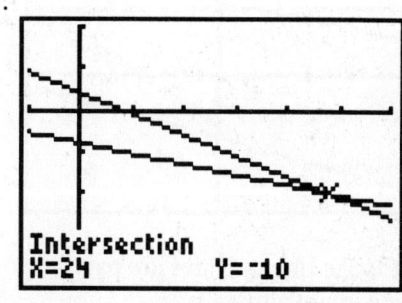

$x = 24$

83.

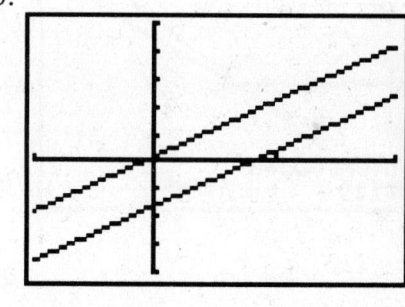

∅

85.

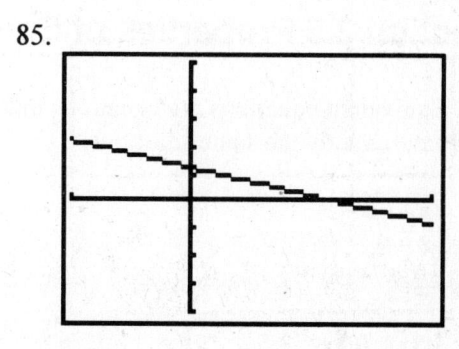

ℝ

87. (a) $2.6x + 21.6 = 58$
   $2.6x = 36.4$
   $x = 14$ years after 1990.
   In 2004, tickets should cost $58.

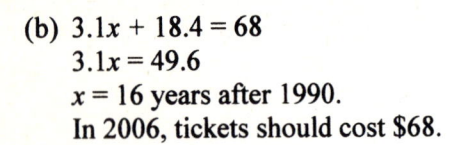

   (b) $3.1x + 18.4 = 68$
   $3.1x = 49.6$
   $x = 16$ years after 1990.
   In 2006, tickets should cost $68.

89.

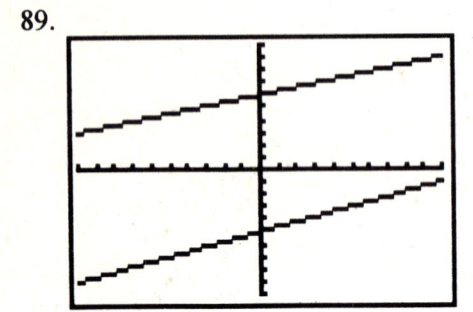

It looks like the lines are parallel so the apparent solution is    .

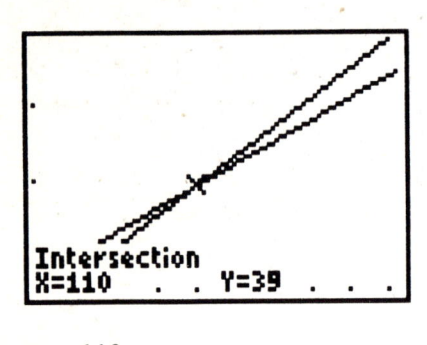

$x = 110$

91.

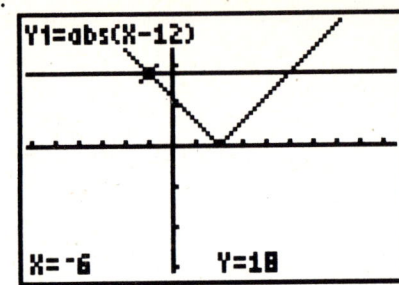

$x = -6, 30$

93.

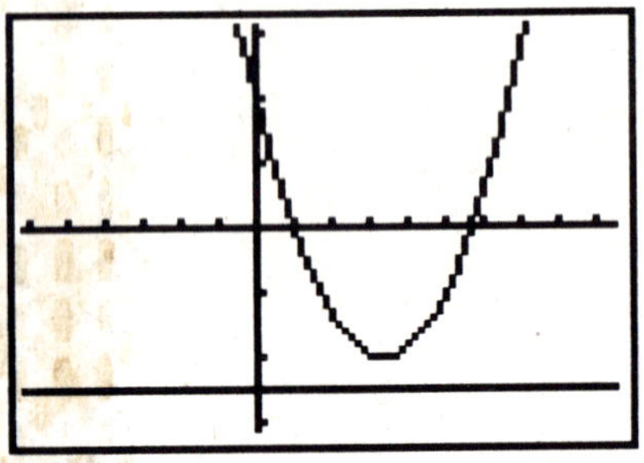

## Section 2.5 Properties of Equations

1. Equivalent equations are equations that have exactly the same solutions.

3. $7x - 6x = 9$
   $1x = 9$
   $x = 9$

5. $1.8x - 0.8x = 3.5$
   $1x = 3.5$
   $x = 3.5$

7. $x - 5 = 0$
   $x - 5 + 5 = 0 + 5$
   $x + 0 = 5$
   $x = 5$
   Step (i) was the best choice.

Algebra Basics, Equations, and Inequalities

9. $5x = 4x - 6$
$5x + (-4x) = -4x + 4x - 6$
$x = 0x - 6$
$x = -6$
Step (ii) was the best choice.

11. $x + 5 = 9$
$x + 5 + (-5) = 9 + (-5)$
$x = 4$

13. $-4 = x - 6$
$-4 + 6 = x - 6 + 6$
$2 = x$
$x = 2$

15. $25 + x = 39$
$25 + x + (-25) = 39 + (-25)$
$x = 14$

17. $x + 27.6 = 8.9$
$x + 27.6 + (-27.6) = 8.9 + (-27.6)$
$x = -18.7$

19. $12 = x + 12$
$12 + (-12) = x + 12 + (-12)$
$0 = x$

21. $3x = 2x - 1$
$3x + (-2x) = 2x - 1 + (-2x)$
$x = -1$

23. $7x = 5 + 6x$
$7x + (-6x) = 5 + 6x + (-6x)$
$x = 5$

25. $-4x + 2 = -3x$
$-4x + 2 + 4x = -3x + 4x$
$2 = x$

27. $9x - 6 = 10x$
$9x - 6 + (-9x) = 10x + (-9x)$
$-6 = x$

29. $-3a + 5a = 9 + a + 6$
$2a = a + 15$
$2a - a = a + 15 - a$
$a = 15$

31. $3x + 3 - 3x + 7 = 6x - 6 - 5x$
$10 = x - 6$
$10 + 6 = x - 6 + 6$
$16 = x$

33. $7x - 9 = 8 + 6x$
$7x - 9 + 9 = 8 + 6x + 9$
$7x = 6x + 17$
$7x + (-6x) = 6x + 17 + (-6x)$
$x = 17$

35. $1 - 3y = 4 - 4y$
$1 - 3y + 4y = 4 - 4y + 4y$
$1 + y = 4$
$1 + y + (-1) = 4 + (-1)$
$y = 3$

37. $5x + 3 = 6x + 1$
$5x + 3 + (-5x) = 6x + 1 + (-5x)$
$3 = x + 1$
$3 + (-1) = x + 1 + (-1)$
$2 = x$

39. $1 - 5x = 1 - 4x$
$1 - 5x + 5x = 1 - 4x + 5x$
$1 = 1 + x$
$1 + (-1) = 1 + x + (-1)$
$0 = x$

41. When we add 3 to both sides of $x - 3 = 12$, the left side becomes $x - 3 + 3 = x + 0 = x$. When we divide both sides of $3x = 12$ by 3, the left side becomes $\dfrac{3x}{3} = x$. In both cases, the variable is isolated.

43. $-4x = 20$
$\dfrac{-4x}{-4} = \dfrac{20}{-4}$
$x = -5$
Step (ii) is best.

45. $\dfrac{1}{5}x = 2$
$5\left(\dfrac{1}{5}x\right) = 5 \cdot 2$
$x = 10$
Step (i) is best.

47. All of the given steps are correct. In each case, the result is $1x = 20$ because:

   (i) $\dfrac{4}{3}$ and $\dfrac{3}{4}$ are reciprocals whose product is 1.

   (ii) $\dfrac{3}{4}$ divided by $\dfrac{3}{4}$ is 1.

   (iii) multiplying by 4 and dividing by 3 is equivalent to multiplying by $\dfrac{4}{3}$.

49. $5x = 20$
$\dfrac{1}{5}(5x) = \dfrac{1}{5}(20)$
$x = 4$

51. $48 = -16x$
$\dfrac{48}{-16} = \dfrac{-16x}{-16}$
$-3 = x$

53. $-3x = 0$
$\dfrac{-3x}{-3} = \dfrac{0}{-3}$
$x = 0$

55. $-3x = 0$
$\dfrac{-3x}{-3} = \dfrac{0}{-3}$
$x = 0$

57. $\dfrac{x}{9} = -4$
$9\left(\dfrac{x}{9}\right) = 9(-4)$
$x = -36$

59. $\dfrac{-t}{4} = -5$
$-4\left(\dfrac{-t}{4}\right) = -4(-5)$
$\dfrac{4t}{4} = 20$
$t = 20$

61. $\dfrac{3}{4}x = 9$
$\dfrac{4}{3}\left(\dfrac{3}{4}x\right) = \dfrac{4}{3}(9)$
$x = \dfrac{36}{3}$
$x = 12$

63. $2x - 7x = 0$
$-5x = 0$
$\dfrac{-5x}{-5} = \dfrac{0}{-5}$
$x = 0$

65. $y - 6y = -25$
$-5y = -25$
$\dfrac{-5y}{-5} = \dfrac{-25}{-5}$
$y = 5$

67. $8 = 3t - 4t$
$8 = -t$
$(-1)8 = (-1)(-t)$
$t = -8$

69. $9x = 8x$
$9x - 8x = 8x - 8x$
$x = 0$

Algebra Basics, Equations, and Inequalities

71. $-4x = -28$
$\dfrac{-4x}{-4} = \dfrac{-28}{-4}$
$x = 7$

73. $5 - 7x = -6x$
$5 - 7x + 7x = -6x + 7x$
$x = 5$

75. $-3 + x = 0$
$3 + (-3) + x = 3 + 0$
$x = 3$

77. $\dfrac{1}{4} = -2y$
$-\dfrac{1}{2}\left(\dfrac{1}{4}\right) = \left(-\dfrac{1}{2}\right)(-2y)$
$y = -\dfrac{1}{8}$

79. Distributive Property

81. Multiplication Property of Equations

83. Addition Property of Equations

85. $145x - 263 = 1042$
$145x - 263 + 263 = 1042 + 263$
$145x = 1305$
$x = 9$

87. $2200 = 145x - 263$

89. $x + a = b$
$x + a - a = b - a$
$x = b - a$

91. $5x - a = b$
$5x - a + a = b + a$
$5x = b + a$
$\dfrac{5x}{5} = \dfrac{b+a}{5}$
$x = \dfrac{a+b}{5}$

93. $x - k = 0$
$-4 - k = 0$
$-4k + k = k$
$-4 = k$

95. $kx + 1 = 7$
$k(2) + 1 = 7$
$2k + 1 = 7$
$2k + 1 - 1 = 7 - 1$
$2k = 6$
$\dfrac{2k}{2} = \dfrac{6}{2}$
$k = 3$

97. $2x - 1 = 5$
$2x - 1 + 1 = 5 + 1$
$2x = 6$
$\dfrac{2x}{2} = \dfrac{6}{2}$
$x = 3$

So $3x - 4$
$= 3(3) - 4$
$= 9 - 4$
$= 5$

## Section 2.6 Solving Linear Equations

1. Dividing both sides by 2 would create fractions and make the equation more difficult to solve. The best first step is to add 5 to both sides.

3. $5x + 14 = 4$
$5x + 14 - 14 = 4 - 14$
$5x = -10$
$\dfrac{5x}{5} = \dfrac{-10}{5}$
$x = -2$

5.  $2 - 3x = 0$
    $2 - 3x + 3x = 0 + 3x$
    $2 = 3x$
    $\dfrac{2}{3} = \dfrac{3x}{3}$
    $x = \dfrac{2}{3}$

7.  $0.2 - 0.37t = -0.91$
    $0.2 - 0.37t + 0.91 = -0.91 + 0.91$
    $-0.37t + 1.11 = 0$
    $-0.37t + 1.11 + 0.37t = 0 + 0.37t$
    $1.11 = 0.37t$
    $\dfrac{1.11}{0.37} = t$
    $t = 3$

9.  $7x = 2x - 5$
    $7x - 2x = 2x - 5 - 2x$
    $5x = -5$
    $\dfrac{5x}{5} = \dfrac{-5}{5}$
    $x = -1$

11. $3x = 5x + 14$
    $3x - 14 = 5x + 14 - 14$
    $3x - 14 = 5x$
    $3x - 14 - 3x = 5x - 3x$
    $-14 = 2x$
    $\dfrac{-14}{2} = \dfrac{2x}{2}$
    $x = -7$

13. $4x - 7 = 6x + 3$
    $4x - 7 - 4x = 6x + 3 - 4x$
    $-7 = 2x + 3$
    $-7 - 3 = 2x + 3 - 3$
    $-10 = 2x$
    $\dfrac{-10}{2} = \dfrac{2x}{2}$
    $-5 = x$

15. $5x + 2 = 2 - 3x$
    $5x + 2 + 3x = 2 - 3x + 3x$
    $8x + 2 = 2$
    $8x + 2 - 2 = 2 - 2$
    $8x = 0$
    $\dfrac{8x}{8} = \dfrac{0}{8}$
    $x = 0$

17. $x + 6 = 5x - 10$
    $x + 6 - x = 5x - 10 - x$
    $6 = 4x - 10$
    $6 + 10 = 4x - 10 + 10$
    $16 = 4x$
    $\dfrac{16}{4} = \dfrac{4x}{4}$
    $x = 4$

19. $5 - 9x = 8x + 5$
    $5 - 9x + 9x = 8x + 5 + 9x$
    $5 = 17x + 5$
    $5 - 5 = 17x + 5 - 5$
    $0 = 17x$
    $\dfrac{0}{17} = \dfrac{17x}{17}$
    $x = 0$

Algebra Basics, Equations, and Inequalities

21. $9 - x = x + 15$
$9 - x + x = x + 15 + x$
$9 = 2x + 15$
$9 - 15 = 2x + 15 - 15$
$-6 = 2x$
$\dfrac{-6}{2} = \dfrac{2x}{2}$
$x = -3$

23. $3x + 8 - 5x = 2 - x + 2x - 3$
$-2x + 8 = x - 1$
$-2x + 8 + 2x = x - 1 + 2x$
$8 = 3x - 1$
$8 + 1 = 3x - 1 + 1$
$9 = 3x$
$\dfrac{9}{3} = \dfrac{3x}{3}$
$x = 3$

25. $3 + 5y - 24 + 3y = y - 1 + y$
$8y - 21 = 2y - 1$
$8y - 21 + 21 = 2y - 1 + 21$
$8y = 2y + 20$
$8y - 2y = 2y + 20 - 2y$
$6y = 20$
$\dfrac{6y}{y} = \dfrac{20}{6}$
$y = \dfrac{10}{3}$

27. $5 - 5x + 8 - 4 = -9 + x$
$5 - 5x + 4 = -9 + x$
$-5x + 9 = -9 + x$
$-5x + 9 + 5x = -9 + x + 5x$
$9 = -9 + 6x$
$9 + 9 = -9 + 6x + 9$
$18 = 6x$
$\dfrac{18}{6} = \dfrac{6x}{6}$
$3 = x$

29. $3(x + 1) = 1$
$3x + 3 = 1$
$3x + 3 - 3 = 1 - 3$
$3x = -2$
$\dfrac{3x}{3} = -\dfrac{2}{3}$
$x = -\dfrac{2}{3}$

31. $5(2x - 1) = 3(3x - 1)$
$10x - 5 = 9x - 3$
$10x - 5 - 9x = 9x - 3 - 9x$
$x - 5 = -3$
$x - 5 + 5 = -3 + 5$
$x = 2$

33. $-10 = 3x + 8 + 4(x - 1)$
$-10 = 3x + 8 + 4x - 4$
$10 = 7x + 4$
$-10 - 4 = 7x + 4 - 4$
$-14 = 7x$
$\dfrac{-14}{7} = \dfrac{7x}{7}$
$x = -2$

35. $4(x - 1) = 5 + 3(x - 6)$
$4x - 4 = 5 + 3x - 18$
$4x - 4 = 3x - 13$
$4x - 4 + 4 = 3x - 13 + 4$
$4x = 3x - 9$
$4x - 3x = 3x - 9 - 3x$
$x = -9$

37. $6x - (5x - 3) = 0$
$6x - 5x + 3 = 0$
$x + 3 = 0$
$x + 3 - 3 = 0 - 3$
$x = -3$

39. $2 - 2(3x - 4) = 8 - 2(5 + x)$
$2 - 6x + 8 = 8 - 10 - 2x$
$-6x + 10 = -2 - 2x$
$-6x + 10 + 6x = -2 - 2x + 6x$
$10 = -2 + 4x$
$10 + 2 = -2 + 4x + 2$
$12 = 4x$
$\dfrac{12}{4} = \dfrac{4x}{4}$
$x = 3$

41. The number $n$ is the LCD of all fractions in the equation.

43. $\dfrac{2}{3}x - \dfrac{2}{3} = \dfrac{5}{3}$
$3\left(\dfrac{2}{3}x - \dfrac{2}{3}\right) = 3\left(\dfrac{5}{3}\right)$
$2x - 2 = 5$
$2x - 2 + 2 = 5 + 2$
$2x = 7$
$\dfrac{2x}{2} = \dfrac{7}{2}$
$x = \dfrac{7}{2}$

45. $1 + \dfrac{3t}{4} = \dfrac{1}{4}$
$4\left(1 + \dfrac{3t}{4}\right) = 4\left(\dfrac{1}{4}\right)$
$4 + 3t = 1$
$4 + 3t - 4 = 1 - 4$
$3t = -3$
$\dfrac{3t}{3} = -\dfrac{3}{3}$
$t = -1$

47. $\dfrac{1}{2}x - 2 = 1 + \dfrac{2}{3}x + \dfrac{1}{6}$
$6\left(\dfrac{1}{2}x - 2\right) = 6\left(1 + \dfrac{2}{3}x + \dfrac{1}{6}\right)$
$3x - 12 = 6 + 4x + 1$
$3x - 12 = 4x + 7$
$3x - 12 - 3x = 4x + 7 - 3x$
$-12 = x + 7$
$-12 - 7 = x + 7 - 7$
$x = -19$

49. $\dfrac{x + 5}{3} = \dfrac{-x}{2} + 5$
$6\left(\dfrac{x + 5}{3}\right) = 6\left(\dfrac{-x}{2} + 5\right)$
$2(x + 5) = -3x + 30$
$2x + 10 = -3x + 30$
$2x + 10 + 3x = -3x + 30 + 3x$
$5x + 10 = 30$
$5x + 10 - 10 = 30 - 10$
$5x = 20$
$\dfrac{5x}{5} = \dfrac{20}{5}$
$x = 4$

51. $\dfrac{3}{5}(x - 1) = 9$
$5\left[\dfrac{3}{5}(x - 1)\right] = 5(9)$
$3(x - 1) = 45$
$3x - 3 = 45$
$3x - 3 + 3 = 45 + 3$
$3x = 48$
$\dfrac{3x}{3} = \dfrac{48}{3}$
$x = 16$

53. $\dfrac{x}{3}+2=\dfrac{3}{2}(x+3)-6$

$6\left(\dfrac{x}{3}+2\right)=6\left[\dfrac{3}{2}(x+3)-6\right]$

$2x+12=9(x+3)-36$

$2x+12=9x+27-36$

$2x+12=9x-9$

$2x+12-2x=9x-9-2x$

$12=7x-9$

$21=7x$

$x=3$

55. $\dfrac{3}{7}x=\dfrac{5}{14}(x+1)+\dfrac{1}{7}(8-x)$

$14\left(\dfrac{3}{7}x\right)=14\left[\dfrac{5}{14}(x+1)+\dfrac{1}{7}(8-x)\right]$

$6x=5(x+1)+2(8-x)$

$6x=5x+5+16-2x$

$6x=3x+21$

$6x-3x=3x+21-3x$

$3x=21$

$\dfrac{3x}{3}=\dfrac{21}{3}$

$x=7$

57. $7.35x+5.1=2.6x-0.6$

$7.35x+5.1-5.1=2.6x-0.6-5.1$

$7.35x=2.6x-5.7$

$7.35x-2.6x=2.6x-5.7-2.6x$

$4.75x=-5.7$

$\dfrac{4.75x}{4.75}=-\dfrac{5.7}{4.75}$

$x=-1.2$

59. $2.7(3x-1.5)+6.3=9.1x+6.55$

$8.1x-4.05+6.3=9.1x+6.55$

$8.1x+2.25=9.1x+6.55$

$8.1x+2.25-8.1x=9.1x+6.55-8.1x$

$2.25=x+6.55$

$2.25-6.55=x+6.55-6.55$

$x=-4.3$

61. Identity, $\mathbb{R}$

63. $x=x+3$

$x-x=x+3-x$

$0=3$

Contradiction, $\varnothing$

65. $2x=3x$

$2x-2x=3x-2x$

$x=0$

Conditional

67. $6t+7=7$

$6t=0$

$t=0$

Conditional

69. $0x=3$

$0=3$

Contradiction, $\varnothing$

71. $6x-3-x=6x+5$

$5x-3=6x+5$

$5x-3-5x=6x+5-5x$

$-3=x+5$

$-3-5=x+5-5$

$x=-8$

73. $2+3x+5=6x+7-6x$

$3x+7=7$

$3x=0$

$\dfrac{3x}{3}=\dfrac{0}{3}$

$x=0$

75. $\dfrac{3}{4}x+\dfrac{5}{2}=\dfrac{1}{2}(5+x)+\dfrac{1}{4}x$

$4\left(\dfrac{3}{4}x+\dfrac{5}{2}\right)=4\left[\dfrac{1}{2}(5+x)+\dfrac{1}{4}x\right]$

$3x+10=2(5+x)+x$

$3x+10=10+2x+x$

$3x+10=3x+10$

$\mathbb{R}$

77. $4(x-2)+8=6(x+1)-2x$
    $4x-8+8=6x+6-2x$
    $4x=4x+6$
    $4x-4x=4x+6-4x$
    $0=6$
    $\varnothing$

79. $2(3x-4)=2-6(1-x)$
    $6x-8=2-6+6x$
    $6x-8=6x-4$
    $-8=-4$
    $\varnothing$

81. $3x-1-2(x-4)=2x+7-x$
    $3x-1-2x+8=x+7$
    $x+7=x+7$
    $\mathbb{R}$

83. (a) (i) The cost at a private colleges is $150,000.

    (ii) The cost at a public college is double the 1997 cost.

    (b) $5.88x + 40.58 = 150$
    $5.88x = 109.42$
    $x = 18.61$
    $x \approx 19$ years after 1990

    So, in 2009, private college education costs are expected to be $150,000.

    $2.28x + 15.7 = 2(31.637)$
    $2.28x + 15.7 = 63.274$
    $2.28x = 47.574$
    $x = 20.87$
    $x \approx 21$ years after 1990

    So, in 2011, public college education costs are expected to be double 1997 costs.

85. 1976 is 20 years after 1956, so $t = 20$.

    Men: $-0.014(20) + 10.5$
    $= 10.22$ seconds

    Women: $-0.017(20) + 11.5$
    $= 11.16$ seconds

87. The winning times are projected to be the same 333 years after 1956.

    $1956 + 333 = 2289$

    $-0.014(333) + 10.5$
    $= -4.662 + 10.5$
    $= 5.838$ seconds

89. $x-5\{x-5[x-5(x-5)]\}=1$
    $x-5\{x-5[x-5x+25]\}=1$
    $x-5\{x-5(-4x+25)\}=1$
    $x-5\{x+20x-125\}=1$
    $x-5(21x-125)=1$
    $x-105x+625=1$
    $-104x+625=1$
    $-104x=-624$
    $x=\dfrac{-624}{-104}$
    $x=6$

91. $x-[2(x-3)-(x+1)]=-3(x-1)+4+3x$
    $x-[2x-6-x-1]=-3x+3+4+3x$
    $x-(x-7)=7$
    $x-x+7=7$
    $7=7$
    $\mathbb{R}$

93. $ax+b=c, a \neq 0$
    $ax=c-b$
    $x=\dfrac{c-b}{a}$

95. $2t^2 - 15 = t(t+5) + t^2$
$2t^2 - 15 = t^2 + 5t + t^2$
$2t^2 - 15 = 2t^2 + 5t$
$-15 = 5t$
$t = -3$

## Section 2.7 Inequalities: Graphing Methods

1. The value of x + 3 is less than or equal to 7.

3. $2 + x > 7$
$2 + x - 2 > 7 - 2$
$x > 5$
So, 9 and 11 are solutions.

5. $4 \le 1 - 3x$
$4 - 1 \le 1 - 3x - 1$
$3 \le -3x$
$\dfrac{3}{-3} \ge \dfrac{-3x}{-3}$
$-1 \ge x$

So, $-1, -2,$ and $-4$ are solutions.

7. $-3 \le 2x - 7 < 5$
$-3 + 7 \le 2x - 7 + 7 < 5 + 7$
$4 \le 2x \le 12$
$\dfrac{4}{2} \le \dfrac{2x}{2} \le \dfrac{12}{2}$
$2 \le x \le 6$

So, 2, 5, and 6 are solutions.

9. The symbols [ and ] indicate that an endpoint is included. The symbols ( and ) indicate that an endpoint is not included.

11. [number line showing open circle at 8, shaded left, marks at 7 and 8]

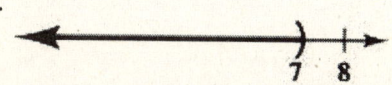

13. [number line showing open circle at -3, shaded right, marks at -4 and -3]

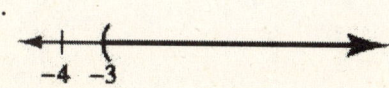

15. [number line showing bracket at 1, shaded right, marks at 0 and 1]

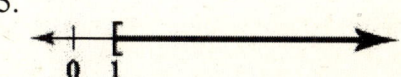

17. [number line showing bracket at -4, shaded left, marks at -5 and -4]

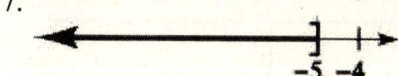

19. [number line showing open circle at -5 and open circle at -1, shaded between, marks at -6, -5, -1, 0]

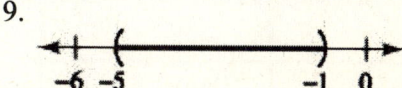

21. [number line showing bracket at 1 and open circle at 5, shaded between, marks at 0, 1, 4, 5]

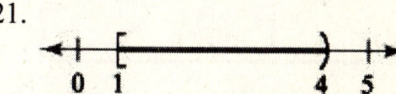

23. $n \ge 0$

25. $x \le 3$

27. $n < 0$

29. $x \le 6$

31. $x < -3$

33. $x \ge 5$

35. $-4 < x \le 7$

37. The point of intersection represents a solution if the inequality symbol is $\le$ or $\ge$.

39. $x > -15$
[number line with open circle at -15, shaded right, marks at -16 and -15]

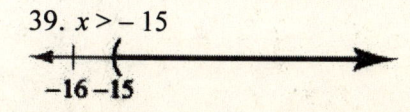

41. $x \leq 0$

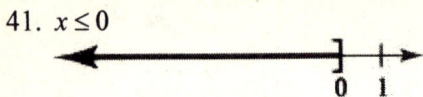

43.

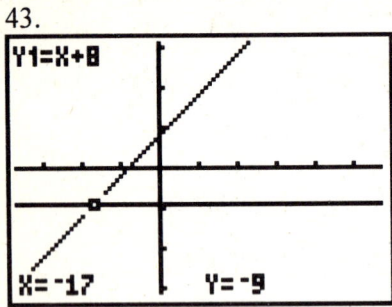

$x > -17$

45.

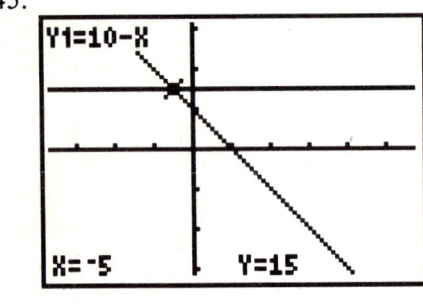

$x \geq -5$

47.

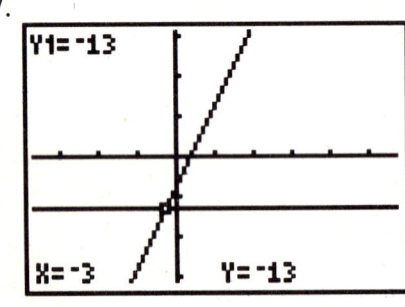

$x < -3$

49.

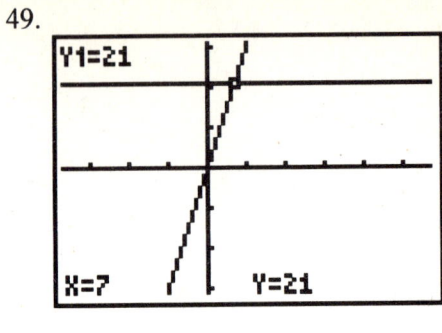

$x \leq 7$

51.

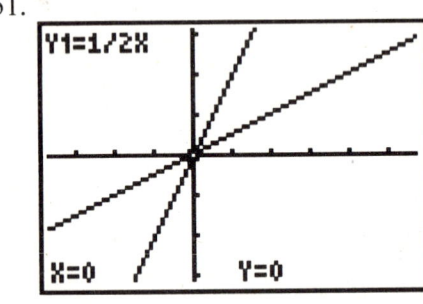

$x < 0$

53.

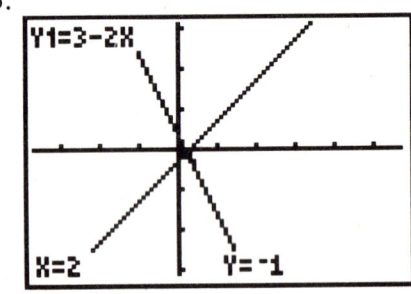

$x < 2$

Algebra Basics, Equations, and Inequalities 59

55.

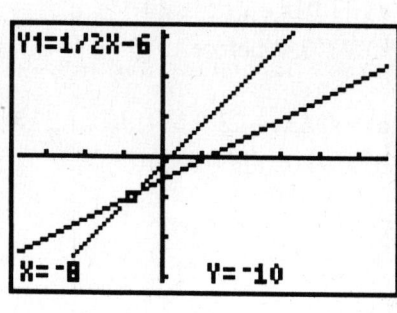

[calculator graph: Y1=1/2X-6, X=-8, Y=-10]

$x \geq -8$

57.

[calculator graph: Y1=7+.8X, X=12, Y=16.6]

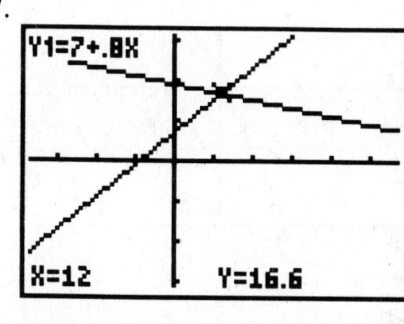

$x < 12$

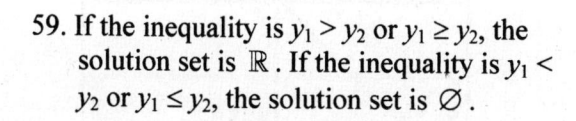

59. If the inequality is $y_1 > y_2$ or $y_1 \geq y_2$, the solution set is $\mathbb{R}$. If the inequality is $y_1 < y_2$ or $y_1 \leq y_2$, the solution set is $\varnothing$.

61. Since $y_1 > y_2$ for all values of $x$, the solution set is $\mathbb{R}$

63.

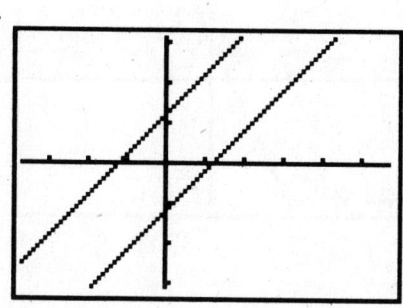

[calculator graph]

(a) $\mathbb{R}$
(b) $\varnothing$

65. $-11 \leq x \leq 23$

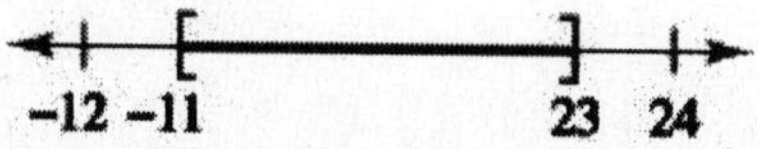

[number line from -12 to 24 with brackets at -11 and 23]

67. $-19 < x \leq 13$

[number line from -20 to 14 with ( at -19 and ] at 13]

69. $-13 < x \leq 9$

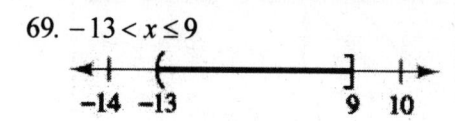

[number line from -14 to 10 with ( at -13 and ] at 9]

71. $-6 < x < 8$

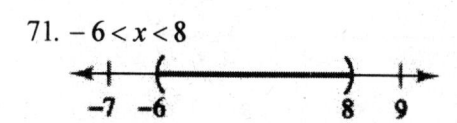

[number line from -7 to 9 with ( at -6 and ) at 8]

73. $x \leq 9$

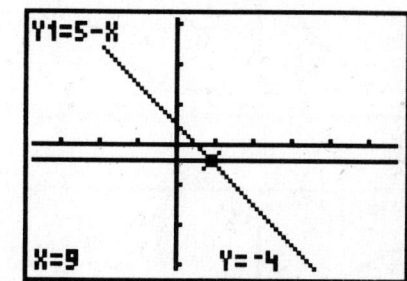

[calculator graph: Y1=5-X, X=9, Y=-4]

75. $\varnothing$

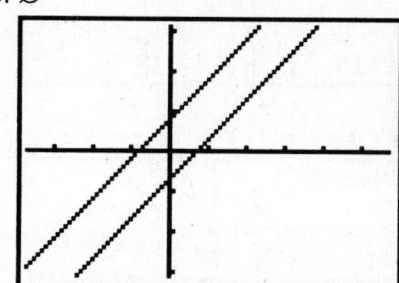

[calculator graph]

77. $x > -9$

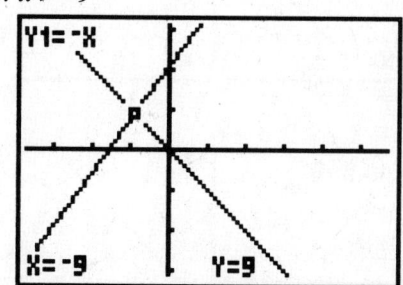

[calculator graph: Y1=-X, X=-9, Y=9]

79. $\mathbb{R}$

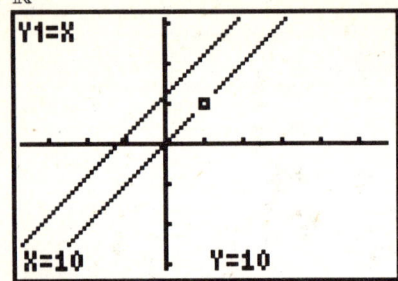

81. $x < 0$

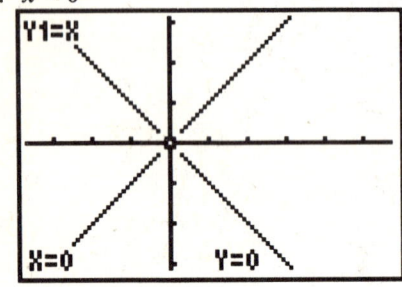

83. $\varnothing$

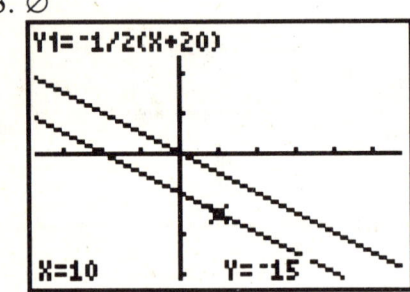

85. $x + 6 \geq 3; x \geq -3$

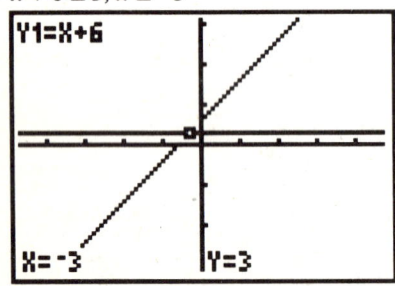

87. $x - 8 > -2; x > 6$

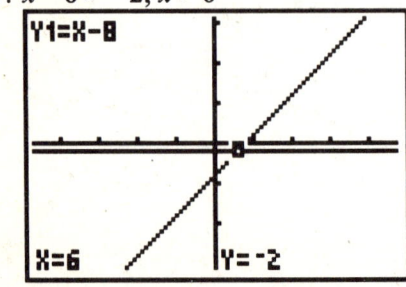

89. (a) $-1.01x + 27.5 \geq 20$
    (b) $x \leq 7.4$, before 1988

91. (a) $-0.83x + 23.9 > -1.01x + 27.5$
    (b) $x > 20$; after 2000

93. $\varnothing$

95. (a) $\varnothing$
    (b) $\mathbb{R}$

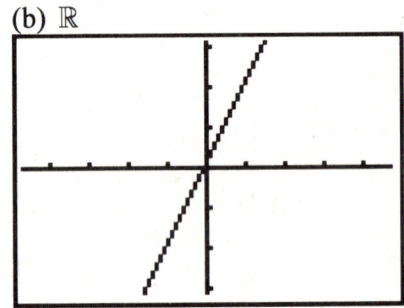

97. $\mathbb{R}$

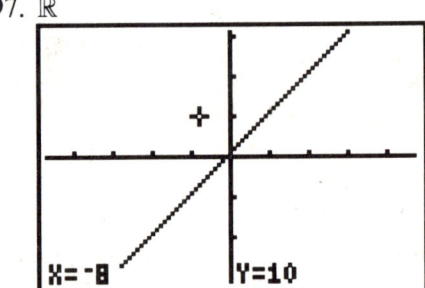

99. $\varnothing$

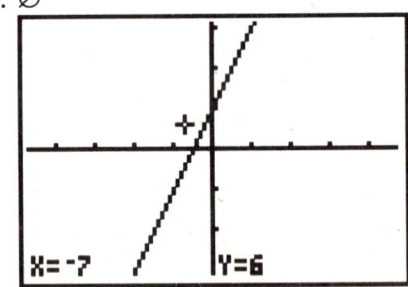

## Section 2.8 Inequalities: Algebraic Methods

1. The properties allow us to add the same number to both sides on an equation or inequality.

3. $>$

5. $>$

7. $>$

9. $<$

11. $x + 5 < 1$
    $x + 5 - 5 < 1 - 5$
    $x < -4$

13. $-3 \leq 2 + x$
    $-3 - 2 \leq 2 + x - 2$
    $-5 \leq x$

15. $5x - 1 \geq -1 + 4x$
    $5x - 1 + 1 \geq -1 + 4x + 1$
    $5x \geq 4x$
    $5x - 4x \geq 4x - 4x$
    $x \geq 0$

17. $x - 2 > 2x + 6$
    $x - 2 + 2 > 2x + 6 + 2$
    $x > 2x + 8$
    $x - 2x > 2x + 8 - 2x$
    $-x > 8$
    $(-1)(-x) < (-1)(8)$
    $x < -8$

19. $3x > 12$
    $\dfrac{3x}{3} > \dfrac{12}{3}$
    $x > 4$

21. $-5x < 15$
    $\dfrac{-5x}{-5} > \dfrac{15}{-5}$
    $x > -3$

23. $0 \leq -x$
    $(-1)(0) \geq (-1)(-x)$
    $0 \geq x$

25. $\dfrac{3}{4}x \leq 12$
    $\dfrac{4}{3}\left(\dfrac{3}{4}x\right) \leq \dfrac{4}{3}(12)$
    $x \leq 16$

27. $\dfrac{x}{-5} \leq 1$
    $-5\left(\dfrac{x}{-5}\right) \geq -5(1)$
    $x \geq -5$

29. No. Because we multiplied both sides by a negative number, the inequality symbol must be reversed: $32x > 89$.

31. $5 + 3x > -4$
    $5 + 3x - 5 > -4 - 5$
    $3x > -9$
    $\dfrac{3x}{3} > \dfrac{-9}{3}$
    $x > -3$

33. $3 - x \leq 5$
    $3 - x - 3 \leq 5 - 3$
    $-x \leq 2$
    $(-1)(-x) \geq (-1)(2)$
    $x \geq -2$

35. $3 - 4x \leq 19$
    $3 - 4x - 3 \leq 19 - 3$
    $-4x \leq 16$
    $\dfrac{-4x}{-4} \geq \dfrac{16}{-4}$
    $x \geq -4$

37. $5x + 5 \leq 4x - 2$
    $5x + 5 - 4x \leq 4x - 2 - 4x$
    $x + 5 \leq -2$
    $x + 5 - 5 \leq -2 - 5$
    $x \leq -7$

39. $6 - 3x > x - 2$
    $6 - 3x + 3x > x - 2 + 3x$
    $6 > 4x - 2$
    $6 + 2 > 4x - 2 + 2$
    $8 > 4x$
    $\dfrac{8}{4} > \dfrac{4x}{4}$
    $2 > x$

41. $7 + x \leq 3(1 - x)$
    $7 + x \leq 3 - 3x$
    $7 + x + 3x \leq 3 - 3x + 3x$
    $7 + 4x \leq 3$
    $7 + 4x - 7 \leq 3 - 7$
    $4x \leq -4$
    $\dfrac{4x}{4} \leq \dfrac{-4}{4}$
    $x \leq -1$

43. $6(2 - 3x) > 11(3 - x)$
    $12 - 18x > 33 - 11x$
    $12 - 18x + 18x > 33 - 11x + 18x$
    $12 > 33 + 7x$
    $12 - 33 > 33 + 7x - 33$
    $\dfrac{-21}{7} > \dfrac{7x}{7}$
    $-3 > x$

45. $3y + 2(2y + 1) > 11 + y$
    $3y + 4y + 2 > 11 + y$
    $7y + 2 > 11 + y$
    $7y + 2 - y > 11 + y - y$
    $6y + 2 > 11$
    $6y + 2 - 2 > 11 - 2$
    $6y > 9$
    $\dfrac{6y}{6} > \dfrac{9}{6}$
    $y > \dfrac{3}{2}$

47. $6(t - 3) + 3 \leq 2(4t + 3) + 5t$
    $6t - 18 + 3 \leq 8t + 6 + 5t$
    $6t - 15 \leq 13t + 6$
    $6t - 15 - 6t \leq 13t + 6 - 6t$
    $-15 \leq 7t + 6$
    $-15 - 6 \leq 7t + 6 - 6$
    $-21 \leq 7t$
    $\dfrac{-21}{7} \leq \dfrac{7t}{7}$
    $-3 \leq t$

49. $2x + \dfrac{1}{2} \geq \dfrac{1}{3}$
    $2x + \dfrac{1}{2} - \dfrac{1}{2} \geq \dfrac{1}{3} - \dfrac{1}{2}$
    $2x \geq \dfrac{2}{6} - \dfrac{3}{6}$
    $2x > -\dfrac{1}{6}$
    $\dfrac{1}{2}(2x) > \dfrac{1}{2}\left(-\dfrac{1}{6}\right)$
    $x > -\dfrac{1}{12}$

Algebra Basics, Equations, and Inequalities

51. $\dfrac{1}{2}t - \dfrac{3}{4} < -\dfrac{1}{3}t$

$\dfrac{1}{2}t - \dfrac{3}{4} + \dfrac{1}{3}t < -\dfrac{1}{3}t + \dfrac{1}{3}t$

$\dfrac{3}{6}t - \dfrac{3}{4} + \dfrac{2}{6}t < 0$

$\dfrac{5}{6}t - \dfrac{3}{4} < 0$

$\dfrac{5}{6}t - \dfrac{3}{4} + \dfrac{3}{4} < 0 + \dfrac{3}{4}$

$\dfrac{5}{6}t < \dfrac{3}{4}$

$\dfrac{6}{5}\left(\dfrac{5}{6}t\right) < \left(\dfrac{3}{4}\right)\left(\dfrac{6}{5}\right)$

$t < \dfrac{9}{10}$

53. $-1.6t - 1.4 \geq 4.2 - 3t$

$-1.6t - 1.4 + 3t \geq 4.2 - 3t + 3t$

$1.4t - 1.4 \geq 4.2$

$1.4t - 1.4 + 1.4 \geq 4.2 + 1.4$

$1.4t \geq 5.6$

$\dfrac{1.4t}{1.4} \geq \dfrac{5.6}{1.4}$

$t \geq 4$

55. If the resulting inequality is true, the solution set is $\mathbb{R}$. If the resulting inequality is false, the solution set is $\varnothing$.

57. $3(x-2) \geq 2x + x$

$3x - 6 \geq 3x$

$-6 \geq 0$

$\varnothing$

59. $2(3-x) + x \leq 6 - x$

$6 - 2x + x \leq 6 - x$

$6 - x \leq 6 - x$

$\mathbb{R}$

61. $2x - 5 + x \leq -3(2-x)$

$3x - 5 \leq -6 + 3x$

$-5 \leq -6$

$\varnothing$

63. $4 < x + 2 < 11$

$4 - 2 < x + 2 - 2 < 11 - 2$

$2 < x < 9$

65. $10 \leq -5x < 35$

$\dfrac{10}{-5} \geq \dfrac{-5x}{-5} > \dfrac{35}{-5}$

$-2 \geq x \geq -7$

67. $0 \leq \dfrac{y}{3} \leq 2$

$3 \cdot 0 \leq 3\left(\dfrac{y}{3}\right) \leq 3(2)$

$0 \leq y \leq 6$

69. $-5 < 2x - 1 < -1$

$-5 + 1 < 2x - 1 + 1 < -1 + 1$

$-4 < 2x < 0$

$\dfrac{-4}{2} < \dfrac{2x}{2} < \dfrac{0}{2}$

$-2 < x < 0$

71. $-2 \leq 4 - 3x \leq 4$

$-2 - 4 \leq 4 - 3x - 4 \leq 4 - 4$

$-6 \leq -3x \leq 0$

$\dfrac{-6}{-3} \geq \dfrac{-3x}{-3} \geq \dfrac{0}{-3}$

$2 \geq x \geq 0$

73. $-3x + 5 \geq x - 1$

$-3x + 5 + 3x \geq x - 1 + 3x$

$5 \geq 4x - 1$

$5 + 1 \geq 4x - 1 + 1$

$6 \geq 4x$

$\dfrac{6}{4} \geq \dfrac{4x}{4}$

$\dfrac{3}{2} \geq x$

75. $5x - 6 \geq 5x - 1$
$5x - 6 - 5x \geq 5x - 1 - 5x$
$-6 \geq -1$
$\varnothing$

77. $x + 5 \leq x + 7$
$x + 5 - x \leq x + 7 - x$
$5 \leq 7$
$\mathbb{R}$

79. $1 - 3(x + 2) < 7$
$1 - 3x - 6 < 7$
$-3x - 5 < 7$
$-3x - 5 + 5 < 7 + 5$
$-3x < 12$
$\dfrac{-3x}{-3} > \dfrac{12}{-3}$
$x > -4$

81. $2x + 1 > \dfrac{3x - 1}{2}$
$2(2x + 1) > 2\left(\dfrac{3x - 1}{2}\right)$
$4x + 2 > 3x - 1$
$4x + 2 - 3x > 3x - 1 - 3x$
$x + 2 > -1$
$x + 2 - 2 > -1 - 2$
$x > -3$

83. $2(x + 3) < 2x + 1$
$2x + 6 < 2x + 1$
$2x + 6 - 2x < 2x + 1 - 2x$
$6 < 1$
$\varnothing$

85. $2.3y + 0.75 \geq 4.2$
$2.3y + 0.75 - 0.75 \geq 4.2 - 0.75$
$2.3y \geq 3.45$
$\dfrac{2.3y}{2.3} \geq \dfrac{3.45}{2.3}$
$y \geq 1.5$

87. $0.57x + 27.4 > 65$
$0.57x + 27.4 - 27.4 > 65 - 27.4$
$0.57x > 37.6$
$\dfrac{0.57x}{0.57} > \dfrac{37.6}{0.57}$
$x > 65.96$ years from 1940. So, in 2006 and later, the percentage of employed women will exceed 65%.

89. $0.57x + 27.4 < -0.57x + 72.6$
$0.57x + 27.4 - 27.4 < -0.57x + 72.6 - 27.4$
$0.57x < -0.57x + 45.2$
$0.57x + 0.57x < -0.57x + 45.2 + 0.57x$
$1.14x < 45.02$
$\dfrac{1.14x}{1.14} < \dfrac{45.02}{1.14}$
$x < 39.49$ years after 1940.

91. $x + a < b$
$x + a - a < b - a$
$x < b - a$

93. $ax + b \leq 0, 9 < 0$
$ax + b - b \leq 0 - b$
$ax \leq -b$
$\dfrac{ax}{a} \leq \dfrac{-b}{a}$ since $a < 0$
$x \geq \dfrac{-b}{a}$

95. $x(2x - 5) > 2x^2 + 5x + 20$
$2x^2 - 5x > 2x^2 + 5x + 20$
$2x^2 - 5x - 2x^2 > 2x^2 + 5x + 20 - 2x^2$
$-5x > 5x + 20$
$-5x - 5x > 5x + 20 - 5x$
$-10x > 20$
$\dfrac{-10x}{-10} < \dfrac{20}{-10}$
$x < -2$

Algebra Basics, Equations, and Inequalities

97. $3 + x < 2x - 4 < x + 6$
$3 + x - x < 2x - 4 - x < 6 - x$
$3 < 2x - 4 < 6$
$3 + 4 < x - 4 + 4 < 6 + 4$
$7 < x < 10$

# Chapter 2 Review Exercises

1. (a) $-4(-6) + 7$
   $= 24 + 7$
   $= 31$

   (b) $-(1)^2 + 1$
   $= -1 + 1$
   $= 0$

3. $(-3, -8)$

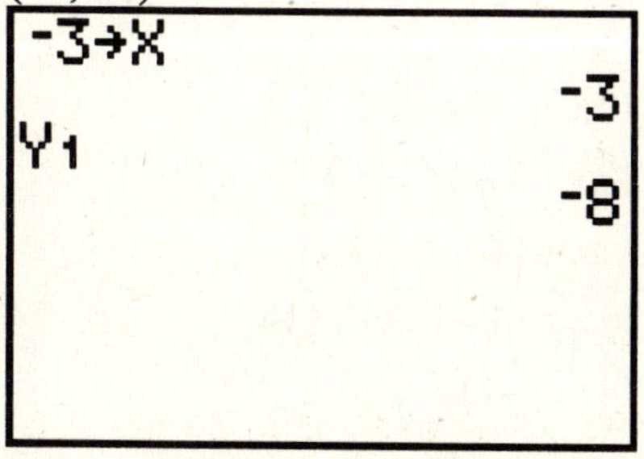

5. (a) $3xy + x - 5xy + 4x$
   $= (3xy - 5xy) + (y + 4x)$
   $= -2xy + 5x$

   (b) $8C^2 - 6C^2 + 5C^2$
   $= 2C^2 + 5C^2$
   $= 7C^2$

7. $-3(2a - b + 4)$
   $= -6a - 3(-b) + (-3)4$
   $= -6a + 3b - 12$

9. $-(x + 2y) - 4(3x - y + 1)$
   $= -x - 2y - 12x - 4(-y) - 4$
   $= -x - 2y - 12x + 4y - 4$
   $= (-x - 12x) + (-2y + 4y) - 4$
   $= -13x + 2y - 4$

11. $R(4, -1)$

13. Q

15. The set of all points with a y-coordinate of 0 is the $x$ – axis.

17. $+ (0, 0)$ is called the origin.

19. 

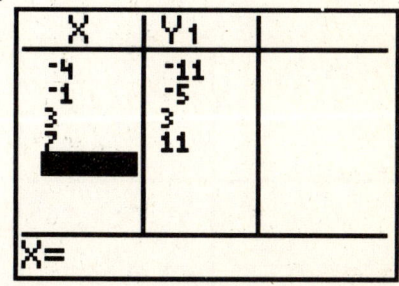

21. 

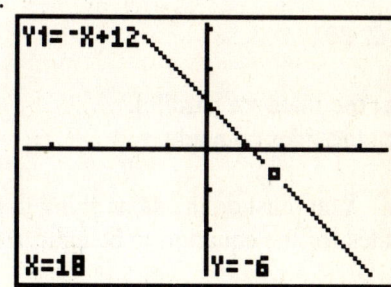

    $x = 18$

23. The y-coordinate will be $-3$ because the y-coordinate is the value of the expression when $x = 5$ and $7 - 2(5) = 7 - 10 = -3$

25. (ii) $x - 3 = 3 - x$
    $x - 3 + x = 3 - x + x$
    $2x - 3 = 3$
    $2x - 3 + 3 = 3 + 3$
    $2x = 6$
    $x = 3$
    conditional

27. $(-3)^2 + 1 = -3(-3) + 1$
    $9 + 1 = 9 + 1$
    $10 = 10$
    So $-3$ is a solution to the equation.

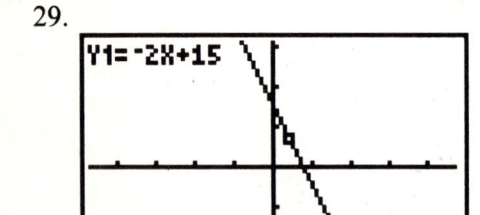

29.
    [Graphing calculator screen: Y1=-2X+15, X=4, Y=7]

31.
    [Graphing calculator screen: Y1=14-3X, X=-2, Y=20]

    $(-2, 20)$

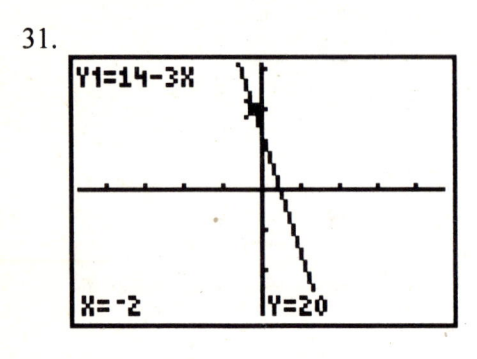

33. (a) the lines are parallel.
    (b) the lines coincide.

35. iii. You must do the same thing to both sides for the equation to be equivalent.

37. (ii) $2x + 3 = 8$
    $2x = 5$
    The variable term, not the variable is isolated.

39. $7 = 9x - 8x$
    $7 = x$

41. $-5x + 8 = -4x$
    $-5x + 8 + 5x = -4x + 5x$
    $8 = x$

43. $-\dfrac{x}{3} = -6$
    $-3\left(-\dfrac{x}{3}\right) = -3(-6)$
    $x = 18$

45. $9 = 2x - 6$
    $9 + 6 = 2x - 6 + 6$
    $15 = 2x$
    $\dfrac{15}{2} = \dfrac{2x}{2}$
    $x = \dfrac{15}{2}$

47. $6 + 3x - 4 - 7x - 2 = 6 - 5x - 10$
    $(3x - 7x) + (6 - 4 - 2) = -5x + 6 - 10$
    $-4x + 0 = -5x - 4$
    $-4x + 5x = -5x - 4 + 5x$
    $x = 4$

49. $\dfrac{4}{3}x + \dfrac{14}{9} = \dfrac{x}{9} + \dfrac{1}{3}$
    $9\left(\dfrac{4}{3}x + \dfrac{14}{9}\right) = 9\left(\dfrac{x}{9} + \dfrac{1}{3}\right)$
    $12x + 14 = x + 3$
    $12x - x + 14 = x + 3 - x$
    $11x + 14 = 3$
    $11x + 14 - 14 = 3 - 14$
    $11x = -11$
    $\dfrac{11x}{11} = \dfrac{-11}{11}$
    $x = -1$

51. $2(x - 3) - 3(x + 1) = 0$
    $2x - 6 - 3x - 3 = 0$
    $-x - 9 = 0$
    $-x - 9 + x = 0 + x$
    $-9 = x$

53. $3x + 7 - x = 4 + 2x + 3$
    $2x + 7 = 2x + 7$
    $\mathbb{R}$

55. $-3(x - 4) = 4x - (7x - 13)$
    $-3x + 12 = 4x - 7x + 13$
    $-3x + 12 = -3x + 13$
    $-3x + 12 + 3x = -3x + 13 + 3x$
    $12 = 13$
    $\varnothing$

57. (iii)

Algebra Basics, Equations, and Inequalities     67

59. The point of intersection represents a solution for $\leq$ or $\geq$ inequalities.

61. (a) $x \leq 2$

    (b) $-1 < x \leq 3$

63. (a) $8 + x > x - 6$
$8 + x - x > -6 - x$
$8 > -6$
This inequality is always true.

    (b) $8 + x < x - 6$
This inequality is always false.

65. (iii) since $7 - x > 0$
$7 - x + x > 0 + x$
$7 > x$
$x < 7$

67. $-8 \geq 3 - x$
$-8 - 3 \geq 3 - x - 3$
$-11 \geq -x$
$(-1)(-11) \leq (-1)(-x)$
$11 \leq x$

69. $-\dfrac{3}{4} x \leq 15$
$-\dfrac{4}{3}\left(-\dfrac{3}{4} x\right) \geq -\dfrac{4}{3}(15)$
$x \geq -20$

71. $-3(x + 1) < -(x - 1)$
$-3x - 3 < -x + 1$
$-3x - 3 + 3 < -x + 1 + 3$
$-3x < -x + 4$
$-3x + x < -x + x + 4$
$-2x < 4$
$\dfrac{-2x}{-2} > \dfrac{4}{-2}$
$x > -2$

73. $-12 < -4x \leq 16$
$\dfrac{-12}{-4} > \dfrac{-4x}{-4} \geq \dfrac{16}{-4}$
$3 > x \geq -4$

75. $3(x-1) - (x+1) \leq 2(x+4)$
$3x - 3 - x - 1 \leq 2x + 8$
$2x - 4 \leq 2x + 8$
$2x - 4 - 2x \leq 2x + 8 - 2x$
$-4 \leq 8$
$\mathbb{R}$

# Chapter 2 Test

1. $\dfrac{2x}{y} - (x + y); x = 2, y = -4$

$= \dfrac{2(2)}{-4} - \left[2 + (-4)\right]$

$= \dfrac{4}{-4} - (-2)$

$= -1 + 2$

$= 1$

3. $-(x - 1) + 4(2x + 3)$
$= -x + 1 + 8x + 12$
$= 7x + 13$

5. In Quadrant II, $x < 0$ and $y > 0$.

7. $3(-4) - 2$
$= -12 - 2$
$= -14$
The y-coordinate of a point of the graph of an expression corresponds to the value of the expression. If $x$ is replaced with $-4$, the value of the expression is $-14$, so $y = -14$.

9.

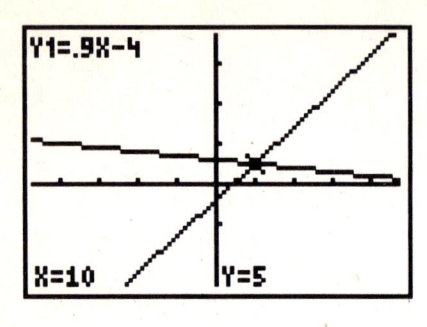

[Graphing calculator screen showing Y1=.9X-4, X=10, Y=5]

$x = 10$

11. (a) identity

(b) contradiction

13. $14x - 20 = 1 + 13x$
$14x - 20 + 20 = 1 + 13x + 20$
$14x = 13x + 21$
$14x - 13x = 13x + 21 - 13x$
$x = 21$

15. $\frac{2}{3}(x-5) = x + 2$
$\frac{3}{2}\left[\frac{2}{3}(x-5)\right] = \frac{3}{2}(x+2)$
$x - 5 = \frac{3}{2}x + 3$
$x - 5 - x = \frac{3}{2}x + 3 - x$
$-5 = \frac{1}{2}x + 3$
$-5 - 3 = \frac{1}{2}x + 3 - 3$
$-8 = \frac{1}{2}x$
$2(-8) = 2\left(\frac{1}{2}x\right)$
$-16 = x$

17. $5 - 5(1-x) = 3 + 5x$
$5 - 5 + 5x = 3 + 5x$
$5x = 3 + 5x$
$0 = 3$
$\emptyset$

19. $x - 3 \geq -2$

21. The line $y = 9-x$ is always above the graph of $y = -11 - x$.

(a) So $9 - x \leq -11 - x$ is never true. The solution set is $\phi$

(b) $9 - x > -11 - x$ is true for all values of $x$. The solution set is $\mathbb{R}$

23. $x + \frac{1}{4} \geq \frac{-4x}{5} - 1.55$
$20\left(x + \frac{1}{4}\right) \geq 20\left(\frac{-4x}{5} - 1.55\right)$
$20x + 5 \geq -16x - 31$
$20x + 5 + 16x \geq -16x - 31 + 16x$
$36x + 5 \geq -31$
$36x + 5 - 5 \geq -31 - 5$
$36x \geq -36$
$\frac{36x}{36} \geq \frac{-36}{36}$
$x \geq -1$

25. $-4 \leq 2(x-7) < 0$
$\frac{-4}{2} \leq \frac{2(x-7)}{2} < 0$
$-2 \leq x - 7 < 0$
$-2 + 7 \leq x - 7 + 7 < 0 + 7$
$5 \leq x < 7$

# Chapter 3

# Modeling and Applications

## Section 3.1 Ratio and Proportion

1. A ratio is a quotient used to compare two numbers.

3. $\dfrac{12}{8} = \dfrac{3}{2}$

5. $\dfrac{600}{780} = \dfrac{10}{13}$

7. $\dfrac{25}{5000} = \dfrac{1}{200}$

9. $\dfrac{173}{42}$

11. Strikes = 120 − 55 = 65

    Balls to strikes: $\dfrac{55}{65} = \dfrac{11}{13}$

13. Area $= \pi r^2 = \pi(16)^2 = 256\pi$

    Circumference $= 2\pi r = 2\pi(16) = 32\pi$

    Area to circumference: $\dfrac{256\pi}{32\pi} = 8$

15. (a) $\dfrac{350}{500+250} = \dfrac{350}{750} = \dfrac{7}{15}$

    (b) $\dfrac{500}{350} = \dfrac{10}{7}$

    (c) $\dfrac{250}{500} = \dfrac{1}{2}$

17. (a) $\dfrac{3+5}{10} = \dfrac{8}{10} = \dfrac{4}{5}$

    (b) $\dfrac{10}{3+2} = \dfrac{10}{5} = 2$

    (c) $\dfrac{3+5+10}{3+5+10+3+2} = \dfrac{18}{23}$

19. $\dfrac{70}{50} = \dfrac{7}{5}$

21. 2 pints = 1 quart so 4 pints = 2 quarts

    $\dfrac{3 \text{ pints}}{4 \text{ pints}} = \dfrac{3}{4}$

23. 1 yard = 3 feet so 5 yards = 15 feet

    $\dfrac{15 \text{ feet}}{20 \text{ feet}} = \dfrac{3}{4}$

25. 1 dollar = 4 quarters so 3 dollars = 12 quarters

    $\dfrac{12 \text{ quarters}}{15 \text{ quarters}} = \dfrac{4}{5}$

27. The unit price is the price charged for each unit, such as weight, volume, or length, in which the item is sold.

29. $\dfrac{55 \text{ cents}}{8 \text{ ounces}} = 6.88$ per ounce

31. $\dfrac{\$1.99}{24 \text{ plates}} = \$0.829$ per plate

    $\approx 8$ cents per plate

33. $\dfrac{\$223.44}{12 \text{ sq yards}} = \$18.62 \text{ per yard}$

35. $\dfrac{\$7.60}{60 \text{ tablets}}$
 $= \$0.1267 \text{ per tablet}$
 $\approx 13 \text{ cents per tablet}$

37. (i) $\dfrac{\$0.81}{14 \text{ ounces}} = \$0.058 \text{ per ounce}$

 (ii) $\dfrac{\$1.59}{28 \text{ ounces}} = \$0.057 \text{ per ounce}$

 (iii) $\dfrac{\$3.46}{64 \text{ ounces}} = \$0.054 \text{ per ounce}$

 The 64-ounce size is most economical. It costs 5.4 cents per ounce.

39. (i) $\dfrac{\$2.97}{7 \text{ ounces}} = \$0.424 \text{ per ounce}$

 (ii) $\dfrac{\$3.88}{11 \text{ ounces}} = \$0.353 \text{ per ounce}$

 (iii) $\dfrac{\$5.69}{15 \text{ ounces}} = \$0.379 \text{ per ounce}$

 The 11-ounce size is the most economical with a unit price of 35.3 cent per ounce.

41. A proportion is an equation stating two ratios are equal.

43. $\dfrac{x}{32} = \dfrac{7}{8}$
 $x \cdot 8 = 32 \cdot 7$
 $8x = 224$

 $x = \dfrac{224}{8}$
 $= 28$

45. $\dfrac{16}{9} = \dfrac{4}{x}$
 $16 \cdot x = 9 \cdot 4$
 $16x = 36$
 $x = \dfrac{36}{16}$
 $x = \dfrac{9}{4}$

47. $\dfrac{x+5}{3} = \dfrac{5}{2}$
 $(x+5) \cdot 2 = 3 \cdot 5$
 $2x + 10 = 15$
 $2x = 5$
 $x = \dfrac{5}{2}$

49. $\dfrac{2-x}{2} = \dfrac{x+3}{4}$
 $(2-x) \cdot 4 = 2(x+3)$
 $8 - 4x = 2x + 6$
 $-4x = 2x - 2$
 $-6x = -2$
 $x = \dfrac{-2}{-6}$
 $x = \dfrac{1}{3}$

51. $\dfrac{4}{r-3} = \dfrac{5}{r-2}$
 $4(r-2) = 5(r-3)$
 $4r - 8 = 5r - 15$
 $-8 = r - 15$
 $7 = r$

53. $\dfrac{r+3}{r-2} = \dfrac{3}{4}$
 $(r+3) \cdot 4 = 3(r-2)$
 $4r + 12 = 3r - 6$
 $r + 12 = -6$
 $r = -18$

Modeling and Applications

55. $\dfrac{3}{2} = \dfrac{2x+1}{x-5}$
$3(x-5) = 2(2x+1)$
$3x - 15 = 4x + 2$
$-15 = x + 2$
$-17 = x$

57. $\dfrac{5}{8} = \dfrac{x}{6}$
$5 \cdot 6 = 8x$
$30 = 8x$
$\dfrac{30}{8} = x$
$x = \dfrac{15}{4} = 3.75$

59. $\dfrac{5}{6} = \dfrac{x-1}{x+1}$
$5(x+1) = 6(x-1)$
$5x + 5 = 6x - 6$
$5 = x - 6$
$x = 11$

The desired lengths are $x - 1 = 11 - 1 = 10$ and $x + 1 = 11 + 1 = 12$.

61. $\dfrac{95}{5} = \dfrac{x}{3}$
$95 \cdot 3 = 5 \cdot x$
$285 = 5x$
$x = \dfrac{285}{5}$
$x = 57¢$

63. $\dfrac{2.25}{4} = \dfrac{5}{x}$
$2.25x = 4 \cdot 5$
$x = \dfrac{20}{2.25}$
$= 8.8$ ounces

65. $\dfrac{8 \text{ oz}}{110 \text{ calories}} = \dfrac{6 \text{ oz}}{x \text{ calories}}$
$8x = 110(6)$
$8x = 660$
$x = \dfrac{660}{8}$
$= 82.5$ calories

67. $\dfrac{2}{24} = \dfrac{x}{45}$ since $\dfrac{3}{4}$ hour = 45 minutes
$8(45) = 24x$
$360 = 24x$
$x = \dfrac{360}{24}$
$= 15$ miles

$6(1200) = 1000x$
$x = 7.2$ hours.

69. (a) $\dfrac{1}{3} = \dfrac{2.2}{x}$
$1 \cdot x = 3(2 \cdot 2)$
$x = 6.6$ feet

(b) $\dfrac{1}{3} = \dfrac{x}{21}$
$1 \cdot 21 = 3x$
$21 = 3x$
$x = 7$ inches

71. $\dfrac{6}{1} = \dfrac{120}{x}$
$6x = 120$
$x = 20$ pounds

73. Let $x$ = # of people who voted for the winner. Then $1320-x$ people voted for the loser.

$$\frac{3}{2} = \frac{x}{1320-x}$$
$$3(1320-x) = 2 \cdot x$$
$$3960 - 3x = 2x$$
$$3960 = 5x$$
$$x = 792 \text{ votes}$$

75. $\dfrac{4}{3} = \dfrac{x}{140-x}$ where $x$ is the number of doctors who recommend aspirin and $140-x$ is the number who do not.

$$4(140-x) = 3x$$
$$560 - 4x = 3x$$
$$560 = 7x$$
$$\frac{560}{7} = 7x$$
$$x = 80 \text{ doctors}$$

77. $\dfrac{250}{175} = \dfrac{20,000}{x}$

$$250x = 175(20,000)$$
$$250x = 3,500,000$$
$$x = \frac{3,500,000}{250}$$
$$x = 14,000 \text{ households}$$

79. (i) $\dfrac{x}{y} = \dfrac{3}{5}$

$$5x = 3y$$
$$x = \frac{3}{5}y$$

(ii) $x + y = 168$

$$\left(\frac{3}{5}y\right) + \frac{5}{5}y = 168$$
$$\frac{3}{5}y + \frac{5}{5}y = 168$$
$$\left(\frac{8}{5}\right)y = 168$$
$$y = 168 \cdot \frac{5}{8}$$
$$y = 105$$

$$x = \frac{3}{5}y$$
$$= \frac{3}{5}(105)$$
$$= 63$$

So, $x = 63$ and $y = 105$.

81. Let $x$ = the amount of material required for 6 windows.

The amount required for 8 windows is $x + 7$.

$$\frac{x}{6} = \frac{x+7}{8}$$
$$8x = 6(x+7)$$
$$8x = 6x + 42$$
$$2x = 42$$
$$x = 21 \text{ yards}$$

Let $y$ = the amount required for 9 windows.

$$\frac{x}{6} = \frac{21}{6} = \frac{y}{9}$$
$$21(9) = 6y$$
$$189 = 6y$$
$$6y = \frac{189}{6}$$
$$y = 31.5 \text{ yards}$$

83. Let $x$ = amount of paint for 2 rooms. The amount of paint for 5 rooms is $x + 4.5$.

$$\frac{x}{2} = \frac{x+4.5}{5}$$
$$5x = 2(x+4.5)$$
$$5x = 2x + 9$$
$$3x = 9$$
$$x = 3$$
$$\frac{3}{2} = \frac{y}{7}$$
$$3(7) = 2y$$
$$21 = 2y$$
$$y = 10.5$$

So 10.5 gallons are needed to paint 7 rooms.

85. $\frac{17}{20} = \frac{x}{2800}$
$17(2800) = 20x$
$47600 = 20x$
$x = 2380$ students

87. $\frac{104}{100,000} = \frac{1}{x}$
$104x = 100,000$
$x = 962$ golfers

89. $\frac{904}{100,000} = \frac{x}{1000}$
$904(1000) = 100,000x$
$x = 9.04$
$= 9$ serious injuries

91.

| | |
|---|---|
| $\frac{1.25}{16} = 0.078$ | $\frac{0.75}{16} = 0.047$ |
| $\frac{2.39}{32} = 0.075$ | $\frac{1.89}{32} = 0.059$ |
| $\frac{3.49}{48} = 0.073$ | $\frac{2.99}{48} = 0.062$ |

Without the coupon, the 48-ounce size is the best value. With the coupon, the 16-ounce is the best value.

93. Let $x$ be the number of town homes. There are $x$ condominiums and $1000 - (x + x) = 1000 - 2x$ individual homes.

$$\frac{1000 - 2x}{1000} = \frac{3}{5}$$
$$5(1000 - 2x) = 3(1000)$$
$$5000 - 10x = 3000$$
$$2000 - 10x = 0$$
$$2000 = 10x$$
$$x = 200$$

There are 200 townhouses.

95. $\frac{x+y}{x} = \frac{y-x}{y}$

$(x+y)y = x(y-x)$
$xy + y^2 = xy - x^2$
$y^2 = -x^2$

Since both $x^2$ and $y^2$ are nonnegative, $-x^2 = y^2$ only when $x = y = 0$. But both $x$ and $y$ are nonzero, so no values of $x$ and $y$ satisfy the equation.

## Section 3.2 Percents

1. Percent means per hundred

3. (a) 0.28

   (b) $\dfrac{28}{100} = \dfrac{7}{25}$

5. (a) 2.35

   (b) $\dfrac{2.35}{100} = \dfrac{47}{20}$

7. (a) 0172

   (b) $\dfrac{17.2}{100} = \dfrac{43}{250}$

9. (a) 5 ¾ %
   = 5.75 %
   = 0.0575

   (b) $\dfrac{5.75}{100} = \dfrac{575}{10,000} = \dfrac{23}{400}$

11. 37.5 %

13. 55 %

15. 690 %

17. 0.02 %

19. $\dfrac{7}{8} = 0.875$
    = 87.5 %

21. $\dfrac{7}{15} = 0.46\overline{6}$
    = $46.\overline{6}$ %

23. 1 ¾ = 1.75
    = 175 %

25. $\dfrac{12}{100} \cdot .658$
    $\dfrac{7896}{100}$
    = 78.96

27. $0.03\, x = 5.4$
    $\dfrac{3}{100} x = \dfrac{540}{100}$
    $\dfrac{3x}{100} = \dfrac{540}{100}$
    $3x = 540$
    $x = 180$

29. $n \cdot 125 = 85$
    $n = \dfrac{85}{125}$
    = 0.68
    = 68%

31. $0.19\, x = 57$
    $x = \dfrac{57}{0.19}$
    = 300

33. $1.20\,(345) = 414$

35. $n \cdot 1600 = 128$
    $n = \dfrac{128}{1600}$
    = 0.08
    = 8%

37. $0.06\overline{3}\,(9600) = 608$

Modeling and Applications

39. $n \cdot 340 = 425$

$n = \dfrac{425}{340}$

$= 1.25$

$= 125\%$

41. $0.025 x = 15$

$x = \dfrac{15}{0.0025}$

$= 6000$

43. The student's result, 1.2, is the amount of the increase. The answer is found by adding this amount to 10.

45. $1.34 (575) = 770.5$

47. $(1 + 1.40)(150)$
$= 2.4 (150)$
$= 360$

49. $(1 - 0.48)(230)$
$= (0.52)(230)$
$= 119.6$

51. $(1 - 0.005) 1600$
$= (0.995) 1600$
$= 1592$

53. (a) $\dfrac{5}{4} = 1.25$

25 percent increase

(b) $\dfrac{4}{5} = 0.80$

$= 1 - 0.20$

20 % decrease

55. (a) $\dfrac{253}{230} = 1.10$

$= 1 + 0.10$

10 % increase

(b) $\dfrac{230}{253} = 0.9090$

$= 1 - 0.0909$

$9.0\overline{9}$ % decrease

57. Statement (ii) is true. In statement (i), the percentage is correct, but we don't know that the club has 100 members. Statement (iii) is not true because 40 is not 60 % of 70.

59. $\dfrac{35,000}{28,000} = 1.25$

25 % increase

61. $1.15 (14) = 16.1$ inches

63. $(1 - 0.20)x = 2452$
$0.80x = 2452$
$x = \dfrac{2452}{0.80}$
$= 3065$

65. $0.015 (3650) = 54.75$
$54 < 54.75$
The lot should not be rejected.

67. $(0.28 + 0.085) \cdot 24,300$
$= 0.365 (24,300)$
$= \$8869.50$

69. $(1 - 0.2)x = 11,221$
$0.98x = 11,221$
$x = \dfrac{11,221}{0.98}$
$= 11,450$

71. (a) $\dfrac{16.8}{17.5} = 0.96$

$= 1 - 0.04$

4 % decrease

(b) 96 %

73. (a) $\dfrac{3}{5} = 60\%$

United States, Japan, France, Germany

(b) $(0.80)(35 \text{ million})$
$= 28$ million children

75. 100 %

77. No. We know only percentages, not actual numbers of homebuyers. The numbers may be close. For example, if there are 100 homes sold then $0.41(100) = 41$ are in $60k+ households and $0.43(100) = 43$ are in $50k or less households.

However, if there are 100,000 homes sold, $0.41(100,000) = 41,000$ are in $60k+ households and $0.43(100,000) = 43,000$ are in $50k or less households.

$41 \approx 43$ but $41,000 \not\approx 43,000$

79. (i) $1.04(2000)$
    $= \$2080$

    $1.07(2080)$
    $= \$2225.60$

(ii) $1.03(2000)$
    $= \$2060$

    $1.08(2060)$
    $= \$2224.80$
    Option (i) is better.

81. $\frac{1}{4}(3\% x) = 900$

    $0.25(0.03x) = 900$

    $0.0075x = 900$

    $x = \frac{900}{0.0075}$

    $= \$120,000$

## Section 3.3 Formulas

1. A formula is an equation that serves as instructions for calculating a certain quantity.

3. $I = Prt$
   $= 250\,(0.07)\,(1)$
   $= \$175$

5. $P = 2L + 2W$
   $= 2(10) + 2(3)$
   $= 20 + 6$
   $= 26$ feet

   $A = LW$
   $= 10 \cdot 3$
   $= 30$ square feet

7. $C = 2\pi r$
   $= 2\pi(3.2)$
   $\approx 20.11$ meters

   $A = \pi r^2$
   $= \pi(3.2)^2$
   $\approx 32.17$ square meters

9. $F = \frac{9}{5}C + 32$

   (a) $F = \frac{9}{5}(10) + 32$
       $= 18 + 32$
       $= 50$

   (b) $F = \frac{9}{5}(32) + 32$
       $= 57.6 + 32$
       $= 89.6$

11. $d = rt$

    (a) $d = 55 \cdot 3$
        $= 165$

    (b) $d = 16 \cdot 4$
        $= 64$

13. $V = LWH$

    (a) $V = 5 \cdot 3 \cdot 2$
        $= 30$

(b) $V = 7.5 \cdot 2.1 \cdot 1.6$
$= 25.2$

15. $A = \dfrac{1}{2}bh$

(a) $A = \dfrac{1}{2}(12.3)(6.1)$
$= 37.515$

(b) $A = \dfrac{1}{2}(8)(12)$
$= 48$

17. $A = \pi r^2$

(a) $A = \pi(5)^2$
$\approx 78.54$

(b) $A = \pi(3.7)^2$
$\approx 43.01$

19. The hypotenuse is the longest side of a right triangle. It is the side opposite the right angle.

21. $6^2 + 9^2 = 12^2$
$36 + 81 = 144$
$117 = 144$
No

23. $8^2 + 15^2 = 17^2$
$64 + 225 = 289$
$289 = 289$
Yes

25. Yes

```
            28
35→C
            35
Y₁
          1225
Y₂
          1225
```

27. No

```
             4
4.9→C
            4.9
Y₁
          25.61
Y₂
          24.01
```

29. $r = \dfrac{d}{t}$
$d = rt$
$r = \dfrac{50 \text{ km}}{\dfrac{3}{4} \text{ hour}}$
$= \dfrac{50}{0.75}$ km per hour
$= 66.\overline{6}$ km/hr

31. $h = 100t - 16t^2$

(a) $h = 100(1) - 16(1)^2$
$= 100 - 16$
$= 84$ feet

(b) $h = 100(3) - 16(3)^2$
$= 300 - 16(9)$
$= 300 - 144$
$= 156$ feet

(c) $(h = 100(6.25) - 16(6.25)^2$
$= 625 - 625$
$= 0$ feet

33. The visible height of the post is $15 - 3 = 12$ feet. The cable forms a triangle with the post and the ground. Since $5^2 + 12^2 = 25 + 144 = 169 = 13^2$, the triangle is a right triangle and the post is perpendicular to the ground.

35. The procedure is exactly the same.

37. $P = 4s$

$\dfrac{P}{4} = s$

$s = \dfrac{1}{4}P$

39. $A = LW$

$\dfrac{A}{W} = L$

$L = \dfrac{A}{W}$

41. $I = Prt$

$\dfrac{I}{rt} = P$

$P = \dfrac{I}{rt}$

43. $C = \pi d$

$\dfrac{C}{\pi} = d$

$d = \dfrac{1}{\pi}C$

45. $C = \dfrac{5}{9}(F - 32)$

$\dfrac{9}{5}C = \dfrac{9}{5}\left[\dfrac{5}{9}(F - 32)\right]$

$\dfrac{9}{5}C = F - 32$

$\dfrac{9}{5}C + 32 = F$

$F = \dfrac{9}{5}C + 32$

47. $A = \dfrac{1}{2}bh$

$2A = 2\left(\dfrac{1}{2}bh\right)$

$2A = bh$

$\dfrac{2A}{h} = b$

$b = \dfrac{2A}{h}$

49. $A = \dfrac{1}{2}(a + b)$

$2A = a + b$

$2A - a = b$

$b = 2A - a$

51. $x + y = 8$

$y = 8 - x$

$\phantom{y} = -x + 8$

53. $5x - y = -7$

$5x = 7 + y$

$5x - 7 = y$

$y = 5x - 7$

55. $7x + 2y = 10$

$2y = -7x + 10$

$y = \dfrac{1}{2}(-7x + 10)$

$y = \dfrac{-7}{2}x + 5$

57. $4y - 3x + 4 = 0$

$4y + 4 = 3x$

$4y = 3x - 4$

$y = \dfrac{1}{4}(3x - 4)$

$y = \dfrac{3}{4}x - 1$

59. $y - 3 = 2(x - 4)$

$y - 3 = 2x - 8$

$y = 2x - 8 + 3$

$y = 2x - 5$

Modeling and Applications

61. $\dfrac{3}{4}x - \dfrac{2}{5}y = \dfrac{1}{2}$

$20\left(\dfrac{3}{4}x - \dfrac{2}{5}y\right) = 20\left(\dfrac{1}{2}\right)$

$15x - 8y = 10$

$15x = 8y + 10$

$15x - 10 = 8y$

$\dfrac{1}{8}(15x - 10) = y$

$y = \dfrac{15}{8}x - \dfrac{10}{8}$

$y = \dfrac{15}{8}x - \dfrac{5}{4}$

63. $V = LWH$

$\dfrac{V}{WH} = L$

$L = \dfrac{V}{WH}$

(a) $L = \dfrac{1200}{15 \cdot 8}$

$= 10$ feet

(b) $L = \dfrac{1200}{10 \cdot 10}$

$= 12$ fee

(c) $L = \dfrac{1200}{\left(8 + \dfrac{3}{12}\right)\left(9 + \dfrac{2}{12}\right)}$

$= \dfrac{1200}{(8.25)(9.1\overline{6})}$

$= 15 \cdot 87$ feet

65. (a) $R = W + rW$

$= 78 + 0.32(78)$

$= \$102.96$

(b) $R = W + rW$

$R - W = rW$

$\dfrac{R - W}{W} = r$

$r = \dfrac{R - W}{W}$

$r = \dfrac{448.50 - 390}{390}$

$= \dfrac{58.50}{390}$

$= 0.15$

$= 15\%$

67. $F = 1.8C + 32$

(a) $F = 1.8(31) + 32$

$= 88°$

(b) $F = 1.8(17) + 32$

$= 63°$

(c) $F = 1.8(-6) + 32$

$= 21°$

(d) $F = 1.8(-13) + 32$

$= 9°$

(e) $F = 1.8(26) + 32$

$= 79°$

(f) $F = 1.8(18) + 32$

$= 64°$

69. $ax + by = c$

$by = -ax + c$

$y = \dfrac{-ax + c}{b}$

71. $m = \dfrac{y-b}{x}$

$mx = y - b$

$mx + b = y$

$y = mx + b$

73. $y - a = m(x - b)$

$\dfrac{y-a}{x-b} = m$

$m = \dfrac{y-a}{x-b}$

75. $A = P + Prt$

$A = P(1 + rt)$

$\dfrac{A}{1+rt} = P$

$P = \dfrac{A}{1+rt}$

77. $E = IR + Ir$

$E = I(R + r)$

$\dfrac{E}{R+r} = I$

$I = \dfrac{E}{R+r}$

## Section 3.4 Translation

1. The equation may not be correct. The answer should be checked against the wording of the problem.

3. $x - (-7)$
$= x + 7$

5. $\dfrac{x}{2}$

7. $2x - 2$

9. $\dfrac{1}{4}x = 8$

11. $6x - 1 = 11$

13. $2x - 1 = x + 5$

15. $3x - 4$

17. $x - 5 = -8$

19. $2 - 3x$

21. $2x + 5 = 9$

23. $x - \dfrac{1}{2}$ since 6 inches $= \dfrac{1}{2}$ foot

25. $3W$

27. $x, x + 8$

29. $x, x + 7$

31. $x, \dfrac{1}{2}x + 5$

33. The angles are supplementary angles because the sum of their measures is $x + (180 - x) = 180$.

35. $180 - x$

37. $130 - m$

39. $x, 90 - x$

41. $x, 90 - x$

43. $n + 1$

45. $x + 1, x + 3$  Since $x$ is odd, $x + 1$ is even.

47. $x + (x + 1) + (x + 2)$
$= 3x + 3$

49. $x + (x + 2)$
$= 2x + 2$

51. $0.3x$

53. $0.92q$

Modeling and Applications

55. $18d$

57. $60t$

59. $2500y$

61. $x + 0.25x$
$= 1.25x$

63. $x - 0.34x$
$= 0.66x$

65. $x + 0.05x$
$= 1.05x$

67. $2x - 40$ miles driven in the morning
$x$ miles driven in the afternoon
$2x - 40 + x$
$= 3x - 40$

69. $x =$ length
$x - 7 =$ width
$2(x) + 2(x - 7)$
$= 2x + 2x - 14$
$= 4x - 14$

71. $35 + 0.3x$, $x =$ number of miles

73. $x =$ part invested at 7%
$5000 - x =$ part invested at 8%

$0.07x + 0.08(5000 - x)$
$= 0.07x + 400 - 0.08x$
$= -0.01x + 400$

75. $n =$ number of nickels
$30 - n =$ number of dimes

$5n + 10(30 - n)$
$= 5n + 300 - 10n$
$= -5n + 300$

77. $x =$ number with 25% acid
$50 - x$ number with 40% acid

$0.25x + 0.4(50 - x)$
$= 0.25x + 20 - 0.4x$
$= -0.15x + 20$

79. $x =$ pounds of grapes
$20 - x =$ pounds of strawberries

$1.20x + 1.50(20 - x)$
$= 1.20x + 30 - 1.5x$
$= -0.3x + 30$

81. $x =$ number of hours at \$6/hour
$40 - x =$ number of hours at \$8/hour

$6x + 8(40 - x)$
$= 6x + 320 - 8x$
$= -2x + 320$

83. (a) $16 + \dfrac{1}{3}x$

(b) $20 = 16 + \dfrac{1}{3}x$

85. $x^2 - 7^2$
$= x^2 - 49$

87. $(n+3)^2$

89. $5.5x + 1.25y$

## Section 3.5 Modeling and Problem Solving

1. The equation may not be correct. The answer should be checked against the wording of the problem.

3. $7 - x = -3$
$7 = -3 + x$
$7 + 3 = -3 + x + 3$
$10 = x$

5. $3x - 5 = 4$
   $3x = 9$
   $x = 3$

7. $3x + 4 \leq 37$
   $3x \leq 33$
   $x \leq 11$

9. $2(x - 7) = 14$
   $x - 7 = 7$
   $x = 14$

11. $\dfrac{2}{3}x - 5 = x - 9$
    $\dfrac{2}{3}x = x - 4$
    $-\dfrac{1}{3}x = -4$
    $-3\left(-\dfrac{1}{3}x\right) = -3(-4)$
    $x = 12$

13. $x + (x + 7) \leq 25$
    $2x + 7 \leq 25$
    $2x \leq 18$
    $x \leq 9$
    If $x \leq 9$ then $x + 7 \leq 16$.

15. $x =$ first number (smaller)
    $36 - x =$ second number (larger)
    $(36 - x) - 3x = 4$
    $36 - 4x = 4$
    $-4x = -32$
    $x = 8$ and
    $36 - x = 36 - 8 = 28$

    **The two numbers are 8 and 28.**

17. $x =$ larger number
    $x - 5 =$ smaller number
    $(x - 5) + 4x = -10$
    $4x - 5 = -10$
    $5x = -5$
    $x = -1$ and
    $x - 5 = -1 - 5 = 6$

    **The two numbers are −1 and −6.**

19. $x + (x + 2) = -32$
    $2x + 2 = -32$
    $2x = -32$
    $x = -17$ and
    $x + 2 = -17 + 2 = -15$

    **The two numbers are −15 and −17.**

21. $x + (x + 2) + (x + 4) = 102$
    $3x + 6 = 102$
    $3x = 96$
    $x = 32$ and
    $x + 2 = 34$ and
    $x + 4 = 36$

    **The numbers are 32, 34, and 36.**

23. $x =$ first number
    $x + 1 =$ second number
    $x + 2 =$ third number
    $x + (x + 1) = 3(x + 2) - 5$
    $2x + 1 = 3x + 6 - 5$
    $2x + 1 = 3x + 1$
    $2x = 3x$
    $x = 0$

    **The numbers are 0, 1, and 2.**

Modeling and Applications

25. $x$ = first number
$x + 2$ = second number
$x + 4$ = third number
$2x + [(x+2)+(x+4)] = 114$
$2x + 2x + 6 = 114$
$2x + 6 = 114$
$4x = 108$
$x = 27$

The numbers are 27, 29, and 31.

27. $x$ = first piece
$4x$ = second piece
$2x$ = third piece
$x + 4x + 2x = 35$
$7x = 35$
$x = 5$

The pieces are 5 feet, 20 feet, and 10 feet long.

29. $x$ = first piece
$x + 15$ = second piece
$2x + 5$ = third piece
$x + (x+15) + (2x+5) = 100$
$4x + 20 = 100$
$4x = 80$
$x = 20$

The pieces are 20 yards, 35 yards, and 45 yards long.

31. Because the sum of the measures of the three angles is 180° and the measure of the right angle is 90°, the sum of the measures of the other two is 90°. The other two angles are the same size so each has a measure of 45°.

33. $x$ = first angle measure
$2x$ = second angle measure
$2x + 30$ = third angle measure

$x + 2x + (2x+30) = 180$
$5x + 30 = 180$
$5x = 150$
$x = 30$

The angles are 30°, 60°, and 90°.

35. $2x$ = measure of first angle
$x$ = measure of second angle
$2x - 20$ = measure of third angle

$2x + x + (2x-20) = 180$
$5x - 20 = 180$
$5x = 200$
$x = 40$

So, the angles are 80°, 40°, and 60°.

37. $30 = 2l + 2w$ and $l = 2w + 3$
$30 = 2(2w+3) + 2w$
$30 = 4w + 6 + 2w$
$30 = 6w + 6$
$24 = 6w$
$w = 4$ meters
$l = 2(4) + 3 = 11$ meters

39. $w = \dfrac{3}{4}l$

$2w + 2l \leq 70$

$2\left(\dfrac{3}{4}l\right) + 2l \leq 70$

$\dfrac{3}{2}l + 2l \leq 70$

$3.5l \leq 70$

$l \leq 20$

$w \leq \dfrac{3}{4}(20)$

$w \leq 15$

The width is less than 15 feet and the length is less than 20 feet.

41. $90 - x = \dfrac{1}{4}(180 - x)$

$90 - x = 45 - \dfrac{1}{4}x$

$45 - x = -\dfrac{1}{4}x$

$45 = \dfrac{3}{4}x$

$\dfrac{4}{3}(45) = \dfrac{4}{3}\left(\dfrac{3}{4}x\right)$

$60 = x$

$x = 60°$

43. $(90 - x) + (180 - x) = 170$

$270 - 2x = 170$

$-2x = -100$

$x = 50°$

45. $x =$ first angle measure
$180 - x =$ second angle measure

$180 - x = 5x$

$180 = 6x$

$x = 30°$

$180 - x = 180° - 30° = 150°$

47. $x =$ first angle measure
$90 - x =$ second angle measure

$x = 2(90 - x) + 30$

$x = 180 - 2x + 30$

$3x = 210$

$x = 70°$

$90 - x = 90 - 70 = 20°$

49. $x =$ votes for
$993 - x =$ votes against

$x - (993 - x) = 137$

$x - 993 + x = 137$

$2x - 993 = 137$

$2x = 1130$

$x = 565$

$993 - x = 993 - 565 = 428$

There were 428 votes against.

51. $x =$ shortstop hits
$2x =$ right fielder hits
$(2x + x) + 3 =$ catcher hits

$x + 2x + (2x + x) + 3 = 27$

$6x + 3 = 27$

$6x = 24$

$x = 4$

$2x = 8$

The right fielder has 8 hits.

Modeling and Applications

53. $w = $ width
$2w - 6 = $ length

$w + w + 2w - 6 = 30$
$4w - 6 = 30$
$4w = 36$
$w = 9$ feet
$2w - 6 = 2(9) - 6 = 12$ feet

55. $x = $ number of women
$x - 6 = $ number of men

$x + (x - 6) \leq 30$
$2x - 6 \leq 30$
$2x \leq 36$
$x \leq 18$

There can be at most 18 women in the class.

She made 4 three–point shots and 8 two–point shots.

57. $9700 - 230x = 7860$
$-230x = 7860 - 9700$
$230x = -1840$
$x = 8$ years

59. $x = $ number of fiction books
$3x = $ number of nonfiction
$x + 3x = 64$
$4x = 64$

$x = 16$

There are 16 fiction books and 48 nonfiction books.

61. $x = $ length of stick
$x - \left(\dfrac{1}{6}x + 9\right) = \dfrac{1}{3}x$

$\dfrac{5}{6}x - 9 = \dfrac{1}{3}x$

$\dfrac{5}{6}x - \dfrac{2}{6}x = 9$

$\dfrac{3}{6}x = 9$

$\dfrac{1}{2}x = 9$

$x = 18$ inches

63. $38 + 0.36t < 29 + 0.51t$
$9 + 0.36t < 0.51t$
$9 < 0.15t$
$60 < t$

If the person uses more than 60 minutes, the $29-plan is more expensive.

65. $x = $ decaffeinated
$2x = $ regular
$\dfrac{3}{4}x = $ premium

$x + 2x + \dfrac{3}{4}x = 90$
$3.75x = 90$
$x = \dfrac{90}{3.75}$
$= 24$
$2x = 2(24) = 48$
$\dfrac{3}{4}x = \dfrac{3}{4}(24) = 18$

There are 24 pounds of decaffeinated, 48 pounds of regular, and 18 pounds of premium.

85

67. $49 + 0.02x < 0.09x$
$49 < 0.07x$
$700 < x$

They must spend over $700 to make the membership worthwhile.

69. $\frac{2}{3}x =$ first side
$x =$ second side
$x - 1 =$ third side

$\frac{2}{3}x + x + x - 1 = 7$
$\frac{8}{3}x - 1 = 7$
$\frac{8}{3}x = 8$
$x = 3$

The first side is 2 feet long, the second is 3 feet long, and the third is 2 feet long.

71. $x =$ low temperature
$\frac{x + 79}{2} < 72$
$x + 79 < 144$
$x < 65$

The low temperature was less than 65°.

73. 0.54 billion dollars a year. The slope of the line is the average change in revenue.

75.

| $t$ | $0.54t - 0.88$ |
|---|---|
| 3 | 0.74 |
| 4 | 1.28 |
| 5 | 1.82 |
| 6 | 2.36 |

$0.74 - 0.7 = 0.04$
$1.28 - 1.4 = -0.12$
$1.82 - 1.7 = 0.12$
$2.36 - 2.4 = -0.04$

The model is most accurate for 1998 and 2001.

77. Any two consecutive even integers since $(n + 2) - n = 2$ for all $n$.

79. $[x + (x + 2)] + 6 = 2(x + 4)$
$(2x + 2) + 6 = 2x + 8$
$2x + 8 = 2x + 8$

This is an identity, so the rule holds for all values of $x$.

Modeling and Applications

## Section 3.6 Applications

1. The unit price of the mixture must be between $1.20 and $1.80.

3. $x =$ the number of quarters
$28 - x =$ the number of dimes

$$0.25x + 0.10(28 - x) = 4.15$$
$$0.25x + 2.8 - 0.10x = 4.15$$
$$0.15x + 2.8 = 4.15$$
$$0.15x = 1.35$$
$$x = 9$$
$$28 - x = 28 - 9 = 19$$

There are 9 quarters and 19 dimes.

5. $x =$ number of dimes
$x - 2 =$ number of nickels
$3x + 4 =$ number of pennies

$$0.10x + 0.05(x - 2) + 0.01(3x + 4) = 1.38$$
$$0.10x + 0.05x - 0.10 + 0.03x + 0.04 = 1.38$$
$$0.18x - 0.06 = 1.38$$
$$0.18x = 1.44$$
$$x = 8$$
$$x - 2 = 8 - 2 = 6$$
$$3x + 4 = 3(8) + 4 = 28$$

There are 8 dimes, 6 nickels, and 28 pennies.

7. $x =$ number of ten-dollar bills
$x - 2 =$ number of twenty-dollar bills
$2x =$ number of five-dollar bills

$$10x + 20(x - 2) + 5(2x) = 160$$
$$10x + 20x - 40 + 10x = 160$$
$$40x - 40 = 160$$
$$40x = 200$$
$$x = 5$$
$$x - 2 = 5 - 2 = 3$$
$$2x = 2(5) = 10$$

There are 10 five-dollar bills, 5 ten-dollar bills, and 3 twenty-dollar bills.

9. $x =$ amount of first-class mail
$2020 - x =$ amount of second-class mail

$$0.34x + 0.25(2020 - x) = 638.20$$
$$0.34x + 505 - 0.25x = 638.20$$
$$0.09x + 505 = 638.20$$
$$0.09x = 133.20$$
$$x = 1480$$
$$2020 - x = 2020 - 1480 = 540$$

There are 1480 pieces of first-class mail and 540 pieces of second-class mail.

11. $x =$ number of tapes
    $2x =$ number of CDs

    $7x + 12(2x) = 4030$
    $7x + 24x = 4030$
    $31x = 4030$
    $x = 130$
    $2x = 260$

    There were 130 tapes and 260 CDs sold.

13. $x =$ amount invested at 6%
    $8500 - x =$ amount invested at 7%

    $x(0.06)(1) + (8500 - x)(0.07)(1) = 558$
    $0.06x + 595 - 0.07x = 558$
    $-0.01x + 595 = 558$
    $-0.01x = -37$
    $x = 3700$
    $8500 - x = 8500 - 3700 = 4800$

    So, $3700 is invested at 6%, and $4800 is invested at 7%.

15. $x =$ amount invested at 12%
    $x + 900 =$ amount invested at 9%

    $x(0.12)(1) + (x + 900)(0.09)(1) = 984$
    $0.12x + 0.09x + 81 = 984$
    $0.21x + 81 = 984$
    $0.21x = 903$
    $x = 4300$
    $x + 900 = 4300 + 900 = 5200$

    So, $4300 is invested at 12%, and $5200 is invested at 9%.

17. $x =$ amount of sales
    $0.05x =$ amount of tax

    $x + 0.05x = 2482.20$
    $1.05x = 2482.20$
    $x = 2364$
    $0.05x = 0.05(2364) = \$118.20$

    The amount of sales tax is $118.20, and the amount of sales is $2364.

19. $x =$ amount of 4% solution

    $0.04x + 0.01(7) = 0.03(x + 7)$
    $0.04x + 0.07 = 0.03x + 0.21$
    $0.01x + 0.07 = 0.21$
    $0.01x = 0.14$
    $x = 14$

    Fourteen milliliters of the 4% solution should be added.

21. $x =$ gallons of water
    $x(0.00) + 6(0.80) = (x + 6)(0.15)$
    $4.80 = 0.15x + 0.90$
    $3.90 = 0.15x$
    $x = 26$

    Twenty-six gallons of water should be added.

23. (a) If the cars travel in opposite directions, then the distance between them is the sum of their distances: $55t + 60t = 115t$.

    (b) If the cars travel in the same direction, then the distance between them is the positive difference of their distances: $60t - 55t = 5t$.

Modeling and Applications

25. $x =$ speed of first cyclist
$x + 7 =$ speed of second cyclist

$$x(3) + (x+7)(3) = 147$$
$$3x + 3x + 21 = 147$$
$$6x + 21 = 147$$
$$6x = 126$$
$$x = 21$$
$$3x = 3(21) = 63$$
$$3x + 21 = 3(21) + 21 = 84$$

One cyclist traveled 63 miles and the other cyclist traveled 84 miles.

27. $r =$ speed of first person
$r + 2 =$ speed of second person
$r(1.5) =$ distance traveled by first person
$(r+2)(1.5) =$ distance traveled by second person

$$1.5r + 1.5(r+2) = 15$$
$$1.5r + 1.5r + 3 = 15$$
$$3r + 3 = 15$$
$$3r = 12$$
$$r = 4$$
$$r + 2 = 4 + 2 = 6$$

One person walks 4 mph, and the other walks 6 mph

29. $r =$ walking rate
$r + 2 =$ jogging rate

$$r(0.75) + (r+2)(0.25) = 4.5$$
$$0.75r + 0.25r + 0.50 = 4.5$$
$$1r + 0.5 = 4.5$$
$$r = 4$$
$$r + 2 = 4 + 2 = 6$$

She walks at 4 mph and jogs at 6 mph.

31. $x =$ age of stucco house
$2x =$ age of brick house

$$3(x - 10) = 2x - 10$$
$$3x - 30 = 2x - 10$$
$$x - 30 = -10$$
$$x = 20$$
$$2x = 2(20) = 40$$

The stucco house is 20 years old, and the brick house is 40 years old.

33. $x =$ left fielder games
$4x =$ short stop games

$$2(x + 10) = 4x + 10$$
$$2x + 20 = 4x + 10$$
$$-2x + 20 = 10$$
$$-2x = -10$$
$$x = 5$$
$$4x = 4(5) = 20$$

The left fielder has played 5 games, and the short stop has played 20 games.

35. $x =$ acres in wheat
    $3x =$ acres in corn

    $2(x + 200) = 3x + 200$
    $2x + 400 = 3x + 200$
    $-x + 400 = 200$
    $-x = -200$
    $x = 200$
    $3x = 3(200) = 600$

    Six hundred acres are planted in corn, currently.

37. $120 =$ first person rate
    $120 + 100 =$ second person rate
    $t =$ time walking

    $220t - 120t = 40$
    $100t = 40$
    $t = \dfrac{40}{100}$
    $t = 0.4$ minutes
    $= 24$ seconds

39. $x =$ amount invested at 8%
    $7000-x=$ amount invested at 11%

    $0.11(7000 - x) - 0.08x = 314$
    $770 - 0.11x - 0.08x = 314$
    $770 - 0.19x = 314$
    $-0.19x = 456$
    $x = 2400$
    $7000 - x = 7000 - 2400 = 4600$

    So, $2400 was invested at 8%, and $4600 was invested at 11%.

41. $x =$ sales price
    $0.08x =$ commission

    $x - 0.08x = 103,500$
    $0.92x = 103,500$
    $x = 112,500$

    The purchase price must be $112,500.

43. $x =$ amount invested at 15%
    $3700 - x =$ amount invested at 18%

    $x(0.15) - (3700 - x)(0.18) = 192$
    $0.15x - 666 + 0.18x = 192$
    $0.33x - 666 = 192$
    $0.33x = 858$
    $x = 2600$
    $3700 - x = 3700 - 2600 = 1100$

    So, $2600 is invested at 15%, and $1100 is invested at 18%.

45. $x =$ rate of first printer
    $x + 2 =$ rate of second printer

    $7 \cdot x + 7(x + 2) = 70$
    $7x + 7x + 14 = 70$
    $14x + 14 = 70$
    $14x = 56$
    $x = 4$

    The first printer prints 4 pages per minute, and the second printer prints 6 pages per minute.

## Modeling and Applications

**47.** $x =$ original price
$0.25x =$ discount

$x - 0.25x = 6900$
$0.75x = 6900$
$x = 9200$

The dealer should charge $9200.

**49.** $x =$ amount in student lot
$\frac{1}{3}x =$ amount in faculty lot

$\frac{1}{5}(x - 120) = \frac{1}{3}x - 120$
$\frac{1}{5}x - 24 = \frac{1}{3}x - 120$
$\frac{1}{5}x + 96 = \frac{1}{3}x$
$96 = \frac{1}{3}x - \frac{1}{5}x$
$96 = \frac{5}{15}x - \frac{3}{15}x$
$96 = \frac{2}{15}x$
$x = 720$
$\frac{1}{3}x = 240$

There are 720 student and 240 faculty parking spaces.

**51.** $x =$ amount at 6%
$6200 - x =$ amount at 14%

$x(0.06) - (6200 - x)(0.14) = 272$
$0.06x - 868 + 0.14x = 272$
$0.2x - 868 = 272$
$0.2x = 1140$
$x = 5700$
$6200 - x = 6200 - 5700 = 500$

So $5700 was borrowed at 6 % and $500 was borrowed at 14%.

**53** $x =$ original price
$0.15x =$ student discount

$x - 0.15x - 20 = 82$
$0.85x - 20 = 82$
$0.85x = 102$
$x = 120$

The original price is $120.00.

**55.** $1.12x = 63,280$
$x = 56,500$

The original price was $56,500.

**57.** $x =$ AT & T shares
$x + 15 =$ Coca-Cola shares

$52x + 40(x + 15) = 6120$
$52x + 40x + 600 = 6120$
$92x + 600 = 6120$
$92x = 5520$
$x = 60$
$x + 15 = 60 + 15 = 75$

There are 60 AT&T shares and 75 Coca-Cola shares.

59. $x$ = amount on 18% card
    $1480 - x$ = amount on 21% card

    $x(0.18) - (1480 - x)(0.21) = 9$
    $0.18x - 310.80 + 0.21x = 9$
    $0.39x - 310.80 = 9$
    $0.39x = 319.80$
    $x = 820$
    $1480 - x = 1420 - 820 = 600$

    So $820 was charged on the 18% card
    and $600 was charged on the 21% card.

61. $57t$ = distance traveled by slow car
    $65t$ = distance traveled by fast car

    $65t - 57t = 20$
    $8t = 20$
    $t = 2.5$ hours

    In 2.5 hours they will be 20 miles apart.

63. $\dfrac{26.3 - 22.3 \text{ quadrillion BTUs}}{20 \text{ years}}$

    $= \dfrac{4}{20}$ quadrillion BTUs per year

    $= 0.2$ quadrillion BTUs per year

65. $0.2(4) + 22.3$
    $= 0.8 + 22.3$
    $= 23.1$ quadrillion BTUs

Modeling and Applications

67. $x$ = number of pennies
$y$ = number of nickels
$z$ = number of quarters
$x + 3 = 2y - 1$ = number of dimes

$x + 3 = 2y - 1$
$x + 4 = 2y$
$0.5x + 2 = y$

$y + z + 1 = x + y$
$z + 1 = x$
$z = x - 1$

$0.01x + 0.05y + 0.25z + 0.10(x+3) = 3.23$
$0.01x + 0.05(0.5x + 2) + 0.25(x - 1) + 0.10(x + 3) = 3.23$
$0.01x + 0.025x + 0.10 + 0.25x - 0.25 + 0.1x + 0.3 = 3.23$
$0.385x + 0.15 = 3.23$
$0.385x = 3.08$
$x = 8$
$y = 0.5x + 2 = 0.5(8) + 2 = 6$
$z = x - 1 = 8 - 1 = 7$
$x + 3 = 8 + 3 = 11$

There are 8 pennies, 6 nickels, 11 dimes, and 7 quarters.

69. $x$ = brother's age
$x + 4$ = woman's age

$(x + 4) - 6 = \dfrac{2}{3}(x + 8)$

$x - 2 = \dfrac{2}{3}x + \dfrac{16}{3}$

$3(x - 2) = 3\left(\dfrac{2}{3}x + \dfrac{16}{3}\right)$

$3x = 6 = 2x + 16$
$x - 6 = 16$
$x = 22$

The brother is 22 years old now and will be 25 in 3 years.

## Chapter 3 Review Exercises

1. (a) $\dfrac{43}{57}$

   (b) $\dfrac{57}{57+43}$
   $= \dfrac{57}{100}$

3. 1 pound = 16 ounces so 2 pounds 4 ounces = 36 ounces

   $\dfrac{\$5.40}{36 \text{ ounces}}$
   $= \$0.15$ per ounce

5. $\dfrac{38}{38+22}$
   $= \dfrac{38}{60}$
   $= \dfrac{19}{30}$ in favor of bond

   $\dfrac{19}{30} = \dfrac{x}{2880}$
   $19(2880) = 30x$
   $54,720 = 30x$
   $x = 1824$ voters were in favor of the bond

7. $\dfrac{2}{3+2}$
   $= \dfrac{2}{5}$ of the books are nonfiction

   $\dfrac{2}{5} = \dfrac{x}{10,000}$
   $20,000 = 5x$
   $x = 4000$ nonfiction books

9. $\dfrac{540}{2} = \dfrac{x}{5}$
   $5(540) = 2x$
   $x = 1350$ mg

11. All of the proportions are valid because the proportions are between corresponding sides.

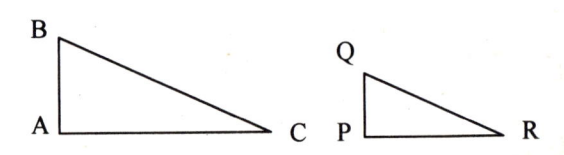

13. (a) 47%
    $= \dfrac{47}{100}$
    $= 0.47$

    (b) 110%
    $= \dfrac{110}{100}$
    $= \dfrac{11}{10}$
    $= 1.1$

    (c) 0.02%
    $= \dfrac{.02}{100}$
    $= \dfrac{2}{10000}$
    $= \dfrac{1}{5000}$
    $= 0.0002$

15. $0.32(80)$
    $= 25.6$

17. $0.98x = 147$
    $x = \dfrac{147}{0.98}$
    $x = 150$

Modeling and Applications

19. $\dfrac{24}{30}$
 $= 80\%$
 The percent decrease is
 $100\% - 80\% = 20\%$.

21. $45 - 0.12(45)$
 $= 45 - 5.4$
 $= 39.6°F$

23. $\dfrac{20}{16}$
 $= 1.25$
 So, 20 is a 25% increase over 16.

 $\dfrac{16}{20}$
 $= 0.80$
 $= 1.00 - 0.20$
 So, 16 is a 20% decrease from 20.

25. $I = \Pr t$
 $= 240(0.18)(0.5)$
 $= \$21.60$

27. (a) $(24)^2 + (45)^2 = 51^2$
 $576 + 2025 = 2601$
 $2601 = 2601$
 Yes, it is a right triangle.

 (b) $(27)^2 + (36)^2 = (47)^2$
 $729 + 1296 = 2209$
 $2025 = 2209$
 No, it is not a right triangle.

29. $A = \dfrac{1}{2}h(b_1 + b_2)$
 $2A = 2\left[\dfrac{1}{2}h(b_1 + b_2)\right]$
 $2A = h(b_1 + b_2)$
 $\dfrac{2A}{h} = b_1 + b_2$
 $\dfrac{2A}{h} - b_1 = b_2$

31. $Q = \dfrac{a+b+c}{3}$
 $3Q = a + b + c$
 $3Q - a - c = b$

33. $\dfrac{x}{2} + \dfrac{y}{6} = \dfrac{1}{3}$
 $6\left(\dfrac{x}{2} + \dfrac{y}{6}\right) = 6\left(\dfrac{1}{3}\right)$
 $3x + y = 2$
 $y = -3x + 2$

35. $C = \dfrac{5}{9}(F - 32)$
 $\dfrac{9}{5}C = \dfrac{9}{5}\left[\dfrac{5}{9}(F - 32)\right]$
 $\dfrac{9}{5}C = F - 32$
 $F = \dfrac{9}{5}C + 32$
 $F = \dfrac{9}{5}(15) + 32$
 $= 27 + 32$
 $= 59°F$

37. (a) $2x - 5$

 (b) $2x = x + 4$

39. $10 - c$ feet

41. $x - 1, x - 2$

43. (a) $p = 2l + 2w$

   $= 2l + 2(l-3)$
   $= 2l + 2l - 6$
   $= 4l - 6$, where $l$ is the length.

   (b) $x - 0.30x = 0.70x$, where $x$ is the original price.

45. $\frac{2}{3}x - 7 = -3$

   $\frac{2}{3}x = 4$

   $\frac{3}{2}\left(\frac{2}{3}x\right) = \frac{3}{2}(4)$

   $x = 6$

47. $x =$ first integer
   $x + 1 =$ second integer
   $x + 2 =$ third integer

   $x + (x+2) = x + 1$
   $2x + 2 = x + 1$
   $x + 2 = 1$
   $x = -1$
   $x + 2 = -1 + 2 = 1$
   The largest integer is 1.

49. $\frac{1}{5}x =$ first angle

   $x =$ second angle

   $\frac{1}{2}\left(\frac{1}{5}x + x\right) =$ third angle

   $\frac{1}{5}x + x + \frac{1}{2}\left(\frac{1}{5}x + x\right) = 180$

   $\frac{1}{5}x + x + \frac{1}{10}x + \frac{1}{2}x = 180$

   $10\left(\frac{1}{5}x + x + \frac{1}{10}x + \frac{1}{2}x\right) = 10(180)$

   $2x + 10x + x + 5x = 1800$
   $18x = 1800$
   $x = 100°$

   $\frac{1}{2}\left(\frac{1}{5}x + x\right) = \frac{1}{2}\left[\frac{1}{5}(100) + 100\right] = 60°$

   $\frac{1}{5}(100°) = 20°$

   The largest angle is 100°.

51. $x =$ angle
   $3(90 - x) - 10 =$ supplement of angle
   $90 - x =$ complement of angle

   $3(90-x) - 10 + x = 180$
   $270 - 3x - 10 + x = 180$
   $-2x + 260 = 180$
   $-2x = -80$
   $x = 40°$

Modeling and Applications

53. $x$ = number of quarters
$x - 3$ = number of dimes
$2(x - 3)$ = number of nickels

$0.25x + 0.1(x - 3) + 0.05[2(x - 3)] = 6.15$
$0.25x + 0.1x - 0.3 + 0.05(2x - 6) = 6.15$
$0.35x - 0.3 + 0.1x - 0.3 = 6.15$
$0.45x - 0.6 = 6.15$
$0.45x = 6.75$
$x = 15$

There are 15 quarters.

55. $x$ = amount at 7%
$12,000 - x$ = amount at 4%

$0.07x + 0.04(12,000 - x) = 750$
$0.07x + 480 - 0.04x = 750$
$0.03x + 480 = 750$
$0.03x = 270$
$x = 9000$

The amount invested in the mutual fund is $9000.

57. $x$ = quarts of water

$0.00x + 0.20(1) = 0.15(x + 1)$
$0.20 = 0.15x + 0.15$
$0.05 = 0.15x$
$x = \dfrac{1}{3}$ quart

59. $x$ = nurses
$3x$ = nurse's aids

$2(x + 1) = 3x$
$2x + 2 = 3x$
$2 = x$
$3x = 3(2) = 6$ nurse's aids

## Chapter 3 Test

1. 1 pound 6 ounces = 22 ounces
$\dfrac{\$0.99}{22 \text{ ounces}}$
= $0.045 per ounce
= 4.5¢ per ounce

3. $\dfrac{x + 3}{10} = \dfrac{x - 1}{6}$
$6(x + 3) = 10(x - 1)$
$6x + 18 = 10x - 10$
$18 = 4x - 10$
$28 = 4x$
$x = 7$

$AB = x + 3 = 7 + 3 = 10$

$DE = x - 1 = 7 - 1 = 6$

5. $x - 0.8x = 184$
   $0.92x = 184$
   $x = \dfrac{184}{0.92}$
   $x = 200$ employees originally
   200−184=16 people lost their jobs.

7. $\dfrac{8\%}{\text{year}} = \dfrac{0.08}{\text{year}}$
   $\dfrac{0.08}{\text{year}} \cdot \dfrac{1 \text{ year}}{12 \text{ months}}$
   $= \dfrac{0.08}{12}$ per month
   $= 0.00\overline{6}$ per month interest rate

   $I = Prt$
   $= 71,000(0.00\overline{6})(1)$
   $= \$473.33$

9. $W = \dfrac{P}{2} - L$
   $W + L = \dfrac{P}{2}$
   $2(W + L) = P$
   $P = 2W + 2L$

11. The triangle is a right triangle if the square of the longest side equals the sum of the square of the other two sides. Because $4.25^2 = 2^2 + 3.75^2$, the triangle is a right triangle.

13. The complement of $y$

15. $x =$ first number (positive)
    $0-x =$ second number (negative)
    $\dfrac{1}{5}x - 18 = 0 - x$
    $\dfrac{1}{5}x - 18 = -x$
    $-18 = -\dfrac{6}{5}x$
    $-\dfrac{5}{6}(-18) = \dfrac{-5}{6}\left(\dfrac{6}{5}x\right)$
    $15 = x$

17. $x =$ angle
    $90 - x =$ complement
    $180 - x =$ supplement

    $(90 - x) + (180 - x) = 156$
    $270 - 2x = 156$
    $-2x = -114$
    $x = 57°$

19. Let $t =$ hours since 10:00 a.m.
    $50t + 55(t - 1) = 260$
    $50t + 55t - 55 = 260$
    $105t = 315$
    $t = 3$ hours after 10:00 a.m.
    $= 1$ p.m.

# Cumulative Test, Chapters 1 – 3

1. Rational numbers have terminating or repeating decimal names. Irrational numbers have decimal names that do not terminate and do not repeat.

3. Associative Property of Multiplication

5. $6 - \left(-\dfrac{2}{3}\right)$
   $= \dfrac{18}{3} + \dfrac{2}{3}$
   $= \dfrac{20}{3}$

Modeling and Applications

7. $-7$

```
-2→X
            -2
Y1
            -7
■
```

9. $+, -$

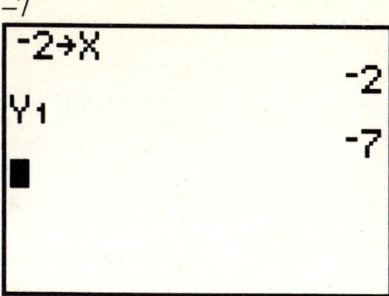

11. (a) Identity

(b) Contradiction, since
$2(x+1) = 2x+1$
$2x+2 = 2x+1$
$2 = 1$
is a false statement.

13. $x > -2$

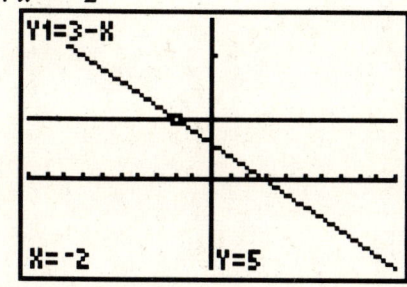

```
Y1=3-X



X=-2    Y=5
```

15. $1.06x = 521,520$
$x = \dfrac{521,520}{1.06}$
$x = \$492,000$

17. $W = \dfrac{1}{2}L - 4$
$P = 2W + 2L$
$= 2\left(\dfrac{1}{2}L - 4\right) + 2L$
$= L - 8 + 2L$
$= 3L - 8$

19. $x =$ first angle
$2x + 6 =$ second angle
$x + (2x+6) + 90 = 180$
$3x + 6 + 90 = 180$
$3x + 6 = 90$
$3x = 84$
$x = 28$
$2x + 6 = 2(28) + 6 = 62$
The angles are $90°, 28°,$ and $62°$.

# Chapter 4

# Properties of Lines

## Section 4.1 Linear Equations in Two Variables

1. If both were 0, the equation would not have a variable term.

3. $x + y = 24$ where $x$ = number of women, and $y$ = number of men.

5. $x + y = 180$ where $x$ = measure of one angle and $y$ = measure of the other angle.

7. $x + y = 12$ where $x$ = length of one piece and $y$ = length of the other piece.

9. $x + y = 12$ where $x$ = number of pounds of Brazilian coffee and $y$ = number of pounds of Turkish coffee.

11. $x + 3y = -3$ is a linear equation in two variables, since there are two variables in the equation and each term consists of a number, or the product of a number and a variable.

13. $\dfrac{2}{x} = y + 1$ is not a linear function since rewriting the equation yields $2 = xy + x$ and $xy$ is the product of two variables.

15. $y = 3$ may be written as $0x + y = 3$, so it is a linear equation in two variables.

17. $xy = 12$ is not a linear equation, since $xy$ is the product of two variables.

19. The value of $x$ must be written first because ordered pairs are of the form $(x, y)$. The solution is $(1, 3)$.

21. $y = -2x - 5$
$-15 = -2(\ ) - 5$
$-15 = -10 - 5$
$15 = -15$
$(5, -15)$ is a solution.

$-7 = -2(-1) - 5$
$-7 = 2 - 5$
$-7 = -3$
$(-1, -7)$ is not a solution.

$-5 = -2(0) - 5$
$-5 = -5$
$(0, -5)$ is a solution.

$7 = -2(-6) - 5$
$7 = 12 - 5$
$7 = 7$
$(-6, 7)$ is a solution.

23. $3 - y = 2x$

$3 - 1 = 2(-1)$
$2 = -2$
$(-1, 1)$ is not a solution.

$3 - (-9) = 2(6)$
$3 + 9 = 12$
$12 = 12$
$(6, -9)$ is a solution.

$3 - (-1) = 2(2)$
$3 + 1 = 4$
$4 = 4$
$(2, -1)$ is a solution.

$3 - 11 = 2(4)$
$-8 = -8$
$(-4, 11)$ is a solution.

25. $3y - 4x = 6$

$3(2) - 4(0) = 6$
$6 - 0 = 6$
$6 = 6$
$(0, 2)$ is a solution.

$3(6) - 4(3) = 6$
$18 - 12 = 6$
$6 = 6$
$(3, 6)$ is a solution.

$3(0) - 4(2) = 6$
$0 - 8 = 6$
$-8 = 6$
$(2, 0)$ is not a solution.

$3\left(\dfrac{10}{3}\right) - 4(1) = 6$
$10 - 4 = 6$
$6 = 6$
$\left(1, \dfrac{10}{3}\right)$ is a solution.

27. Solve the equation for $y$; then enter $y = 3x - 4$.

29.

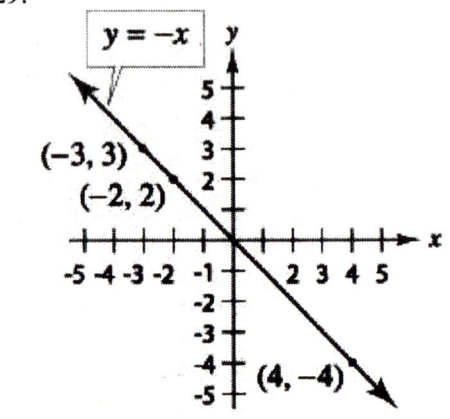

31.

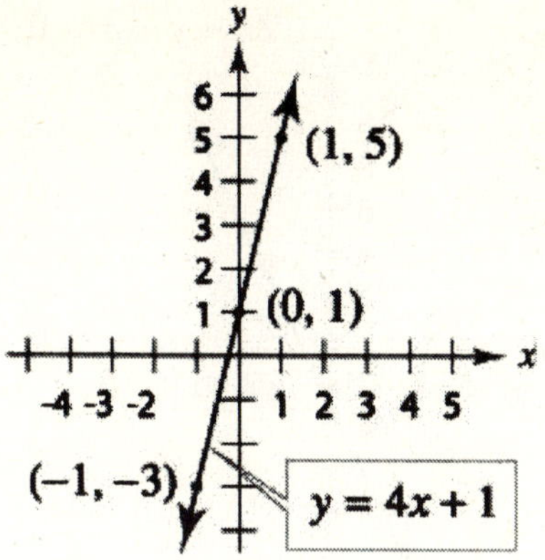


33.

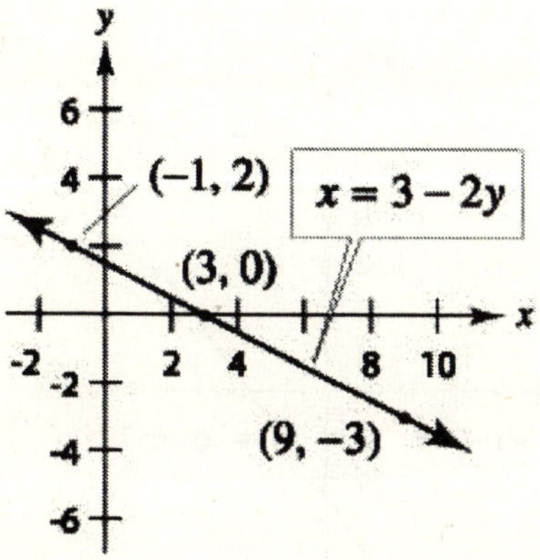


35.

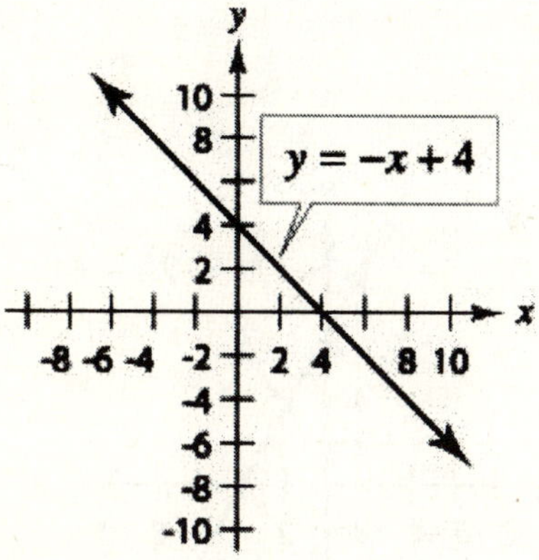


37.

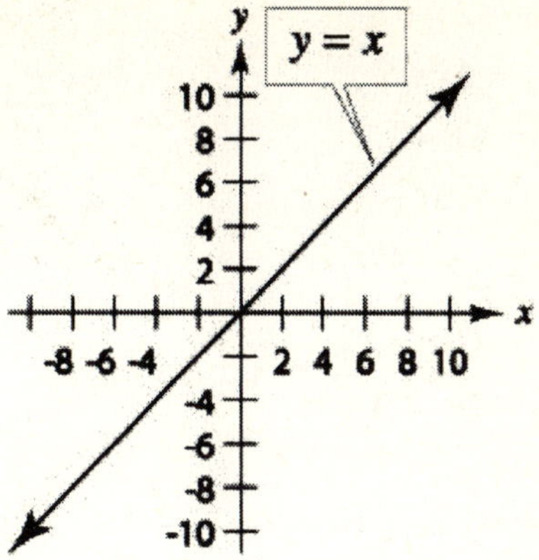


39.

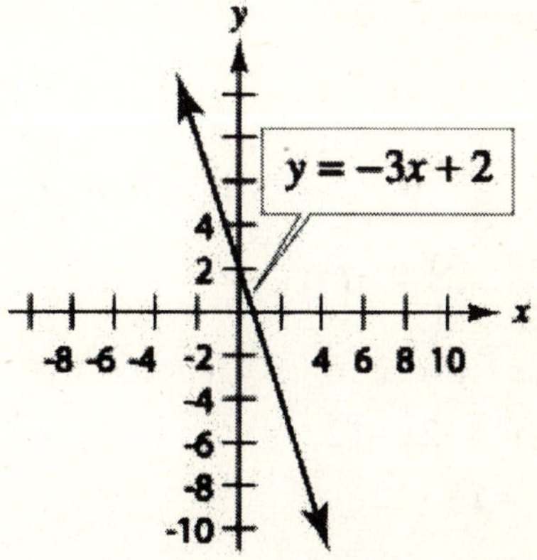


41.

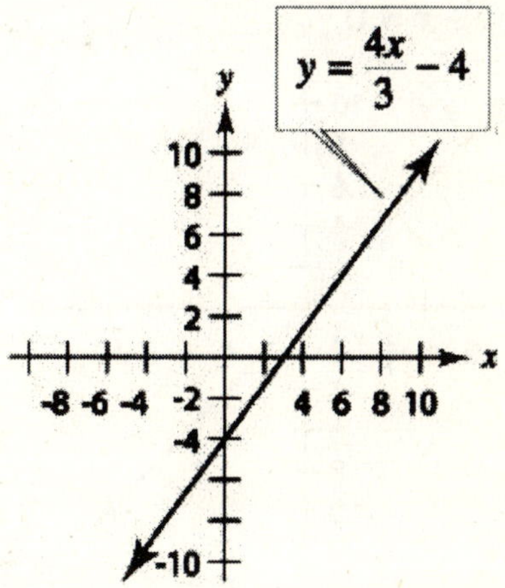


43. $2x + y = 3$

$y = -2x + 3$

45. $2 = 6x - 3y$
$3y + 2 = 6x$
$3y = 6x - 2$
$y = \dfrac{6x - 2}{3}$
$y = 2x - \dfrac{2}{3}$

47. $5y + 2x - 10 = 0$
$5y + 2x = 10$
$5y = -2x + 10$
$y = \dfrac{-2x + 10}{5}$
$y = \dfrac{-2}{5}x + 2$

49. $\dfrac{x}{2} - \dfrac{y}{4} = 3$
$\dfrac{4x}{2} - \dfrac{4y}{4} = 4(3)$
$2x - y = 12$
$-y = -2x + 12$
$y = 2x - 12$

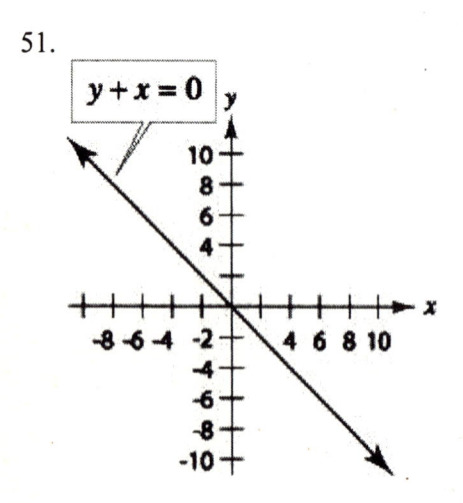

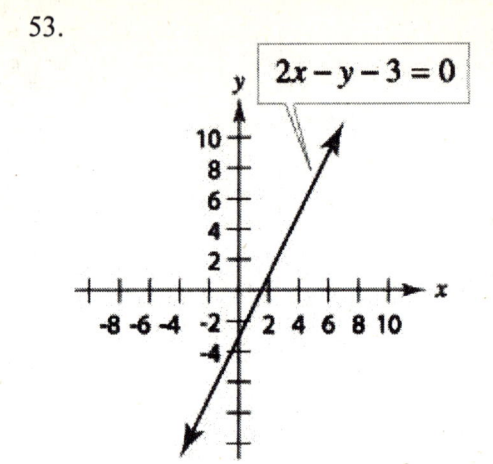

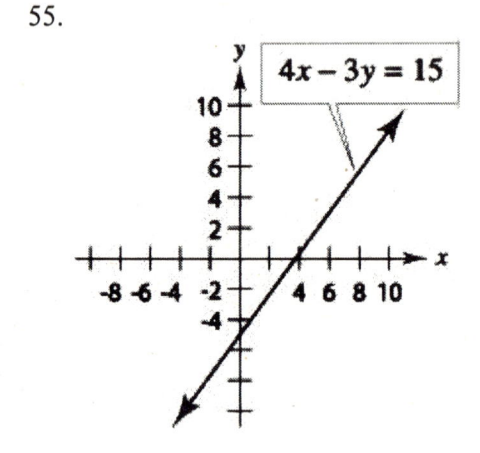

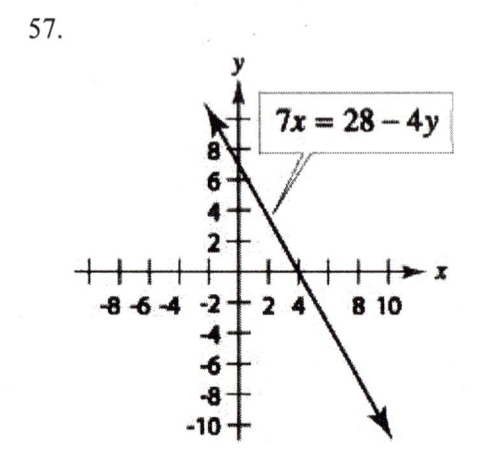

51.

53.

55.

57.

Properties of Lines          105

59.

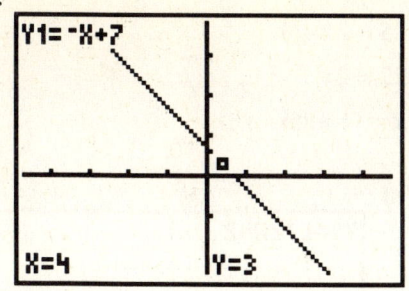

[Graphing calculator screen: Y1=-X+7, X=4, Y=3]

(a) (4, 3)

(b) (0, 7)

(c) (−2, 9)

(d) (7, 0)

61.

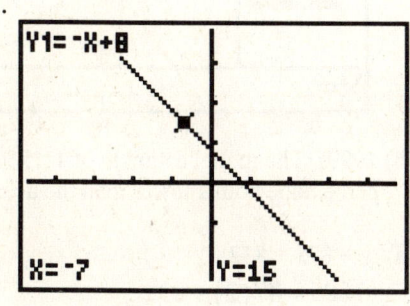

[Graphing calculator screen: Y1=-X+8, X=-7, Y=15]

(a) (−7, 15)

(b) (−2, 10)

(c) (0, 8)

(d) (8, 0)

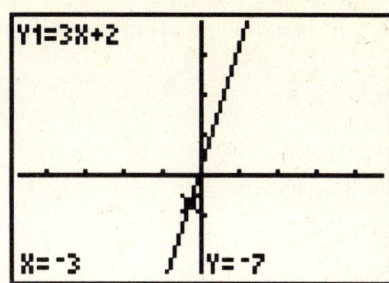

63.

[Graphing calculator screen: Y1=3X+2, X=-3, Y=-7]

(a) (−3, −7)

(b) (5, 17)

(c) (−4, −10)

(d) (2, 8)

65. Equation (ii) is graphed. Plug (0, 2) and (−4, 0) into the equation to verify.

$2y - x = 4$
$2(2) - 0 = 4$
$4 = 4$
$2(0) - (-4) = 4$
$4 = 4$

67. Equation (ii) is graphed. Plug $(-3, 15)$ and $(4, -6)$ into the equation to verify.

    $3x + y = 6$

    $3(-3) + 15 = 6$
    $-9 + 15 = 6$
    $6 = 6$

    $3(4) + (-6) = 6$
    $12 - 6 = 6$
    $6 = 6$

    To find $a$, set $x = 7$ and $y = a$.
    $3(7) + a = 6$
    $21 + a = 6$
    $a = -15$

    To find $b$, set $y = 3$ and $x = b$.
    $3b + 3 = 6$
    $3b = 3$
    $b = 1$

69. $y = x + 3$
    $(0, 3)$ and $(-3, 0)$ satisfy the condition.

71. $y = 2x$
    $(0, 0)$ and $(1, 2)$ satisfy the equation.

73. $x + y = 7$
    $(1, 6)$ and $(4, 2)$ satisfy the equation.

75. $x + y = 0$
    $(-1, 1)$ and $(1, -1)$ satisfy the equation.

77. $C = 1.20x + 5$
    $C(9) = 1.20(9) + 5$
    $= \$15.80$

79. $y = 2x + 3$
    $25 = 2x + 3$
    $22 = 2x$
    $x = 11$ men

81. (a) 

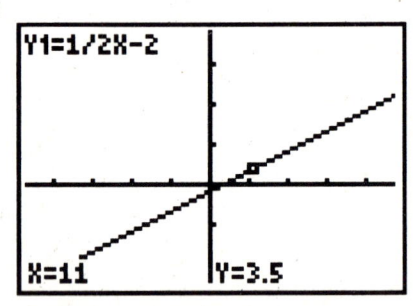

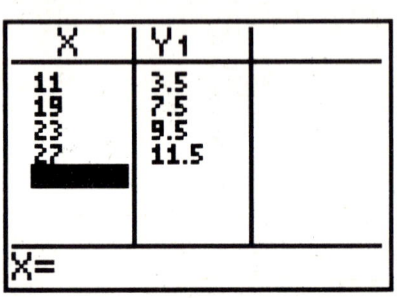

    (b) 1997; The model estimate of 11.5 % is just 0.2 percentage points lower than the actual 11.7%.

83. (a) $y = 87x + 832$
    $1006 = 87(2) + 832$
    $\quad = 174 + 832$
    $\quad = 1006$
    In 1992, the average refund was $1006 (according to the model.)

    (b) $1441 = 87(7) + 832$
    $\quad = 609 + 832$
    $\quad = 1441$
    In 1997, the average refund was $1441 (according to the model.)

85. $y = Kx - 5$
    $-2 = K(1) - 5$
    $3 = K$
    $K = 3$

Properties of Lines

87. $3y - x = K$
$3(K+1) - (K-1) = K$
$3K + 3 - K + 1 = K$
$2K + 4 = K$
$K + 4 = 0$
$K = -4$

89. $\frac{1}{3}x - \frac{2}{5}y = -2$
$\frac{1}{3}(6) - \frac{2}{5}(K) = -2$
$2 - \frac{2}{5}K = -2$
$-\frac{2}{5}K = -4$
$-2K = -20$
$K = 10$

## Section 4.2 Intercepts and Special Cases

1. The points where the graph intersects the x- and y-axes.

3. $(-7, 0), (0, 21)$

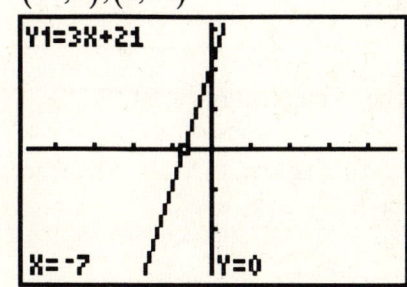

5. $(10, 0), (0, 15)$

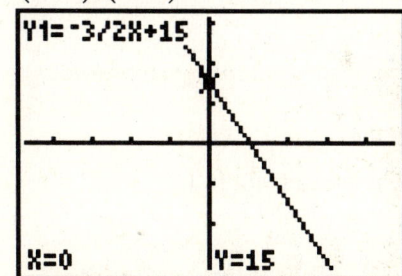

7. $(-18, 0), (0, -9)$

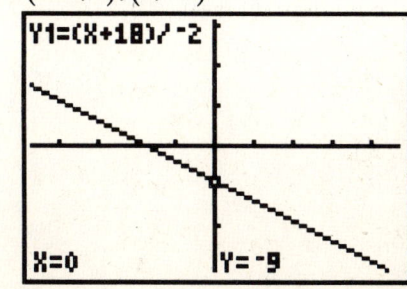

9. $(8, 0), (0, 20)$

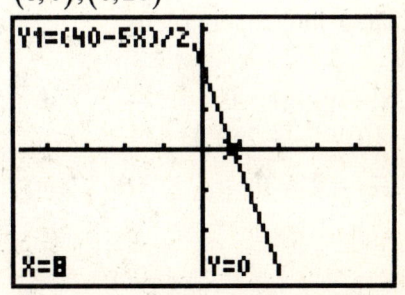

11. Replace $x$ with 0 and solve for $y$. Write the equation in the form $y = ax + b$. The y-intercept is $(0, b)$.

13. $3x = 5y$
$3(0) = 5y$
$0 = 5y$
$y = 0$

The y-intercept is (0, 0) and the x-intercept is (0, 0).

15. $\dfrac{x}{2} - \dfrac{y}{4} = 1$

$\dfrac{0}{2} - \dfrac{y}{4} = 1$

$-\dfrac{y}{4} = 1$

$-4\left(-\dfrac{y}{4}\right) = -4(1)$

$y = -4$

The $y$-intercept is $(0, -4)$.

$\dfrac{x}{2} - \dfrac{0}{4} = 1$

$\dfrac{x}{2} = 1$

$x = 2$

The $x$-intercept is $(2, 0)$.

17. $3x - 7y - 21 = 0$

$3(0) - 7y - 21 = 0$

$-7y - 21 = 0$

$-7y = 21$

$y = -3$

The $y$-intercept is $(0, -3)$.

$3x - 7(0) - 21 = 0$

$3x - 21 = 0$

$3x = 21$

$x = 7$

The $x$-intercept is $(7, 0)$.

19. $8x + 3y = 12$

$8(0) + 3y = 12$

$3y = 12$

$y = 4$

The $y$-intercept is $(0, 4)$.

$8x + 3(0) = 12$

$8x = 12$

$x = \dfrac{12}{8}$

$x = \dfrac{3}{2}$

The $x$-intercept is $\left(\dfrac{3}{2}, 0\right)$.

21. $2y + 3x = 18$

$2y = -3x + 18$

$y = -\dfrac{3}{2}x + 9$

The $y$-intercept is $(0, 9)$.

23. $x - 3y - 18 = 0$

$-3y = -x + 18$

$y = \dfrac{1}{3}x - 6$

The $y$-intercept is $(0, -6)$.

25. $3y - 11 = 4$

$3y = 15$

$y = 5$

The $y$-intercept is $(0, 5)$.

27. $y - 4 = 2(x + 3)$

$y - 4 = 2x + 6$

$y = 2x + 10$

The $y$-intercept is $(0, 10)$.

## Properties of Lines

29. $\dfrac{x}{4}+\dfrac{y}{3}=1$

$12\left(\dfrac{x}{4}+\dfrac{y}{3}\right)=12(1)$

$3x+4y=12$

$4y=-3x+12$

$y=-\dfrac{3}{4}x+3$

The $y$-intercept is $(0, 3)$.

31. (i) $3x-y=2$
    $3x=y+2$
    $y=3x-2$
    Yes

    (ii) $3y+2x=-4$
    $3y=-2x-4$
    $y=-\dfrac{2}{3}x-\dfrac{4}{3}$
    No

    (iii) $4x=3(y+2)$
    $4x=3y+6$
    $4x-6=3y$
    $\dfrac{4}{3}x-2=y$
    $y=\dfrac{4}{3}x-2$
    Yes

33. (i) $-(x-5)=2y$
    $-(5-5)=2(0)$
    $-0=0$
    $0=0$
    Yes

    (ii) $3y-x=-5$
    $3(0)-5=-5$
    $-5=-5$
    Yes

    (iii) $x+y+5=0$
    $5+0+5=0$
    $10=0$
    No

35. (i) $3x-y=-5$
    $-y+5+3x=0$
    $-y+5=-3x$
    $-y=-3x-5$
    $y=3x+5$

    (ii) $x+y=5$
    $y=-x+5$

    (iii) $2x-y=5$
    $2x=y+5$
    $y=2x-5$

    The $y$-intercept of (iii) is $(0, -5)$ while the $y$-intercept of the other equations is $(0, 5)$.

37. (i) $y=3(x-2)$
    $y=3x-6$

    (ii) $y=2(x+3)$
    $y=2x+6$

    (iii) $x-y=6$
    $y=x-6$

    Equation (ii) has $y$-intercept $(0, 6)$ while the other two equations have $y$-intercept $(0, -6)$.

39. The graph is a horizontal line with $y$-intercept $(0, b)$ and no $x$-intercept.

41.

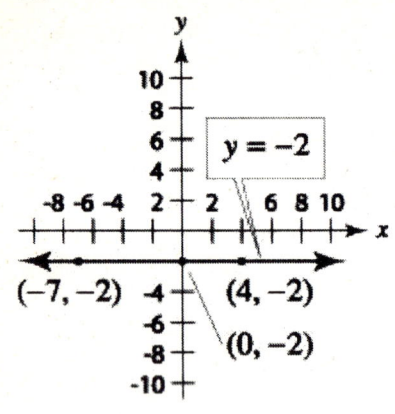

43.

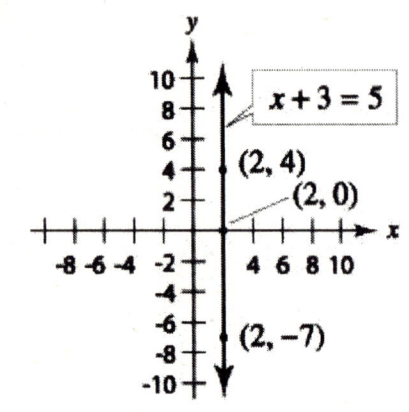

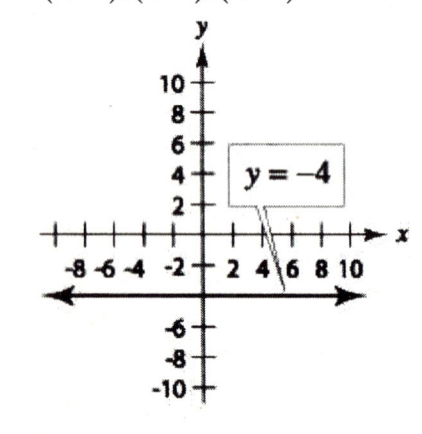

45. $(0,-4), (1,-4), (2,-4)$

47. $(3,1), (3,2), (3,3)$

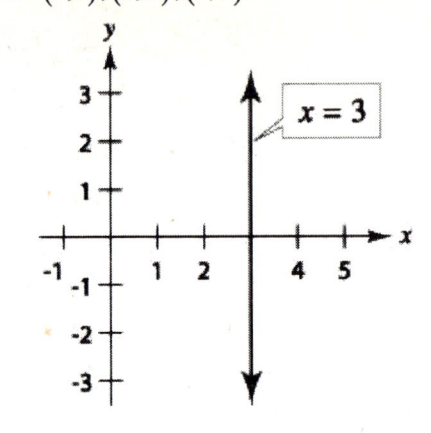

49. $(-7,0), (-7,1), (-7,2)$

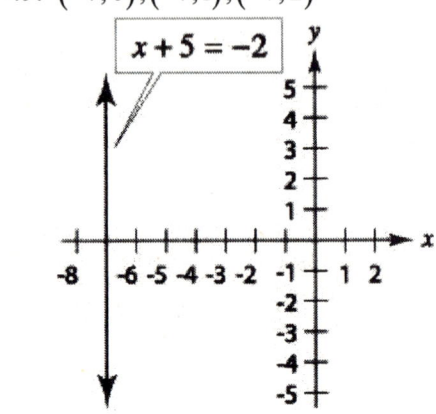

51. $(0,-2), (1,-2), (2,-2)$

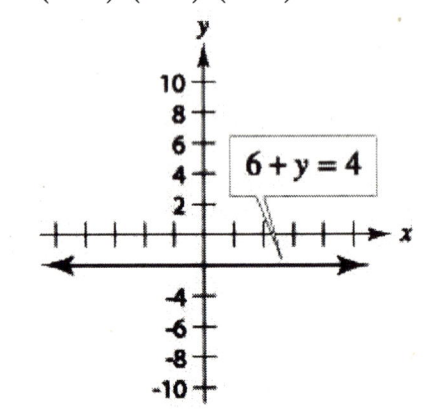

53. Vertical, since the equation contains an $x$-term but no nonzero $y$-term.

55. Neither

57. Horizontal, since the equation contains a $y$-term but no nonzero $x$-term.

## Properties of Lines

59. Horizontal, since the equation contains a y-term but no nonzero x-term.

61. (7, 1), (7, 2)
    $x = 7$

63. (0, 7), (1, 7)
    $2y - 6 = 8$

65. $y = 6$ since all ordered pairs have a y-coordinate of 6.

67. $x = -3$ since all ordered pairs have a x-coordinate of $-3$.

69. $x = 4$ since the line is vertical.

71. $y = 7$ since the line is horizontal.

73. $y = -x + 22$

    (a) $(22, 0), (0, 22)$

    (b) The y-intercept corresponds to "no windows have been washed," and the x-intercept corresponds to "all windows washed".

75. $y = 1.25x + 5$

    (a) (6, 12.5) so the cost is $12.50.

    (b) The y-intercept (0, 5) corresponds to the basic charge for rental.

77. The y-intercept is (0, 37). In year 0 (1995), the number of subscribers was 37 million.

79. The y-intercept is (0, 10.5). In year 0 (1990), there were 10.5 million subscribers.

81. $x = -2$

83. $y = -4$

85. $3y - x - 6 = 2 - y + x$
    $4y = 8 + 2x$
    $y = \dfrac{2}{4}x + \dfrac{8}{4}$
    $y = \dfrac{1}{2}x + 2$

    $0 = \dfrac{1}{2}x + 2$
    $-2 = \dfrac{1}{2}x$
    $x = -4$
    $(-4, 0)$ is the x-intercept.

    $y = \dfrac{1}{2}(0) + 2$
    $y = 2$
    $(0, 2)$ is the y-intercept.

## Section 4.3 Slope of a Line

1. The rise is greater than the run.

3. For each 2 unit rise, the run is 1 so the slope is 2.

5. For each 5 unit fall, the run is 3 so the slope is $\dfrac{-5}{3}$.

7. Vertical lines have an undefined slope.

9. Horizontal lines have a zero slope.

11. Any two points may be used.

13. $\dfrac{9 - 3}{4 - 1}$
    $= \dfrac{6}{3}$
    $= 2$

15. $\dfrac{5-3}{5-3}$
$= \dfrac{2}{2}$
$= 1$

17. $\dfrac{-6-(-4)}{0-(-2)}$
$= \dfrac{-6+4}{0+2}$
$= \dfrac{-2}{2}$
$= -1$

19. $\dfrac{5-5}{3-(-1)}$
$= \dfrac{0}{4}$
$= 0$

21. $\dfrac{2-0}{-6-(-6)}$
$= \dfrac{2}{0}$
$=$ undefined

23. $\dfrac{-1-(-3)}{7-(-5)}$
$\dfrac{-1+3}{12}$
$= \dfrac{2}{12}$
$= \dfrac{1}{6}$

25. $\dfrac{-6-(-2)}{-4-(-8)}$
$= \dfrac{-4}{4}$
$= -1$

27. $\dfrac{-3-(-2)}{1-6}$
$= \dfrac{-1}{-5}$
$= \dfrac{1}{5}$

29. $\dfrac{-11-(-5)}{17-2}$
$= \dfrac{-6}{15}$
$= \dfrac{-2}{5}$

31. $\dfrac{-3-7}{-7-(-3)}$
$= \dfrac{-10}{-4}$
$= \dfrac{5}{2}$

33. $\dfrac{6-(-7.1)}{3.8-5.4}$
$= \dfrac{13.1}{-1.6}$
$= -8.19$

35. $\dfrac{\frac{1}{9}-\frac{5}{9}}{\frac{1}{6}-\frac{5}{6}}$
$= \dfrac{\frac{-4}{9}}{\frac{-4}{6}}$
$= \dfrac{-4}{9} \cdot \left(-\dfrac{6}{4}\right)$
$= \dfrac{6}{9}$
$= \dfrac{2}{3}$

37. No, the ratio must be $\dfrac{3}{5}$.

## Properties of Lines

39. $(0, 0), (-2, 1)$

$m = \dfrac{1-0}{-2-0}$

$= -\dfrac{1}{2}$

41. $(6, 0)\ (6, 1)$

$m = \dfrac{1-0}{6-6}$

$= \dfrac{1}{0}$

$=$ undefined

43. $x - 3y - 9 = 0$

$x - 9 = 3y$

$y = \dfrac{1}{3}x - 3$

$(0, -3), (3, -2)$

$m = \dfrac{-2-(-3)}{3-0}$

$= \dfrac{1}{3}$

45. $(4, 4), (0, 1)$

$m = \dfrac{4-1}{4-0}$

$= \dfrac{3}{4}$

47. For zero slope, the rise is 0. For an undefined slope, the run is 0.

49. The line is horizontal so the slope is 0.

51. The line is vertical so the slope is undefined.

53. The line is vertical so the slope is undefined.

55. The line is horizontal so the slope is 0.

57. The line is horizontal so the slope is 0.

59. The line is vertical and has undefined slope.

61. The line is horizontal and has 0 slope.

63. The line has negative slope.

65. $\dfrac{y-3}{1-0} = 2$

$\dfrac{y-3}{1} = 2$

$y - 3 = 2$

$y = 5$

67. $\dfrac{3-5}{4-x} = \dfrac{-1}{2}$

$\dfrac{-2}{4-x} = \dfrac{-1}{2}$

$-2(2) = -1(4-x)$

$-4 = -4 + x$

$x = 0$

69. $\dfrac{y-3}{5-(-2)} = 0$

$\dfrac{y-3}{7} = 0$

$y - 3 = 0$

$y = 3$

71. (a) $20y - 7x = 36$

$20y = 7x + 36$

$y = \dfrac{7}{20}x + \dfrac{36}{20}$

$y = 0.35x + 1.8$

$(0, 1.8)\ (1, 2, 15)$

$m = \dfrac{2.15 - 1.8}{1 - 0}$

$= 0.35$

$= \dfrac{7}{20}$

(b) $\dfrac{5.4-1.9}{10-0}$
$= \dfrac{3.5}{10}$
$= 0.35$
$= \dfrac{7}{20}$

73. $\dfrac{-1-(-3)}{1-(-1)}$
$= \dfrac{2}{2}$
$= 1$
$\dfrac{y-(-1)}{5-1} = 1$
$\dfrac{y+1}{4} = 1$
$y+1 = 4$
$y = 3$

75. $\dfrac{3k-k}{1-(-1)} = \dfrac{1}{2}$
$\dfrac{2k}{2} = \dfrac{1}{2}$
$k = \dfrac{1}{2}$

77. The point (–2, 3) is in Quadrant II. The slope may be zero or possibly negative. It cannot be positive or undefined, because that would force the line into Quadrant III.

## Section 4.4 Slope and Graphing

1. The coefficient of $x$ is the slope and the constant term is the $y$-coordinate of the $y$-intercept.

3. $3x + y = 4$
$y = -3x + 4$
Slope: –3; $y$-intercept: (0, 4)

5. $y - 4x = 0$
$y = 4x$
Slope: 4; $y$-intercept: (0, 0)

7. $2x - y = 4$
$-y = -2x + 4$
$y = 2x - 4$
Slope: 2; $y$-intercept: (0, –4)

9. $2y + 10 = 0$
$2y = -10$
$y = -5$
Slope: 0; $y$-intercept: (0, –5)

11. $2x + 3y - 12 = 0$
$2x + 3y = 12$
$3y = -2x + 12$
$y = \dfrac{-2}{3}x + 4$
Slope: $\dfrac{-2}{3}$; $y$-intercept: (0, 4)

13. $x - \dfrac{2}{3}y = 2$
$-\dfrac{2}{3}y = -x + 2$
$-\dfrac{3}{2}\left(-\dfrac{2}{3}y\right) = -\dfrac{3}{2}(-x + 2)$
$y = \dfrac{3}{2}x - 3$
Slope: $\dfrac{3}{2}$; $y$-intercept: (0, 3)

## Properties of Lines

15. $\frac{1}{2}x + \frac{2}{3}y = \frac{5}{3}$

$6\left(\frac{1}{2}x + \frac{2}{3}y\right) = 6\left(\frac{5}{3}\right)$

$3x + 4y = 10$

$4y = -3x + 10$

$y = \frac{-3}{4}x + \frac{10}{4}$

$y = \frac{-3}{4}x + \frac{5}{2}$

Slope: $\frac{-3}{4}$; y-intercept: $\left(0, \frac{5}{2}\right)$

17. (i) $2y + 3x = 0$
$2y = -3x$
$y = \frac{-3}{2}x$

(ii) $3x = 10 - 2y$
$2y + 3x = 10$
$2y = -3x + 10$
$y = \frac{3}{2}x + 5$

(iii) $3(y+1) = 2x$
$3y + 3 = 2x$
$3y = 2x - 3$
$y = \frac{2}{3}x - 1$

Equations (i) and (ii) have slope $\frac{-3}{2}$.

19. (i) $y - 3x = 2$
$y = 3x + 2$

(ii) $y - 3 = 2$
$y = 5$

(iii) $y = 3$

Equations (ii) and (iii) have zero slope.

21. Because the slope can be written as $\frac{-2}{1}$ or $\frac{2}{-1}$, move down 2 and right 1 from the given point or move up 2 and left 1.

23.

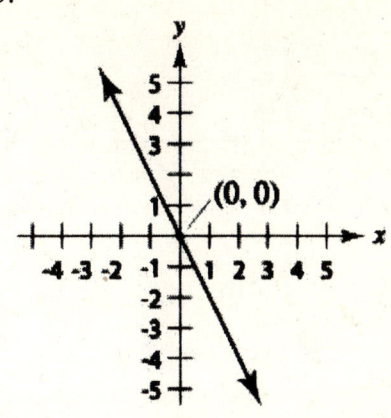

[Graph showing a line with negative slope passing through (0, 0)]

25.

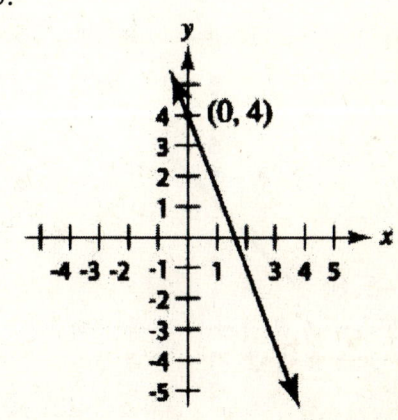

[Graph showing a line with negative slope passing through (0, 4)]

27.

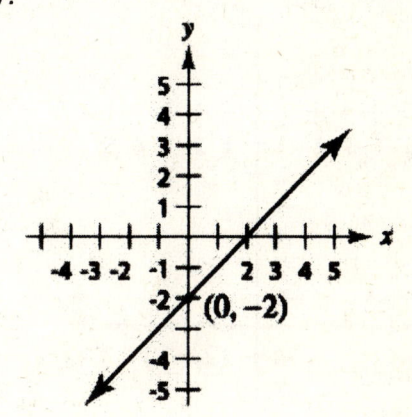

[Graph showing a line with positive slope passing through (0, −2)]

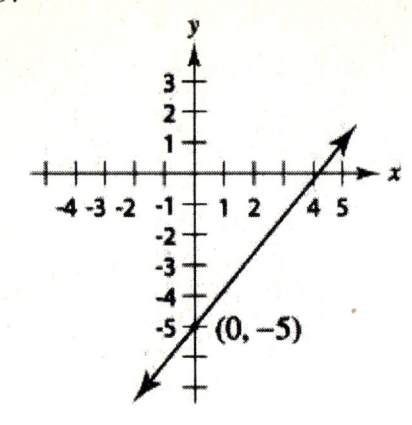

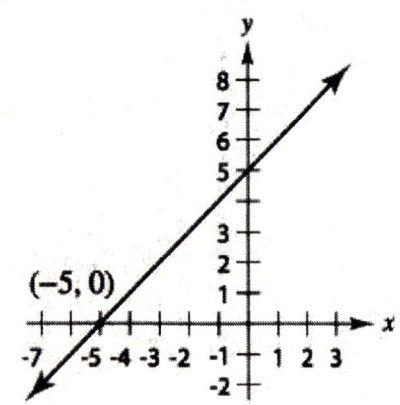

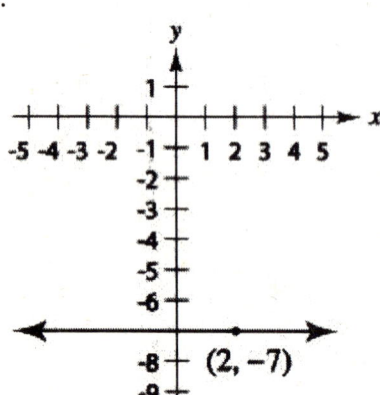

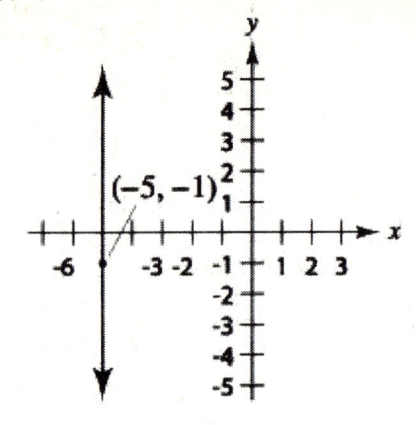

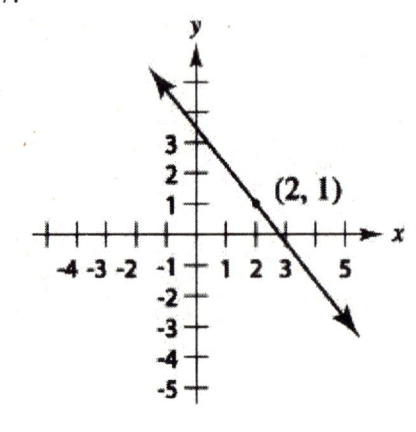

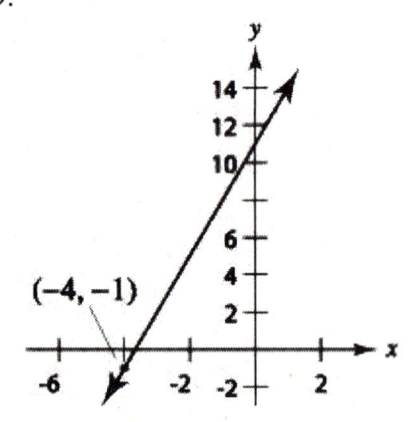

29.

31.

33.

35.

37.

39.

41.

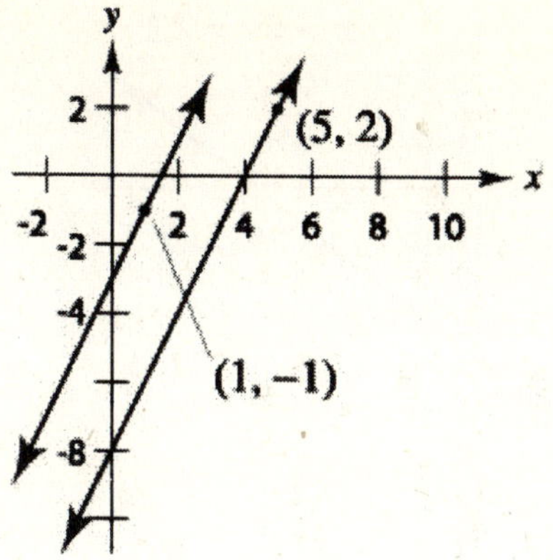


43.

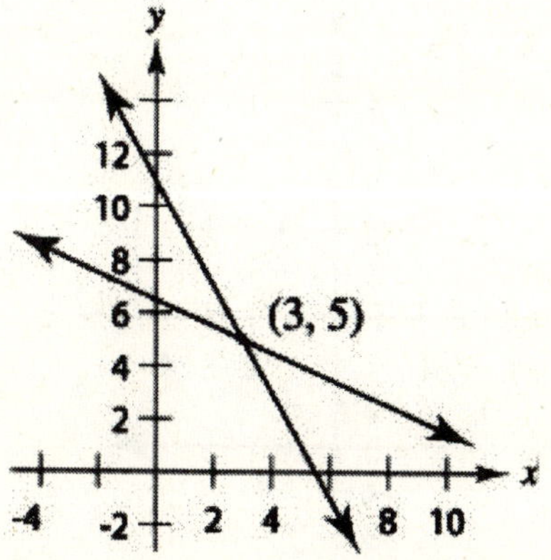


45.

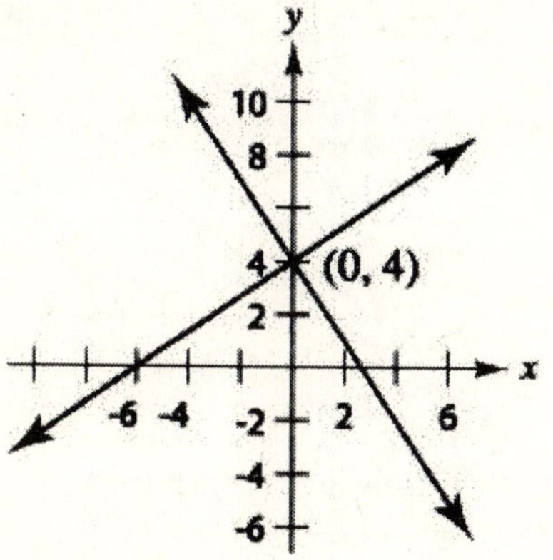


47.

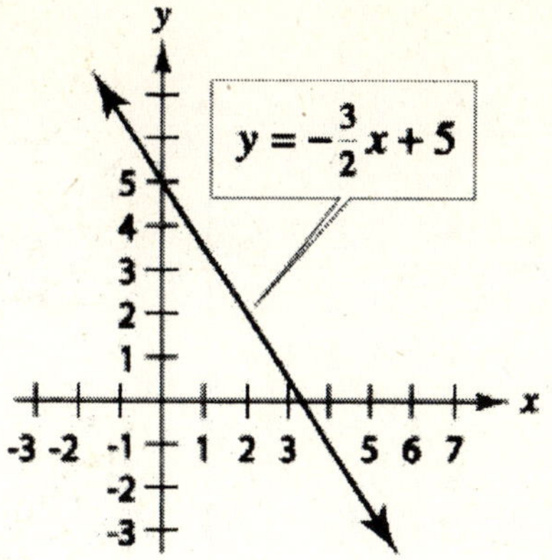


49.

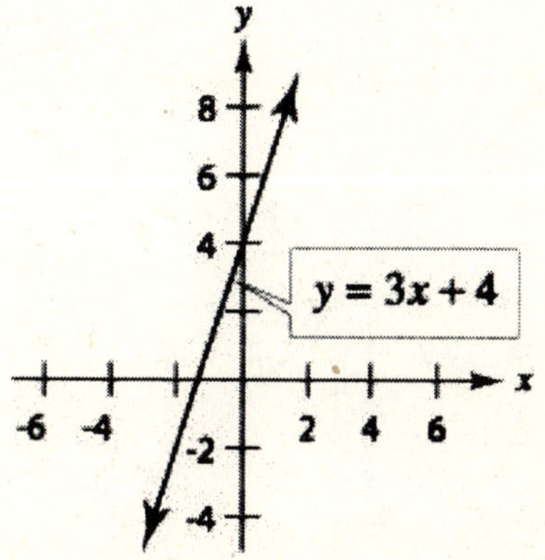


51.

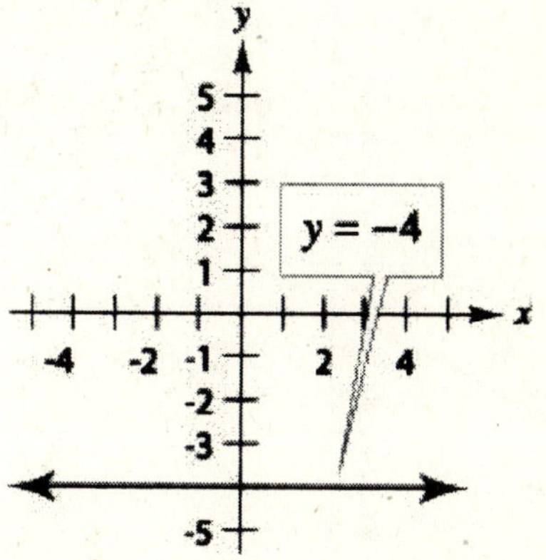

53. $x = 5$

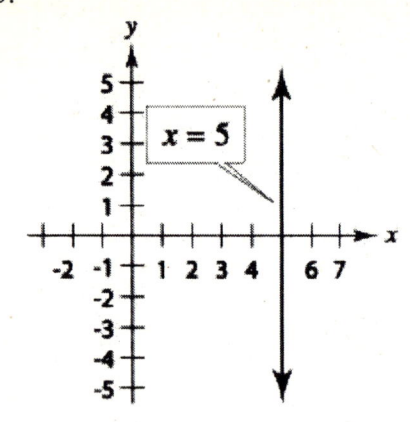

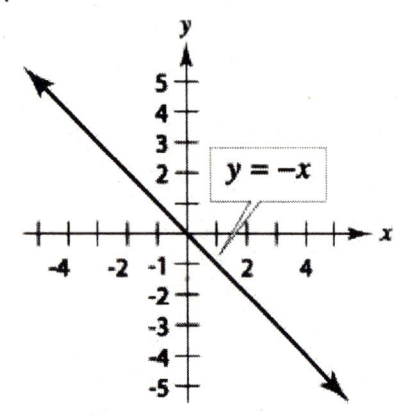

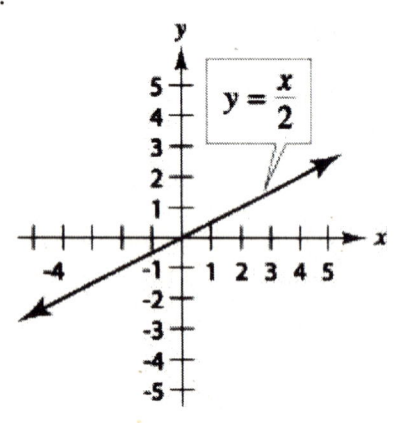

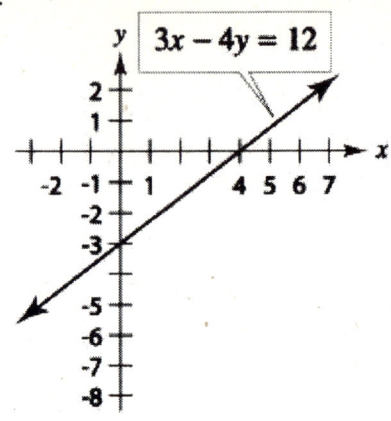

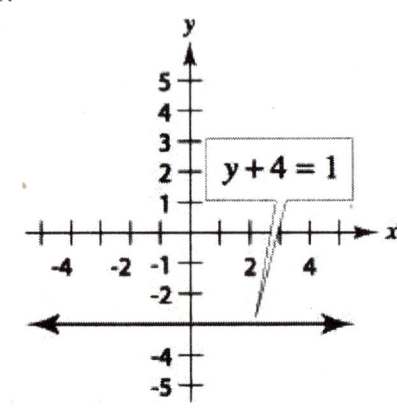

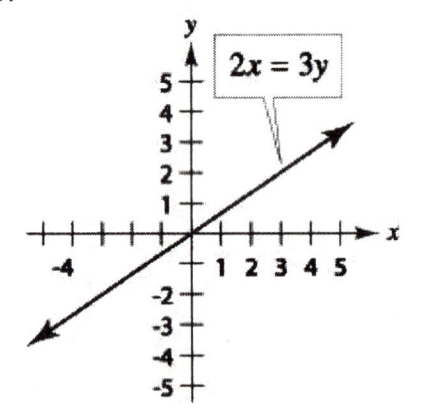

55. $y = -x$

57. $y = \dfrac{x}{2}$

59. $3x - 4y = 12$

61. $y + 4 = 1$

63. $2x = 3y$

65.

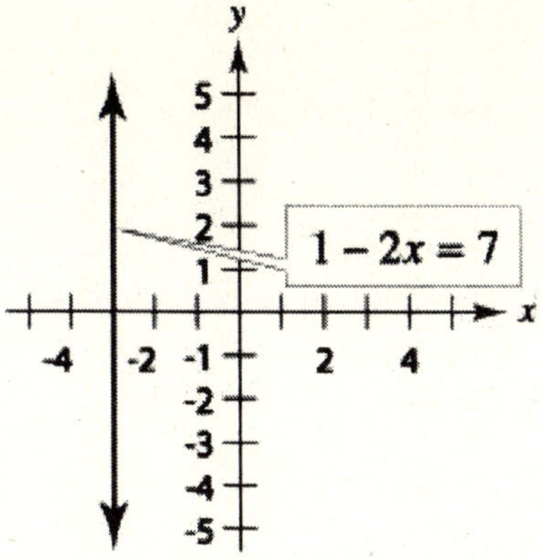
$1 - 2x = 7$

67.

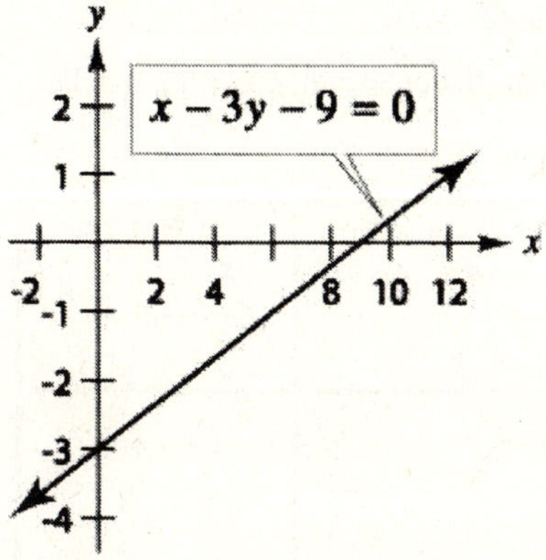
$x - 3y - 9 = 0$

69.

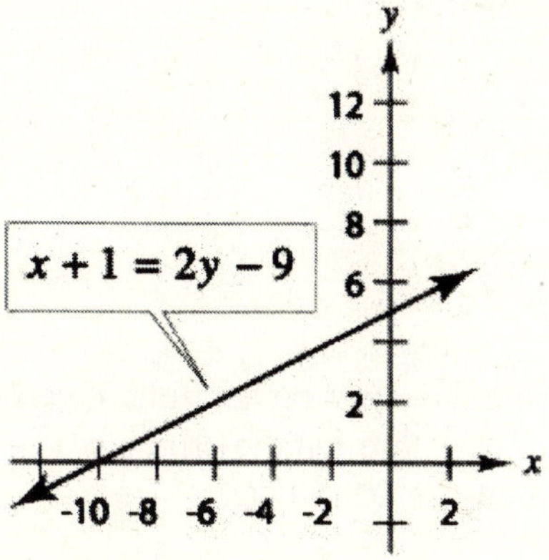
$x + 1 = 2y - 9$

71.

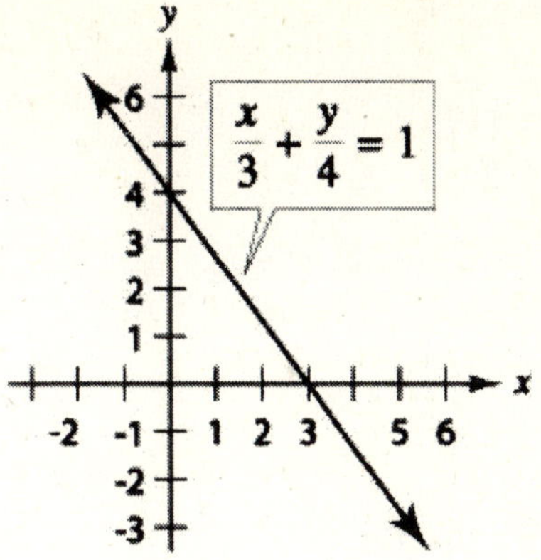
$\frac{x}{3} + \frac{y}{4} = 1$

73. 0

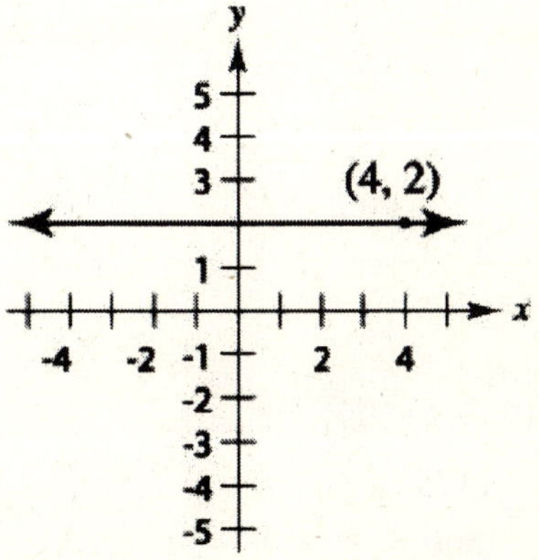
(4, 2)

75. Undefined

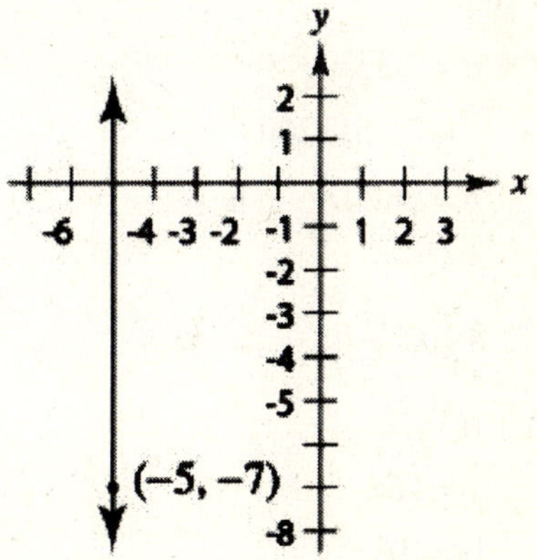
(−5, −7)

77. The slope is $k$ and the graph contains the origin.

79. Yes

81. $\frac{y}{2} = x$

$y = 2x$ ; Yes

83. $y = 5$ ; No

85. $8 = k(4)$
$k = 2$

$y = 2x$
$y = 2(6)$
$y = 12$

87. $-2 = k(-3)$
$k = \dfrac{2}{3}$

$y = \dfrac{2}{3}x$
$y = \dfrac{2}{3}(-18)$
$y = -12$

89. $1175 = k(500)$
$k = \dfrac{1175}{500}$
$k = 2.35$

$y = 2.35x$
$y = 2.35(1200)$
$y = 2820$ pounds

91. $8 = k\left(\dfrac{1}{2}\right)$
$K = 16$
$y = 16x$
$12 = 16x$
$x = \dfrac{12}{16}$
$= \dfrac{3}{4}$ cup

93. (a) $12x - 25y = 100$
$-25y = -12x + 100$
$y = \dfrac{12}{25}x - 4$

(b) Slope: $\dfrac{12}{25}$
The positive slope indicates that the percentage increases with increasing salaries.

95. $166x - 100y = -54$
$-100y = -166x - 54$
$y = 1.66x + 0.54$
Slope: 1.66
Not direct variation since b ≠ 0.

97.

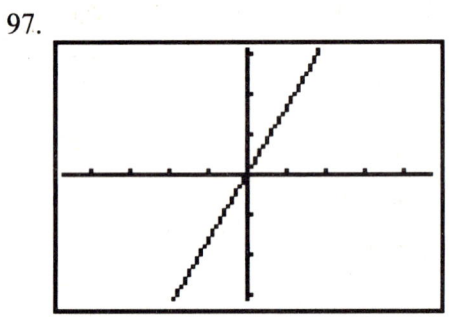

99. $Ax + By = C$
$By = -Ax + C$
$y = \dfrac{-A}{B}x + \dfrac{C}{B}$
Slope $= \dfrac{-A}{B}$

101. The line must be horizontal, vertical, or pass through the origin, so $A = 0$, $B = 0$, or $C = 0$.

103. $A = 0$
Then $y = \dfrac{C}{B}$.

105. For $\dfrac{-A}{B}$ to be negative, $A$ and $B$ must have the same sign.

106. The line must pass through the origin so $C = 0$.

# Section 4.5 Parallel and Perpendicular Lines

1. For nonvertical lines, write the equations in the form $y = mx + b$ and compare the slopes and $y$-intercepts. The lines are parallel if the slopes are the same and the $y$-intercepts are different.

3. $\dfrac{4}{2} = 2$
   $2 = 2$
   parallel

5. $-3 \neq \dfrac{1}{3}$
   not parallel
   $-\left(\dfrac{1}{-3}\right) = \dfrac{1}{3}$
   $\dfrac{1}{3} = \dfrac{1}{3}$
   perpendicular

7. $\dfrac{2}{3} \neq \dfrac{3}{2}$
   not parallel
   $-\left(\dfrac{3}{2}\right) \neq \dfrac{3}{2}$
   not perpendicular

9. $1 \neq -1$
   not parallel
   $-\dfrac{1}{1} = -1$
   perpendicular

11. The lines both have slope $\dfrac{1}{4}$ so they're parallel.

13. The lines are perpendicular, since the negative reciprocal of $\dfrac{4}{5}$ is $-\dfrac{5}{4}$.

15. The lines are neither parallel nor perpendicular.

17. $2y - 3x = 4$
    $2y = 3x + 4$
    $y = \dfrac{3}{2}x + 2$
    $2x + 3y = 6$
    $3y = -2x + 6$
    $y = \dfrac{-2}{3}x + 2$
    The lines are perpendicular.

19. $\dfrac{10 - 3}{-1 - (-4)}$
    $= \dfrac{7}{3}$
    $\dfrac{1 - (-6)}{4 - 1}$
    $= \dfrac{7}{3}$
    The lines are parallel.

21. $\dfrac{7 - 7}{-3 - 2}$
    $= \dfrac{0}{-5}$
    $= 0$
    $\dfrac{5 - 0}{-6 - (-6)}$
    $= \dfrac{5}{0}$
    $=$ undefined
    perpendicular

23. $\dfrac{15-3}{-4-(-8)}$

$=\dfrac{12}{4}$

$=3$

$\dfrac{-2-(-5)}{7-(-2)}$

$=\dfrac{3}{9}$

$=\dfrac{1}{3}$

neither

25. The slope of a vertical line is undefined.

27. (a) 1

(b) $-\left(\dfrac{1}{1}\right)$

$=-1$

29. (a) $-\dfrac{1}{4}$

(b) $-\left(-\dfrac{4}{1}\right)$

$=4$

31. (a) 0

(b) undefined

33. $4y-6x=1$

$4y=6x+1$

$y=\dfrac{6}{4}x+\dfrac{1}{4}$

$y=\dfrac{3}{2}x+\dfrac{1}{4}$

(a) $\dfrac{3}{2}$

(b) $-\left(\dfrac{2}{3}\right)$

$=-\dfrac{2}{3}$

35. $x+5=0$

$x=-5$

(a) undefined

(b) 0

37. $3x-5y-10=0$

$3x-10=5y$

$y=\dfrac{3}{5}x-2$

(a) $\dfrac{3}{5}$

(b) $-\left(\dfrac{5}{3}\right)$

$=-\dfrac{5}{3}$

39. $\dfrac{9-3}{1-(-1)}$

$=\dfrac{6}{2}$

$=3$

parallel

41. $\dfrac{3-1}{-1-(-2)}$

$=\dfrac{2}{1}$

$=2$

neither

43. $\dfrac{0-1}{1-(-4)}$

$=\dfrac{-1}{5}$

$5x-y=-2$

$5x=y-2$

$y=5x+2$

perpendicular

Properties of Lines

45. $\dfrac{7-7}{4-(-2)}$

$= \dfrac{0}{6}$

$= 0$

$x - 3 = 5$

$x = 8$

perpendicular

47. $\dfrac{1-1}{4-(-1)}$

$= \dfrac{0}{5}$

$= 0$

$2y + 1 = 3$

$2y = 2$

$y = 1$

parallel

49. $m$, = undefined

51. $m$, = undefined

53. $2x + 3 = 7$
$2x = 4$
$x = 2$
slope is undefined
A perpendicular line had slope 0, so $k = 0$.

55. $2x - y = -1$
$-y = -2x - 1$
$y = 2x + 1$
The slope is 2 so $k = 2$.

57. (2, 4)

59. $\dfrac{-2-6}{-3-5}$

$= \dfrac{-8}{-8}$

$= 1$

= slope of AC

$\dfrac{6-(-2)}{-3-5}$

$= \dfrac{8}{-8}$

$= -1$

= slope of BD

61. We must show exactly one pair of opposite sides are parallel.

Slope of AB $= \dfrac{-4-2}{-5-11}$

$= \dfrac{-6}{-16}$

$= \dfrac{3}{8}$

Slope of CD $= \dfrac{7-4}{6-(-2)}$

$= \dfrac{3}{8}$

Slope of BC $= \dfrac{4-(-4)}{-2-(-5)}$

$= \dfrac{8}{3}$

Slope of AD $= \dfrac{7-2}{6-11}$

$= \dfrac{5}{-5}$

$= -1$

63. From 1995 to 1998, mail increased by an average of 5.7 billion pieces per year.

65. 1995;
$$\frac{181,000,000,000 \text{ pieces}}{875,000 \text{ employees}}$$
$= 206,857$ pieces per employee

1998;
$$\frac{197,000,000,000 \text{ pieces}}{905,000 \text{ employees}}$$
$= 217,680$ pieces per employee

67. $3x + y = 6$
$y = -3x + 6$
Slope $= -3$

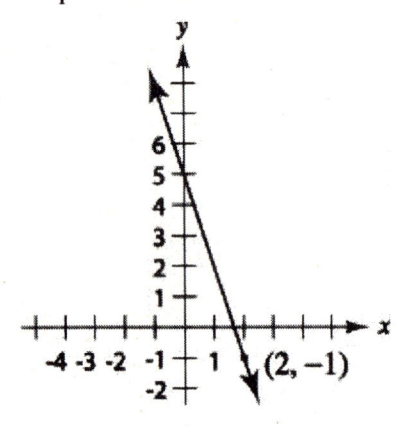

69. $x + y - 3 = 0$
$y = -x + 3$
$y$-intercept: $(0, 3)$

$2x - 3y = 0$
$2x = 3y$
$y = \frac{2}{3}x$

slope $= \frac{-3}{2}$

$y = \frac{-3}{2}x + 3$

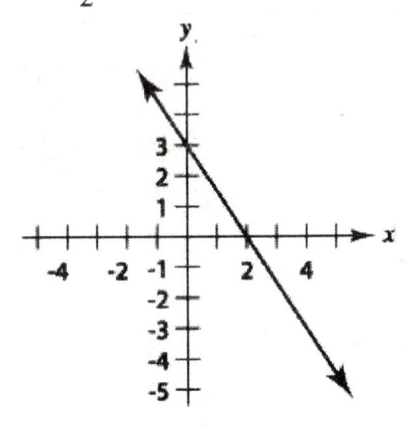

71. It is possible for two lines in space not to intersect and yet not be parallel.

## Section 4.6 Equations of Lines

1. The slope and $y$-intercept are needed.

3. $y = -x + 6$

5. $y = \frac{5}{2}x - 2$

7. $y = 2x + \frac{3}{4}$

9. $(0, 3), (3, -3)$
$m = \frac{3 - (-3)}{0 - 3}$
$= \frac{6}{-3}$
$= -2$
$y = -2x + 3$

Properties of Lines

11. $(0, 1), (3, 3)$
$$m = \frac{3-1}{3-0}$$
$$= \frac{2}{3}$$
$$y = \frac{2}{3}x + 1$$

13. $y = 3$

15. $y = mx + b$
$y = -3x + b$
$0 = -3(-4) + b$
$0 = 12 + b$
$b = -12$
$y = -3x - 12$

17. $y = mx + b$
$y = 1x + b$
$-2 = 1(3) + b$
$-2 = 3 + b$
$b = -5$
$y = x - 5$

19. $y = mx + b$
$y = -\frac{2}{3}x + b$
$2 = -\frac{2}{3}(6) + b$
$2 = -4 + b$
$b = 6$
$y = -\frac{2}{3}x + 6$

21. $y = mx + b$
$y = 0x + b$
$y = b$
$y = -7$

23. $y = mx + b$
$y = 2x + b$
$3 = 2\left(\frac{-3}{2}\right) + b$
$3 = -3 + b$
$b = 6$
$y = 2x + 6$

25. The slope and any point of the line are required.

27. $y - y_1 = m(x - x_1)$
$y - (-3) = 2(x - (-1))$
$y + 3 = 2(x + 1)$
$y + 3 = 2x + 2$
$y = 2x - 1$
$-2x + y = -1$
$2x - y = 1$

29. $y - y_1 = m(x - x_1)$
$y - (-4) = \frac{-5}{2}(x - 3)$
$y + 4 = \frac{-5}{2}x + \frac{15}{2}$
$2y + 8 = -5x + 15$
$2y = -5x + 7$
$5x + 2y = 7$

31. $y - y_1 = m(x - x_1)$
$y - \frac{2}{3} = 3\left(x - \frac{1}{2}\right)$
$y - \frac{2}{3} = 3x - \frac{3}{2}$
$6\left(y - \frac{2}{3}\right) = 6\left(3x - \frac{3}{2}\right)$
$6y - 4 = 18x - 9$
$6y = 18x - 5$
$-18x + 6y = -5$
$18x - 6y = 5$

33. $y - y_1 = m(x - x_1)$

$y - \dfrac{3}{5} = \dfrac{1}{4}(x - 2)$

$y - \dfrac{3}{5} = \dfrac{1}{4}x - \dfrac{2}{4}$

$20\left(y - \dfrac{3}{5}\right) = 20\left(\dfrac{1}{4}x - \dfrac{1}{2}\right)$

$20y - 12 = 5x - 10$

$20y = 5x + 2$

$-5x + 20y = 2$

$5x - 20y = -2$

35. $y = mx + b$

$y = mx + 4$

$0 = m(2) + 4$

$-4 = 2m$

$m = -2$

$y = -2x + 4$

37. $y = mx + b$

$y = mx + (-5)$

$0 = m(-2) - 5$

$5 = -2m$

$m = \dfrac{-5}{2}$

$y = \dfrac{-5}{2}x - 5$

39. $(7, 1), (3, 9)$

$m = \dfrac{9 - 1}{3 - 7}$

$= \dfrac{8}{-4}$

$= -2$

$y = -2x + b$

$1 = -2(7) + b$

$1 = -14 + b$

$b = 15$

$y = -2x + 15$

41. $(-2, 5), (4, 5)$

$m = \dfrac{5 - 5}{4 - (-2)}$

$= \dfrac{0}{6}$

$= 0$

$y = 0x + b$

$y = b$

$y = 5$

43. $(0, 6), (4, 0)$

$m = \dfrac{6 - 0}{0 - 4}$

$= \dfrac{3}{2}$

$y = \dfrac{3}{2}x + b$

$y = \dfrac{3}{2}x + 6$

45. $(-8, 2), (-8, 7)$

The $x$-coordinates are the same, so the line is vertical and has equation $x = -8$.

47. $(1, 2), (5, 8)$

$m = \dfrac{8 - 2}{5 - 1}$

$= \dfrac{6}{4}$

$= \dfrac{3}{2}$

$y = \dfrac{3}{2}x + b$

$2 = \dfrac{3}{2}(1) + b$

$b = \dfrac{1}{2}$

$y = \dfrac{3}{2}x + \dfrac{1}{2}$

## Properties of Lines

49. $(-5, 0), (0, 7)$

$$m = \frac{7-0}{0-(-5)}$$

$$= \frac{7}{5}$$

$$y = \frac{7}{5}x + b$$

$$y = \frac{7}{5}x + 7$$

51. The slope of a vertical line is not defined.

53. $y = 0x + 7$
$y = 7$

55. $x = 4$

57. We need the slope and a point of the line.

59. (a) $m = 1$

$$y = 1 \cdot x + b$$

$$-4 = 1(4) + b$$

$$-8 = b$$

$$y = x - 8$$

(b) $m = -\left(\frac{1}{1}\right)$

$$y = -1x + b$$

$$-4 = -1(4) + b$$

$$b = 0$$

$$y = -x$$

61. (a) $m = \frac{3}{4}$

$$y = mx + b$$

$$y = \frac{3}{4}x + b$$

$$3 = \frac{3}{4}(1) + b$$

$$\frac{9}{4} = b$$

$$y = \frac{3}{4}x + \frac{9}{4}$$

(b) $m = -\frac{4}{3}$

$$y = mx + b$$

$$y = -\frac{4}{3}x + b$$

$$3 = -\frac{4}{3}(1) + b$$

$$b = \frac{13}{3}$$

$$y = -\frac{4}{3}x + \frac{13}{3}$$

62. (a) $m = -\frac{1}{2}$

$$y = mx + b$$

$$y = -\frac{1}{2}x + b$$

$$-4 = -\frac{1}{2}(3) + b$$

$$-4 = -\frac{3}{2} + b$$

$$-\frac{5}{2} = b$$

$$y = -\frac{1}{2}x - \frac{5}{2}$$

(b) $m = \dfrac{2}{1}$

$$y = mx + b$$
$$y = 2x + b$$
$$-4 = 2(3) + b$$
$$-4 = 6 + b$$
$$b = -10$$

$$y = 2x - 10$$

63. $5x - 3y = 7$
$$-3y = -5x + 7$$
$$y = \dfrac{5}{3}x - \dfrac{7}{3}$$

(a) $m = \dfrac{5}{3}$

$$y = mx + b$$
$$y = \dfrac{5}{3}x + b$$
$$-5 = \dfrac{5}{3}(4) + b$$
$$-\dfrac{15}{3} = \dfrac{20}{3} + b$$
$$-\dfrac{35}{3} = b$$

$$y = \dfrac{5}{3}x - \dfrac{35}{3}$$

(b) $m = -\dfrac{3}{5}$

$$y = mx + b$$
$$y = -\dfrac{3}{5}x + b$$
$$-5 = -\dfrac{3}{5}(4) + b$$
$$-\dfrac{25}{5} = -\dfrac{12}{5} + b$$
$$-\dfrac{13}{5} = b$$

$$y = -\dfrac{3}{5}x - \dfrac{13}{5}$$

65. $m = \dfrac{5-1}{1-(-1)}$
$$= \dfrac{4}{2}$$
$$= 2$$

(a) $y = -\dfrac{1}{2}x + b$
$$2 = -\dfrac{1}{2}(3) + b$$
$$\dfrac{4}{2} = -\dfrac{3}{2} + b$$
$$b = \dfrac{7}{2}$$

$$y = -\dfrac{1}{2}x + \dfrac{7}{2}$$

(b) $y = 2x + b$
$$2 = 2(3) + b$$
$$2 = 6 + b$$
$$b = -4$$

$$y = 2x - 4$$

Properties of Lines

67. $m = \dfrac{3-6}{3-2}$

$= \dfrac{3}{1}$

(a) $y = -\dfrac{1}{3}x + b$

$0 = -\dfrac{1}{3}(0) + b$

$b = 0$

$y = -\dfrac{1}{3}x$

(b) $y = 3x + b$

$0 = 3(0) + b$

$b = 0$

$y = 3x$

69. $m = \dfrac{-2-7}{3-3}$

$= -\dfrac{9}{0}$

= undefined

(a) $y = 0x + b$

$2 = 0(-5) + b$

$b = 2$

$y = 2$

(b) $x = -5$

71. $m = \dfrac{4-4}{2-1}$

$= \dfrac{0}{1}$

(a) $x = 3$

(b) $y = 0x + b$

$1 = 0(3) + b$

$b = 1$

$y = 1$

75. $y = x + 7$

77. $y = \dfrac{1}{2}x + 5$

79. $x = -3$, since the line is perpendicular to a line with zero slope.

81. $x = 1$, since the line is parallel to a line with undefined slope.

83. $y = 2x + 7$

85. $(-4, 3), (0, 0)$

$m = \dfrac{3-0}{-4-0}$

$= -\dfrac{3}{4}$

$y = -\dfrac{3}{4}x + b$

$3 = -\dfrac{3}{4}(-4) + b$

$3 = +3 + b$

$b = 0$

$y = -\dfrac{3}{4}x$

87. $y = 120t + 500$ where $t$ is the number of months since July 1 and $y$ is the amount deposited in the account.

89. $(0, 8000), (7, 4859)$

$m = \dfrac{4859 - 8000}{7 - 0}$

$= -\dfrac{3141}{7}$

$y = -\dfrac{3141}{7}x + 8000$ where $x$ is the number of months since logging began and $y$ is the number of trees remaining.

129

91. (a) $(0, 126), (3, 142)$

   (b) $m = \dfrac{142-126}{3-0}$
   $= \dfrac{16}{3}$
   $y-$intercept: $(0, 126)$

   (c) $y = \dfrac{16}{3}x + 126$

93. $(0, 37), (2, 327)$
   $m = \dfrac{327-37}{2-0}$
   $= \dfrac{290}{2}$
   $= 145$
   $y-$intercept: $(0, 37)$

95. $(7, 37), (9, 327)$
   $m = \dfrac{327-37}{9-7}$
   $= \dfrac{290}{2}$
   $= 145$
   The slope is the same.

97. $kx - 2y = 10$
   $kx = 2y + 10$
   $kx - 10 = 2y$
   $y = \dfrac{kx-10}{2}$
   $y = \dfrac{k}{2}x - 5$
   Since the lines are parallel,
   $\dfrac{k}{2} = -5$
   $k = -10$

99. $x + ky + 3 = 0$
   $ky = -x - 3$
   $y = -\dfrac{x}{k} - \dfrac{3}{k}$
   $1 = -\dfrac{3}{k}$ since 1 is the $y$-intercept
   of $y = x + 1$.
   $k = -3$

101. $y = 0x + b$
   $y = b$

102. $y - b = 0(x - 0)$
   $y - b = 0$
   $y = b$
   Note: $(0, b)$ is the $y-$intercept

## Section 4.7 Graphs of Linear Inequalities

1. A solution is a pair $(x, y)$ that satisfies the inequality.

3. $y < -2x + 5$
   $3 < -2(-2) + 5$
   $3 < 4 + 5$
   $3 < 9$
   True, so $(-2, 3)$ is a solution.

   $-8 < -2(6) + 5$
   $-8 < -12 + 5$
   $-8 < -7$
   True, so $(6, -8)$ is a solution.

   $19 < -2(7) + 5$
   $19 < 14 + 5$
   $19 < 19$
   False, so $(-7, 19)$ is not a solution.

## Properties of Lines

$6 < -2(4) + 5$

$6 < -8 + 5$

$6 < -3$

False, so (4, 6) is not a solution.

5. $y \geq \dfrac{x}{2} - 5$

$-3 \geq \dfrac{-4}{2} - 5$

$-3 \geq -7$

True, so $(4, -3)$ is a solution.

$2 \geq \dfrac{8}{2} - 5$

$2 \geq 4 - 5$

$2 \geq -1$

True, so (8, 2) is a solution.

$-7 \geq \dfrac{-3}{2} - 5$

$-7 \geq -6.5$

False, so $(-3, -7)$ is not a solution.

$0 \geq \dfrac{-6}{2} - 5$

$0 \geq -3 - 5$

$0 \geq -8$

True, so $(-6, 0)$ is a solution.

7. $5x > 2y + 10$

$5(4) > 2(3) + 10$

$20 > 6 + 10$

$20 > 16$

True, so (4, 3) is a solution.

$5(0) > 2(-2) + 10$

$0 > -4 + 10$

$0 > 6$

False, so (0, -2) is not a solution.

$5(1) > 2(-4) + 10$

$5 > -8 + 10$

$5 > 2$

True, so (1, -4) is a solution.

$5(4) > 2(5) + 10$

$20 > 10 + 10$

$20 > 20$

False, so (4, 5) is not a solution.

9. It is not correct. The inequality can be rewritten as $y > 2x - 3$, shade above the line.

11. $y \leq 2$

We are looking for a horizontal line at $y = 2$ that is shaded below. (b)

13. $x + 3 > 0$

We are looking for a vertical line that is shaded to the right. (a)

15.

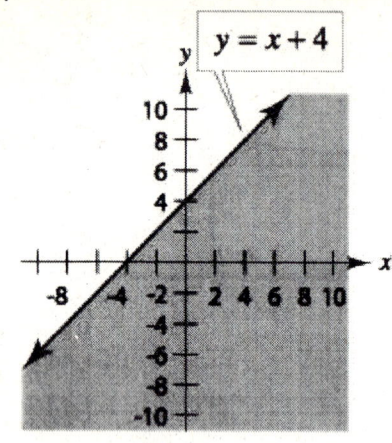

17.

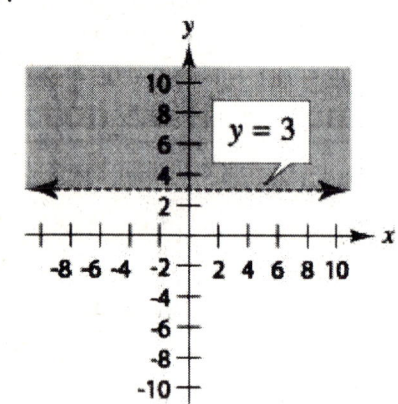

19. For ≤ or ≥, use a solid line, and for < or >, use a dashed line.

21.

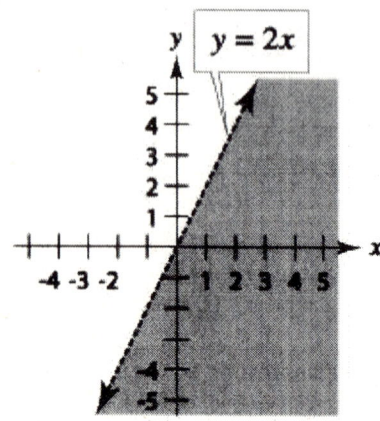

23.

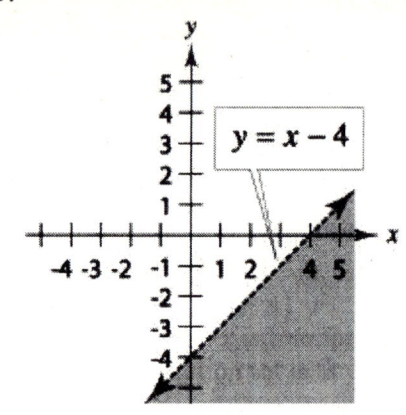

25.

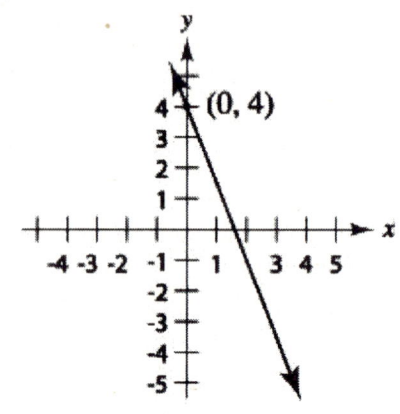

27.

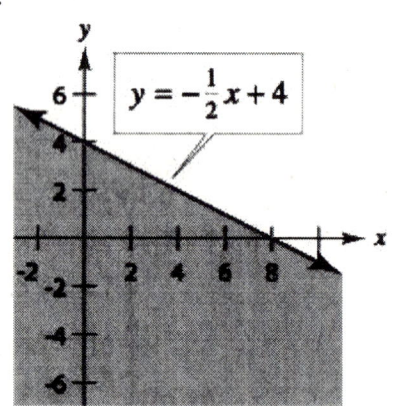

Properties of Lines

29.

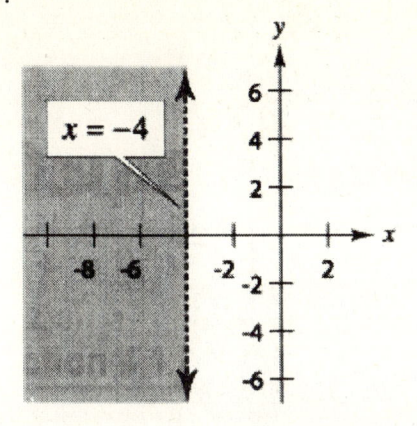

31.

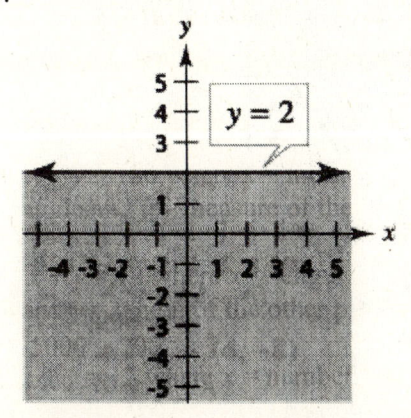

33. $x - y \leq 0$
$x \leq y$
$y \geq x$

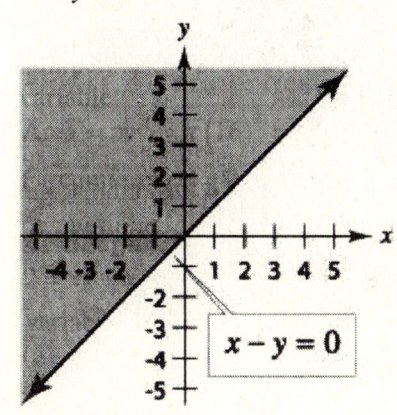

35. $x > 2y$
$\frac{1}{2}x > y$
$y < \frac{1}{2}x$

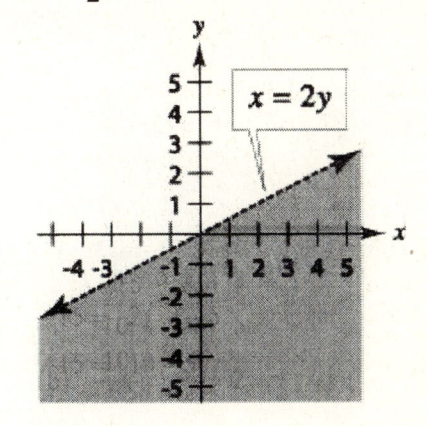

37. $x + 3y < 9$
$3y < -x + 9$
$y < -\frac{1}{3}x + 3$

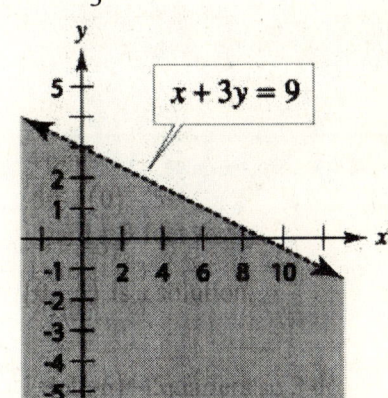

39. $x \geq 6 - 2y$
$2y + x \geq 6$
$2y \geq -x + 6$
$y \geq -\dfrac{1}{2}x + 3$

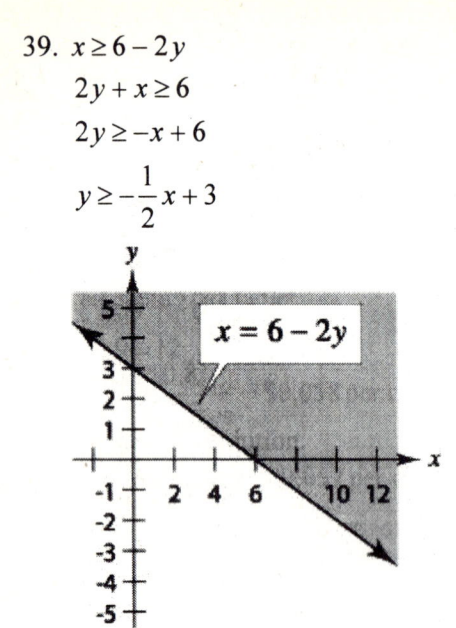

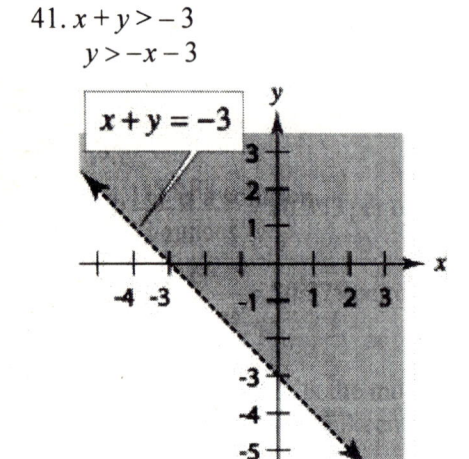

43. $-2 < 2y - 3x$
$-2 + 3x < 2y$
$\dfrac{3x - 2}{2} < y$
$y > \dfrac{3}{2}x - 1$

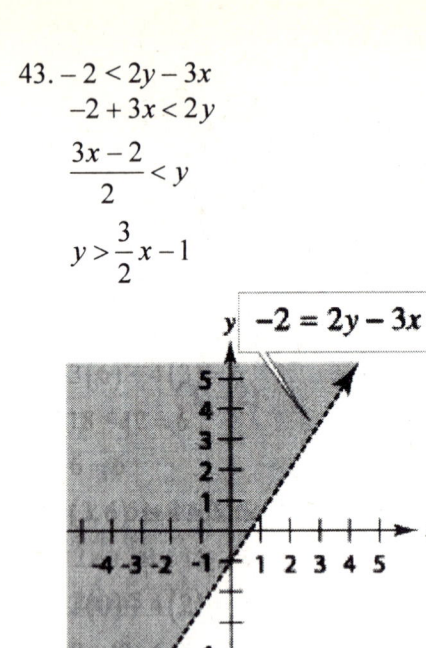

41. $x + y > -3$
$y > -x - 3$

45. $\dfrac{x}{3} - y < \dfrac{1}{4}$
$-y < -\dfrac{1}{3}x + \dfrac{1}{4}$
$y > \dfrac{1}{3}x - \dfrac{1}{4}$

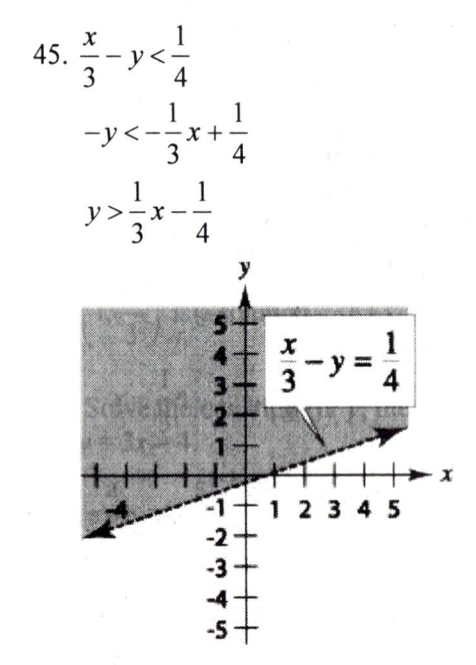

Properties of Lines

47. $2x + 1 > y - 2$
$2x + 3 > y$
$y < 2x + 3$

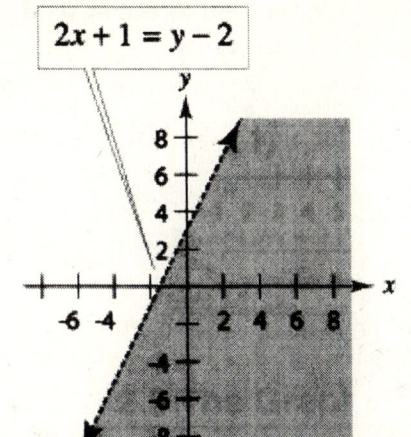

[Graph with boxed label $2x + 1 = y - 2$]

49. $3(x + y) \leq 2(x - y)$
$3x + 3y \leq 2x - 2y$
$x + 3y \leq -2y$
$x \leq -5y$
$-\frac{1}{5}x \geq y$
$y \leq -\frac{1}{5}x$

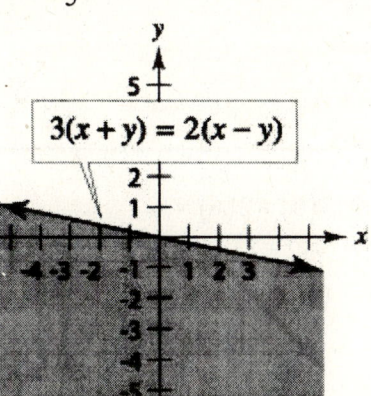

[Graph with boxed label $3(x + y) = 2(x - y)$]

51. $y \geq x + 2$

53. $x + y < 10$

55. $y - 2x > 0$

57. $y \leq \frac{1}{2}x$

59. $y - (x - 3) \geq 0$

61. We must know whether the inequality is being considered as a one-variable or two-variable inequality.

63. (a) [Number line with open parenthesis at 2]

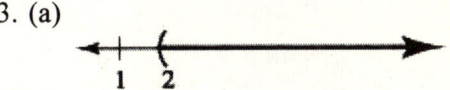

(b) [Graph with boxed label $x = 2$]

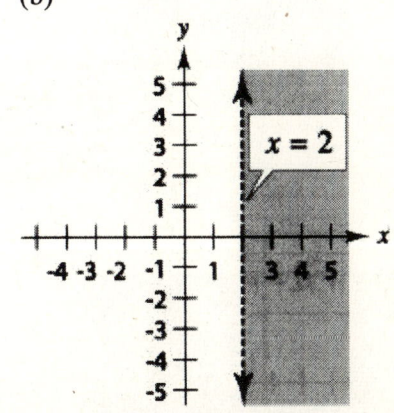

65. (a) [Number line with bracket at $-5$]

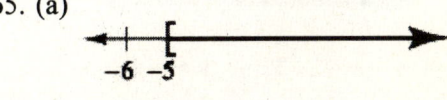

(b) [Graph with boxed label $x = -5$]

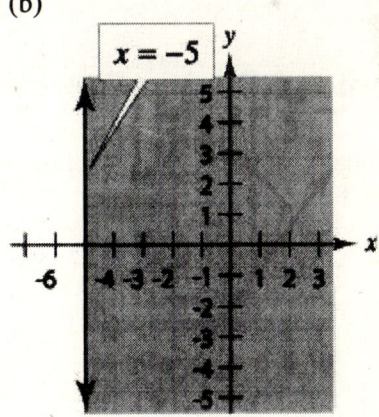

67. Boy: $w > 6.1a + 10.6$
Girl: $w > 6.5a + 8$

69. Boy: $w \geq 1.10(6.1a + 10.6)$
Girl: $w \geq 1.10(6.5a + 8)$
Note: $1.10 = 110\%$

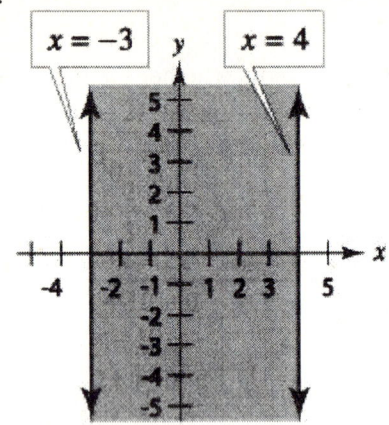

71.

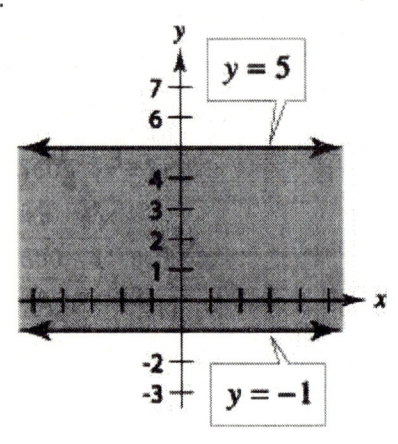

73.

75. $xy \geq 0$

In order for the inequality to be true, $x$ and $y$ must have the same sign. Thus, all points in Quadrants I and III and on the axes, satisfy the inequality.

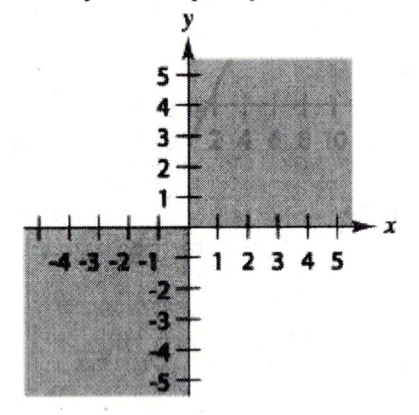

77. (a) $x > 0, y > 0$

(b) $x < 0, y > 0$

(c) $x < 0, y < 0$

(d) $x > 0, y < 0$

## Chapter 4 Review Exercises

1. Equation (ii) and (iii) cannot be written in the form $Ax + By = C$.

3. $3x - 2y = 12$

  (i) $3(2) - 2(-3) = 12$
  $6 + 6 = 12$
  $12 = 12$
  $(2, -3)$ is a solution.

  (ii) $3(-10) - 2(-21) = 12$
  $-30 + 42 = 12$
  $12 = 12$
  $(-10, -21)$ is a solution

  (iii) $3(0) - 2(6) = 12$
  $-12 = 12$
  $(0, 6)$ is not a solution.

  (iv) $3(6) - 2(3) = 12$
  $18 - 6 = 12$
  $12 = 12$
  $(6, 3)$ is a solution.

Properties of Lines

5.

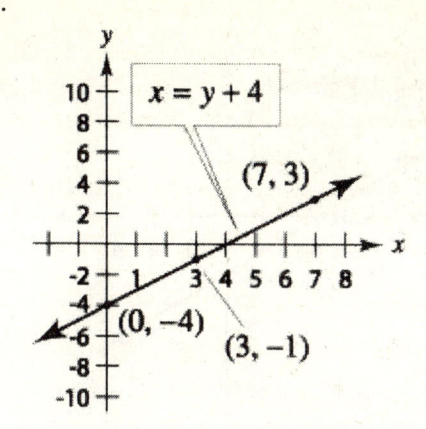

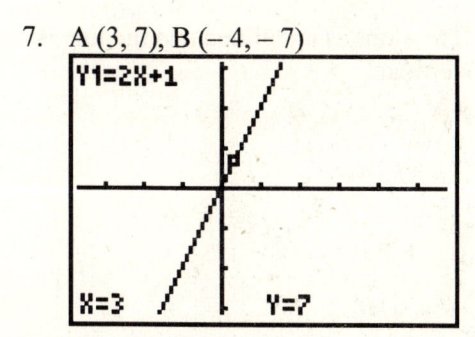

(graph showing line x = y + 4 passing through (7, 3), (3, −1), and (0, −4))

7. A (3, 7), B (−4, −7)

(calculator screen: Y1=2X+1, X=3, Y=7)

9. (i) $x = -2(y + 5)$
$x = -2y - 10$
$2y + x = -10$
$2y = -x - 10$
$y = -\dfrac{1}{2}x - 5$

(ii) $y = -0.5x + 5$

(iii) $x + 2y = 10$
$2y = -x + 10$
$y = -\dfrac{1}{2}x + 5$

Equation (ii) and (iii) have the same graph.

11. (0, 9), (−18, 0)

(calculator screen: Y1=1/2X+9, X=0, Y=9)

13. $2x - 5y - 20 = 0$
$2x - 20 = 5y$
$y = \dfrac{2}{5}x - 4$
(0, −4) is the y-intercept

15. The graph is a vertical line.

17. (a) $x - 7 = 0$
$x = 7$
vertical

(b) $x - 7y = 0$
neither

(c) $7y = 0$
$y = 0$
horizontal

19. $y = 2x + 50$
y-intercept: (0, 50)
$0 = 2x + 50$
$x = -25$
x-intercept: (-25, 0)

21. (iii)

23. Any two points of the line can be used to determine the slope.

25. $2x + 3y = 9$
$3y = -2x + 9$
$y = \dfrac{-2}{3}x + 3$
(0, 3) and (3, 1) are solutions.
$m = \dfrac{3-1}{0-3}$
$= \dfrac{2}{-3}$
$= -\dfrac{2}{3}$

27. (a) $m = \dfrac{y-(-2)}{7-(-8)}$

$m = \dfrac{y+2}{15}$

$0 = y+2$

$0 = -2$

(b) $m = \dfrac{9-(-3)}{0-x}$

$m = \dfrac{12}{-x}$

undefined $= -\dfrac{12}{x}$

$x = 0$

29. The missing point of the triangle is $(1, -7)$ or $(-4, 6)$. The length of the segment between $(1, -7)$ and $(1, 6)$ is 13. The length of the segment between $(1, -7)$ and $(-4, -7)$ is 5. (A similar calculation occurs if you use $(-4, 6)$.) The legs are 5 units and 13 units long.

31. $4x - 5y = 5$

$-5y = -4x + 5$

$y = \dfrac{4}{5}x - 1$

slope: $\dfrac{4}{5}$

$y$ – intercept: $(0, -1)$

33.

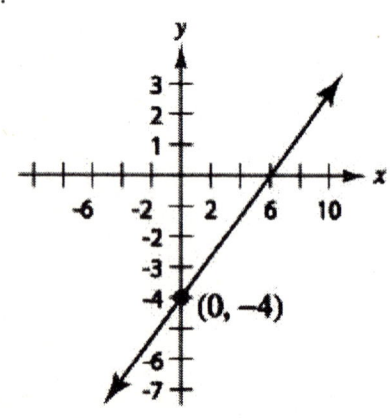

35.

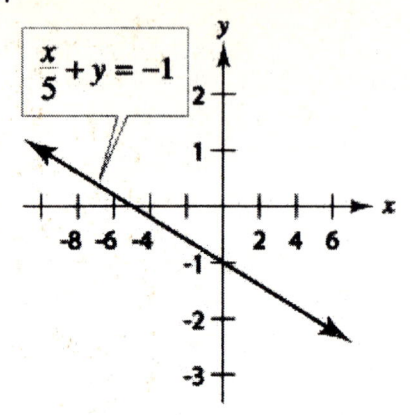

37. The slope is undefined and the line is vertical.

39. $y = kx$

$28 = k \cdot 4$

$k = 7$

$y = 7x$

$y = 7(16)$

$y = 112$

41. The lines are vertical and the slope is undefined.

43. $y = 1 - 3x$

$= -3x + 1$

$x + \dfrac{y}{3} = 2$

$3x + y = 6$

$y = -3x + 6$

The lines both have slope $-3$ so they're parallel

45. $3x + 4y = 0$

$4y = -3x$

$y = -\dfrac{3}{4}x$

(a) $L_2$ has slope $-\dfrac{3}{4}$

(b) $L_2$ has slope $\dfrac{4}{3}$

Properties of Lines

47. $3x + 2y = 5$
$2y = -3x + 5$
$y = -\dfrac{3}{2}x + \dfrac{5}{2}$
$k = -\dfrac{3}{2}$

49. The triangle is a right triangle.

51. $y = mx + b$
$y = \dfrac{3}{5}x - 2$

53. $y = -\dfrac{1}{2}x + b$
$-2 = -\dfrac{1}{2}(3) + b$
$\dfrac{-4}{2} = -\dfrac{3}{2} + b$
$b = -\dfrac{1}{2}$
$y = -\dfrac{1}{2}x - \dfrac{1}{2}$

55. The slope of the line is not defined.

57. $m = \dfrac{1-0}{0-1}$
$= -1$
The slope of $L_1 = -1$.
The slope of $L_2 = 1$.
$y = 1 \cdot x + b$
$4 = 1(-5) + b$
$b = 9$
$y = x + 9$

59. $y = -3x + b$
$7 = -3(3) + b$
$7 = -9 + b$
$b = 16$
$y = -3x + 16$

61. (a) $-2 > 3(1) - 5$
$-2 > -2$
No

(b) $5 > 3(3) - 5$
$5 > 1$
Yes

(c) $-10 > 3(-2) - 5$
$-10 > -11$
Yes

(d) $0 > 3(0) - 5$
$0 > -5$
Yes

63. $x - y < 0$
$x < y$
above

65. Any point other than a point of the line can be used as a test point.

67.

69. (a)

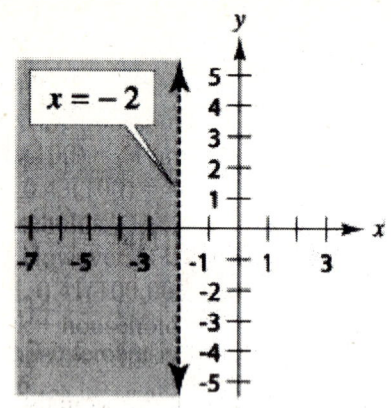

(b)

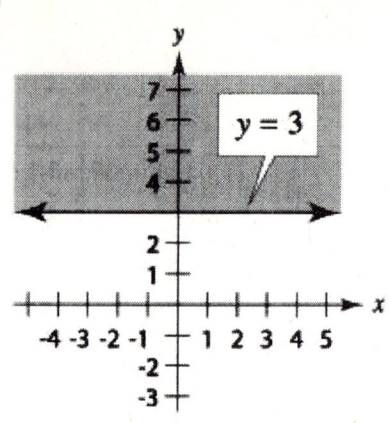

## Chapter 4 Test

1. $x + 4y = 8$
   (i) $4 + 4(1) = 8$
   $8 = 8$

   (ii) $0 + 4(-2) = 8$
   $-8 = 8$

   (iii) $8 + 4(0) = 8$
   $8 = 8$

   (iv) $1 + 4(4) = 8$
   $17 = 8$

   (i) and (iii) are solutions

3. The graph of equation (ii) is different from the other two graphs. All three graphs have the same $y$-intercept, $(0, 1)$. However, the slope in (ii) is $-\dfrac{2}{3}$. Whereas the slope in (i) and (iii) is $+\dfrac{2}{3}$.

5. $4y + 2x + 10 = 0$
   $4y = -2x - 10$
   $y = -\dfrac{2}{4}x - \dfrac{10}{4}$
   $y = -\dfrac{1}{2}x - \dfrac{5}{4}$

   $y$-intercept: $\left(0, -\dfrac{5}{2}\right)$

7. $m = \dfrac{0 - (-8)}{-5 - (-3)}$
   $= \dfrac{8}{-2}$
   $= -4$

9. (a) The line is vertical, so it has undefined slope.

   (b) $m = \dfrac{9 - 0}{-7 - 4}$
   $= \dfrac{9}{-11}$
   negative slope

Properties of Lines

(c) $m = \dfrac{5-0}{0-(-5)}$

$= \dfrac{5}{5}$

$= 1$

positive slope

(d) zero slope

11. Plot the point (–2, –5). From that point, move up 2 units and right 5 units to arrive at another point of the line. Draw a line through these two points.

13. $5y + x = 5$
$5y = -x + 5$
$y = -\dfrac{1}{5}x + 1$

(a) $-\dfrac{1}{5}$

(b) 5

15. Slope of AB $= \dfrac{3-(-5)}{3-(-5)}$

$= \dfrac{8}{8}$

$= 1$

Slope of BC $= \dfrac{3-(-1)}{3-7}$

$= \dfrac{4}{-7}$

$= -1$

Since the slope of AB is the negative reciprocal of the slope of BC, the line segments are perpendicular. So the triangle is a right triangle.

17. $m = \dfrac{4-(-4)}{3-(-3)}$

$= \dfrac{8}{6}$

$= \dfrac{4}{3}$

$y = \dfrac{4}{3}x + b$

$4 = \dfrac{4}{3}(3) + b$

$4 = 4 + b$

$b = 0$

$y = \dfrac{4}{3}x$

$3y = 4x$

$4x - 3y = 0$

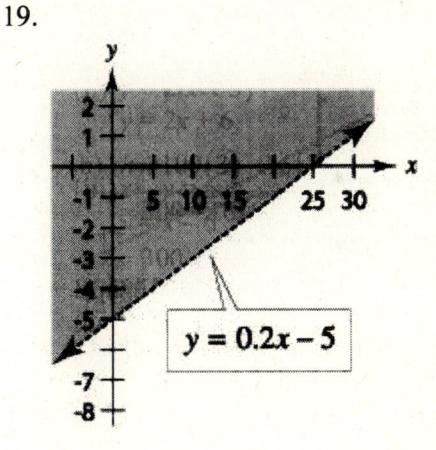

19.

21. $2x + 4y \geq 64$

# Chapter 5

# Systems of Linear Equations

## Section 5.1 The Graphing Method

1. A solution of a system of equations is a pair of numbers that satisfies both equations.

3. $2(-2)-3(-5)=11$
   $-4+15=11$
   $11=11$

   $-2+2(-5)=-12$
   $-2-10=-12$
   $-12=-12$

   $(-2,-5)$ is a solution.

5. $5(3)-2(5)=5$
   $15-0=5$
   $5=5$

   $3=0.4(5)+1$
   $3=2+1$
   $3=3$

   $(5, 3)$ ia a solution.

7. $a(1)-3(-6)=21$
   $a+18=21$
   $a=3$

   $6(1)+b(-6)=36$
   $6-6b=36$
   $-6b=30$
   $b=-5$

9. $2+1=c$
   $c=3$

   $a(2)+1=-5$
   $2a+1=-5$
   $2a=-6$
   $a=-3$

11. $2(-1)+b(3)=-5$
    $-2+3b=-5$
    $3b=-3$
    $b=-1$

    $-1+3(3)=c$
    $-1+9=c$
    $c=8$

13. No, two lines cannot intersect at exactly two points.

15. (0, 0)

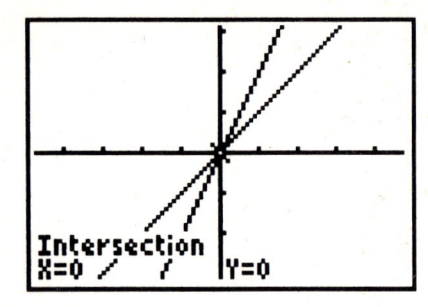

17. (12, −9)

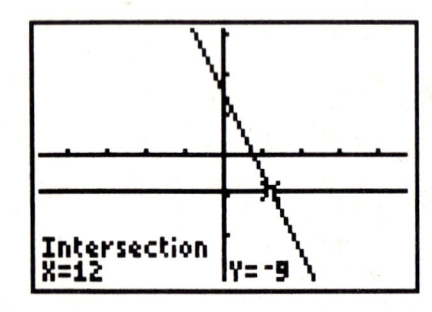

19. (9, 11)

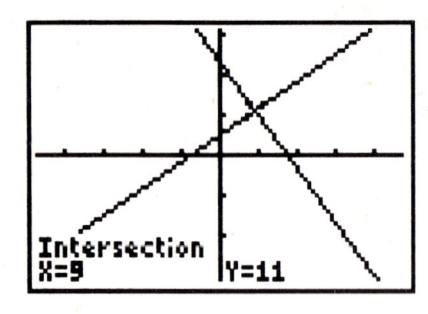

21. (12, −7)

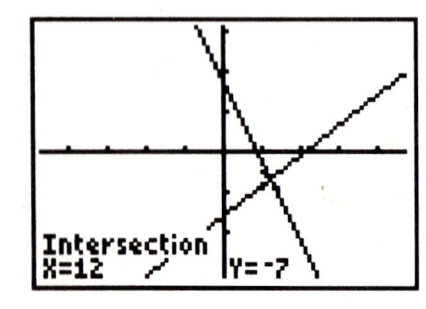

23. (4, −2)

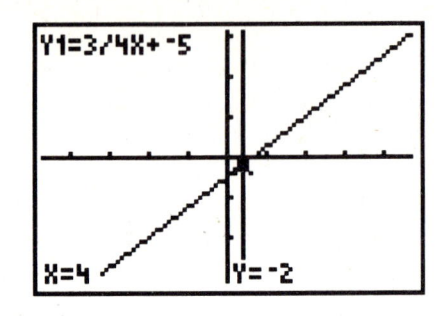

25. (−8, −3)

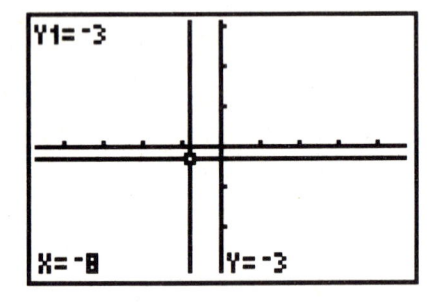

27. (a) The lines intersect at one point.

(b) The lines coincide.

(c) The lines are parallel.

(d) Not possible.

29. Since the lines have different slopes, there is one solution.

31. $3y - x = 15$
$3y = x + 15$
$y = \dfrac{1}{3}x + 15$

$3y - 10 = x + 5$
$3y = x + 15$
$y = \dfrac{1}{3}x + 5$

The lines coincide, so there are infinitely many solutions.

Systems of Linear Equations

33. $y - x = 7$
$y = x + 7$

$2y + 6 = 2x$
$2y = 2x - 6$
$y = x - 3$

The lines are parallel, so there are no solutions.

35. $2y + 8 = x$
$2y = x - 8$
$y = \frac{1}{2}x - 4$

$2x = y + 4$
$2x - 4 = y$
$y = 2x - 4$

The lines have different slopes, so there is one solution.

37. $2x + y = 4$
$y = -2x + 4$

$2x + 3 = -y$
$y = -2x - 3$

inconsistent

39. $3y - x = 6$
$3y = x + 6$
$y = \frac{1}{3}x + 2$

$6y - 12 = 2x$
$6y = 2x + 12$
$y = \frac{1}{3}x + 2$

consistent

41. $x = 4$
$x = 4y$
$y = \frac{1}{4}x$

consistent

43. $y - 2x = 6$
$y = 2x + 6$

$y = 2x + 6$

dependent

45. $y + 2x = 1$
$y = -2x + 1$

$y = 5 - 2x$
$y = -2x + 5$

independent

47. $y = 3x$
$y - 3 = 0$

$y = 3$

independent

49. (5, 4)

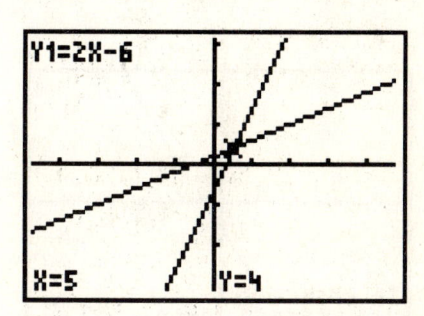

51. No solution

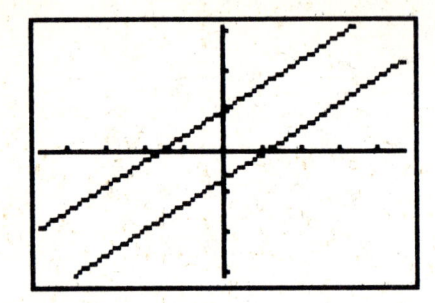

53. $(-5, 3)$

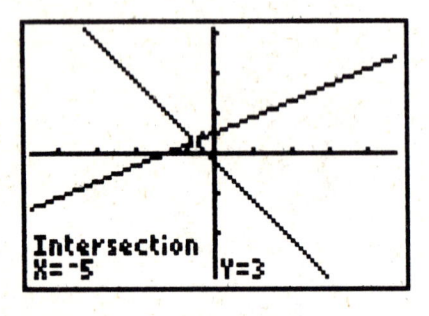

55. Infinitely many solutions

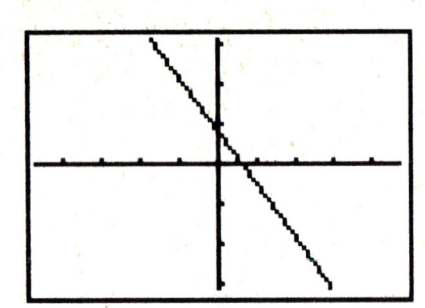

57. $\left(\dfrac{32}{3}, 10\right)$

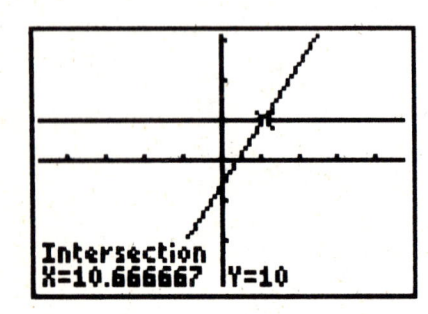

59. $(11, 16)$

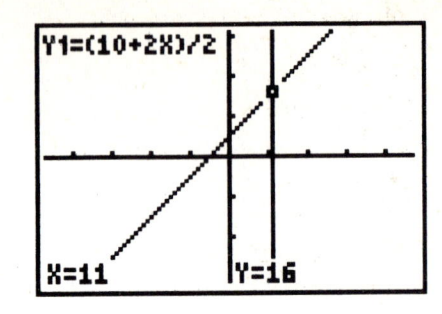

61. $(2.8, -1.6)$

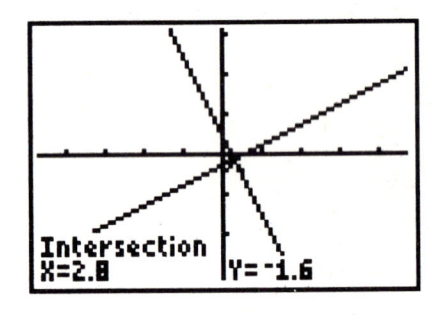

63. $(1.8, -1.4)$

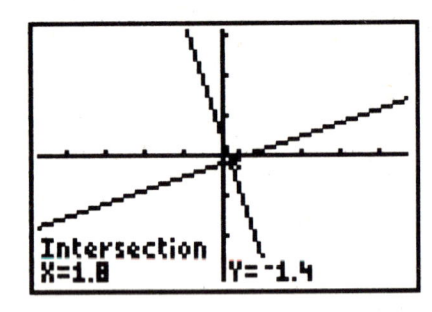

65. No solution.

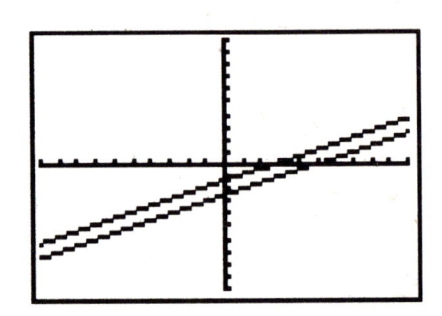

Systems of Linear Equations

67. Infinitely many solutions.

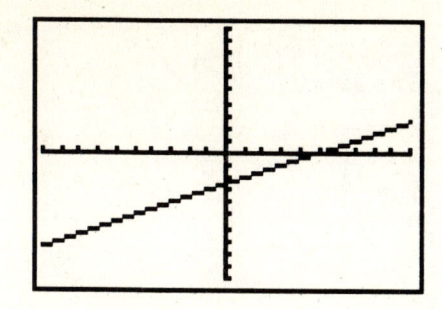

69. $\left(\dfrac{1}{3}, -\dfrac{1}{4}\right)$

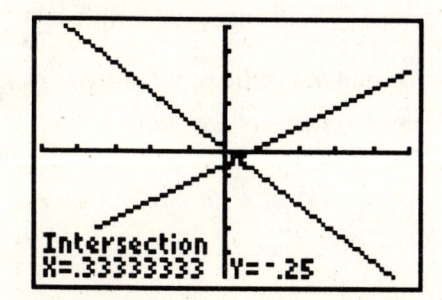

71. $\left(-\dfrac{9}{8}, \dfrac{7}{2}\right)$

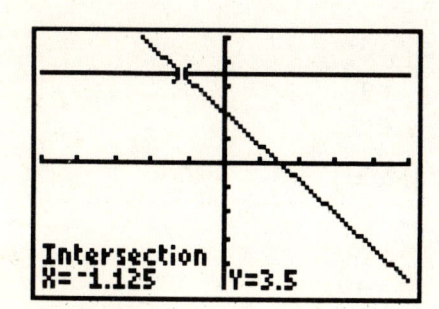

73. $x + y = 69$
    $2x - 9 = y$
    $x = 26$
    $y = 43$

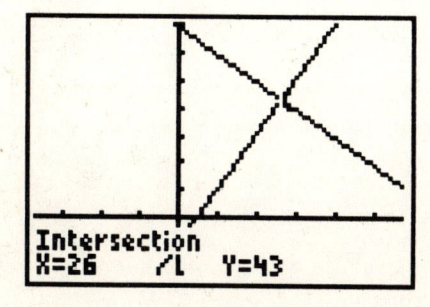

75. $x + y = 180$
    $x + 12 = y$

    $x = 84°$
    $y = 96°$

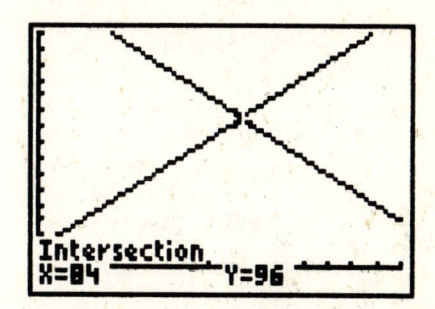

77. (a) $(0, 7.18), (5, 8.30)$
    $$m = \dfrac{8.30 - 7.18}{5 - 0}$$
    $$= \dfrac{1.12}{5}$$
    $$= 0.224$$

    Boys: $y = 0.224x + 7.18$
    $(0, 3.44), (5, 5.35)$
    $$m = \dfrac{5.35 - 3.44}{5 - 0}$$
    $$= 0.382$$
    Girls: $y = 0.382x + 3.44$

    (b) $y = 0.22x + 7.18$
    $y = 0.382x + 3.44$
    $(23.38, 12.32)$

79. (a) $(0, 3034), (8, 2277)$

$$m = \frac{3034 - 2277}{0 - 8}$$

$$= -94.625$$

Cancellations:

$y = -94.625x + 3034$

$y \approx -95x + 3034$

$(0, 624), (8, 1137)$

$$m = \frac{1137 - 624}{8 - 0}$$

$$= 64.125$$

Ticketing:

$y = 64x + 624$

(b) The slopes of the lines are different, so the lines must intersect. Thus, there is a solution.

81. $(8, -4)$

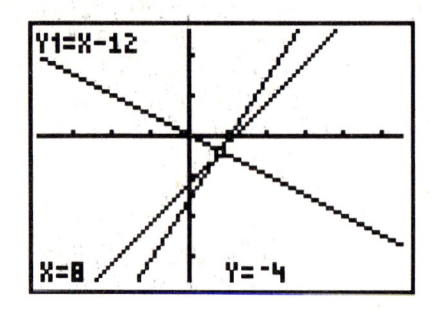

83. No solution.

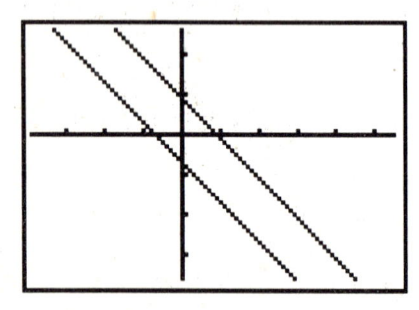

85. $y = \frac{2}{3}x + 13$

$2y + 3x = 0$

$2y = -3x$

$y = -\frac{3}{2}x$

$x = 12$

The triangle is a right triangle, since the slope $\frac{2}{3}$ is the negative reciprocal of $-\frac{3}{2}$. To find the vertices, we must solve three systems of equations.

$y = \frac{2}{3}x + 13$

$y = -\frac{3}{2}x$

$(-6, 9)$

$x = \frac{2}{3}x + 13$

$x = 12$

$(12, 21)$

$y = -\frac{3}{2}x$

$x = 12$

$(12, -18)$

## Section 5.2 The Addition Method

1. The coefficients of one of the variables are opposites.

3. $y = x - 7$
   $-x + y = -7$

   $-x + y = -7$
   $\underline{x + y = 3}$
   $2y = -4$
   $y = -2$

   $-2 = x - 7$
   $x = 5$

   $(5, -2)$ is the solution.

5. $y = x + 6$
   $-x + y = 6$

   $-x + y = 6$
   $2x + y = -15$

   $-x + y = 6$
   $\underline{-2x - y = -15}$ (multiply by $-1$)
   $-3x = 21$
   $x = -7$

   $-(-7) + y = 6$
   $7 + y = 6$
   $y = -1$
   $(-7, -1)$ is the solution.

7. $y = 5 - x$
   $x + y = 5$

   $x + y = 5$
   $2x + 2y = -16$

   $2x + 2y = 10$ (multiply by 5)
   $\underline{2x + 2y = -16}$
   $0 = 26$
   No solution.

9. $2x + y = 2$
   $\underline{x - y = -11}$
   $3x = -9$
   $x = -3$

   $2(-3) + y = 2$
   $-6 + y = 2$
   $y = 8$

   $(-3, 8)$

11. $x + 5y = 13$
    $\underline{-x + 6y = 9}$
    $11y = 22$
    $y = 2$

    $x + 5(2) = 13$
    $x + 10 = 13$
    $x = 3$

    $(3, 2)$

13. $$\begin{aligned} x - 6y &= -11 \\ 3x + 6y &= 7 \\ \hline 4x &= -4 \\ x &= -1 \end{aligned}$$

$$\begin{aligned} (-1) - 6y &= -11 \\ -6y &= -10 \\ y &= \frac{5}{3} \end{aligned}$$

$\left(-1, \dfrac{5}{3}\right)$

15. $$\begin{aligned} 4x - 9y &= -7 \\ -4x + 5y &= -5 \\ \hline -4y &= -12 \\ y &= 3 \end{aligned}$$

$$\begin{aligned} 4x - 9(3) &= -7 \\ 4x - 27 &= -7 \\ 4x &= 20 \\ x &= 5 \end{aligned}$$

$(5, 3)$

17. $$\begin{aligned} \frac{1}{5}x - \frac{3}{4}y &= 4 \\ \frac{-2}{5}x + \frac{3}{4}y &= -2 \\ \hline -\frac{1}{5}x &= 2 \\ x &= -10 \end{aligned}$$

$$\begin{aligned} \frac{1}{5}(-10) - \frac{3}{4}y &= 4 \\ -2 - \frac{3}{4}y &= 4 \\ -\frac{3}{4}y &= 6 \\ -3y &= 24 \\ y &= -8 \end{aligned}$$

$(-10, -8)$

19. When the equations are added, at least one variable is eliminated.

21. $$\begin{aligned} 5x + y &= 12 \\ -3x - y &= -2 \quad \text{(multiply by } -1\text{)} \\ \hline 2x &= 10 \\ x &= 5 \end{aligned}$$

$$\begin{aligned} 5(5) + y &= 12 \\ 25 + y &= 12 \\ y &= -13 \end{aligned}$$

$(5, -13)$

## Systems of Linear Equations

23.    $3x + 7y = 8$
     $\underline{42x - 7y = 7}$   (multiply by 7)
         $45x = 15$
           $x = \dfrac{1}{3}$

    $3\left(\dfrac{1}{3}\right) + 7y = 8$
        $1 + 7y = 8$
           $7y = 7$
            $y = 1$

   $\left(\dfrac{1}{3}, 1\right)$

25.   $-2x + 6y = -10$  (multiply by $-2$)
     $\underline{2x - 6y = -7}$
         $0 = -17$

   No solution

27.   $9x - 2y = 7$
     $\underline{x + 2y = 3}$    $\left(\text{multiply by }\dfrac{1}{4}\right)$
        $10x = 10$
           $x = 1$

    $9(1) - 2y = 7$
      $9 - 2y = 7$
        $-2y = -2$
           $y = 1$

   $(1, 1)$

29.   $4x - y = -3$
     $\underline{-4x + y = 3}$   $\left(\text{multiply by }\dfrac{1}{2}\right)$
         $0 = 0$

   Infinitely many solutions

31.   $-30x + 16y = -16$ (multiply by -2)
     $\underline{30x + 7y = -7}$
         $23y = -23$
           $y = -1$

    $30x + 7(-1) = -7$
       $30x - 7 = -7$
          $30x = 0$
            $x = 0$

   $(0, -1)$

33.   $6x - 8y = 22$ (multiply by 2)
    $\underline{-6x - 9y = 12}$  (multiply by -3)
         $-17y = 34$
            $y = -2$

    $6x - 8(-2) = 22$
       $6x + 16 = 22$
           $6x = 6$
            $x = 1$

   $(1, -2)$

35.   $10x - 4y = 2$ (multiply by 2)
    $\underline{-10x + 15y = -35}$   (multiply by -5)
         $11y = -33$
           $y = -3$

    $10x - 4(-3) = 2$
       $10 + 12 = 2$
          $10x = -10$
            $x = -1$

   $(-1, -3)$

37. $3x - 5y = 6$
$\underline{-3x + 5y = -6}$ (multiply by $-6$)
$0 = 0$

Infinitely many solutions

39. $6x - 4y = 26$ (multiply by 2)
$\underline{-6x - 9y = 0}$ (multiply by -6)
$-13y = 26$
$y = -2$

$6x - 4(-2) = 26$
$6x + 8 = 26$
$6x = 18$
$x = 3$

$(3, -2)$

41. $2x + 5y = -4$
$\underline{-2x - 5y = -4}$ (multiply by $-2$)
$0 = -8$

No solution

43. $24x + 20y = 56$ (multiply by 4)
$\underline{-24x - 42y = -12}$ (multiply by $-6$)
$-22y = 44$
$y = -2$

$24x + 20(-2) = 56$
$24x - 40 = 56$
$24x = 96$
$x = 4$

$(4, -2)$

45. Multiply by the LCD to clear fractions.

47. $2x - 3y = -5$
$3x - 2y = 0$

$6x - 9y = -15$ (multiply by 3)
$\underline{-6x + 4y = 0}$ (multiply by -2)
$-5y = -15$
$y = 3$

$2x - 3(3) = -5$
$2x - 9 = -5$
$2x = 4$
$x = 2$

$(2, 3)$

49. $x + y = -2$
$3x + 3y = 4$

$-3x - 3y = 6$ (multiply by -3)
$\underline{3x + 3y = 4}$
$0 = 10$

No solution

51. $3x - 2y = 4$
$-x + y = 3$

$3x - 2y = 4$
$\underline{-2x + 2y = 6}$ (multiply by 2)
$x = 10$

$3(10) - 2y = 4$
$30 - 2y = 4$
$-2y = -26$
$y = 13$

$(10, 13)$

Systems of Linear Equations

53. $2x - y = 1$
$\underline{-2x + 5y = 1}$
$4y = 2$
$y = \dfrac{1}{2}$

$2x - \dfrac{1}{2} = 1$
$2x = \dfrac{3}{2}$
$x = \dfrac{3}{4}$

$\left(\dfrac{3}{4}, \dfrac{1}{2}\right)$

55. $-3x - y = -10$
$6x + 2y = 5$

$-6x - 2y = -20$ (multiply by 2)
$\underline{6x + 2y = 5}$
$0 = -15$

No solution

57. $8x - 3y = 11$

$4x + 7y = -3$

$8x - 3y = 11$
$\underline{-8x - 14y = 6}$ (multiply by $-2$)
$-17y = 17$
$y = -1$

$8x - 3(-1) = 11$
$8x + 3 = 11$
$8x = 8$
$x = 1$

$(1, -1)$

59. $3x + 5y = 5$
$\underline{4x - 5y = -5}$
$7x = 0$
$x = 0$

$3(0) + 5y = 5$
$5y = 5$
$y = 1$
$(0, 1)$

61. $5x + 2y = 0$ (multiply by 10)
$3x + 5y = 0$ (multiply by 15)

$-25x - 10y = 0$ (multiply by $-5$)
$\underline{6x + 10y = 0}$ (multiply by 2)
$-19x = 0$
$x = 0$

$5(0) + 2y = 0$
$2y = 0$
$y = 0$
$(0, 0)$

63. $6x - y = 6$ (multiply by 3)
$\underline{30x + y = 66}$ (multiply by 6)
$36x = 72$
$x = 2$

$6(2) - y = 6$
$12 - y = 6$
$-y = -6$
$y = 6$
$(2, 6)$

65. $8x+7y=-14$ (multiply by 14)
$4x-25y=50$ (multiply by 5)

$8x+7y=-14$
$\underline{-8x+50y=-100}$ (multiply by -2)
$57y=-114$
$y=-2$

$8x+7(-2)=-14$
$8x-14=-14$
$8x=0$
$x=0$

$(0,-2)$

67. $y=\dfrac{1}{2}x+1$

$\dfrac{y-3}{x-2}=\dfrac{2}{5}$
$5(y-3)=2(x-2)$
$5y-15=2x-4$
$5y=2x+11$
$-2x+5y=11$

$-\dfrac{1}{2}x+y=1$
$-2x+5y=11$

$2x-4y=-4$ (multiply by $-4$)
$\underline{-2x+5y=11}$
$y=7$

$-\dfrac{1}{2}x+7=1$
$-\dfrac{1}{2}x=-6$
$x=12$

So $x=12$ and $y=7$.

69. $x+y=90$
$\underline{3x-y=6}$
$4x=96$
$x=24$

$24+y=90$
$y=66$

The angles are $24°$ and $66°$.

71. (a) $(0,25.3),(1,34.8)$
$m=\dfrac{34.8-25.3}{1-0}$
$=9.5$
men: $y=9.5x+25.3$

$(0,19.6),(1,30.4)$
$m=\dfrac{30.4-19.6}{1-0}$
$=10.8$
women: $y=10.8x+19.6$

(b) $-9.5x+y=25.3$
$-10.8x+y=19.6$

$-9.5x+y=25.3$
$\underline{10.8x-y=-19.6}$ (multiply by $-1$)
$1.3x=5.7$
$x=4.385$
$\approx 4.4$

$-9.5(4.385)+y=25.3$
$-41.655+y=25.3$
$y\approx 67.0$

73. The slope of the first equation indicates that the percentage of male online shoppers is decreasing. The slope of the second equation indicates that the percentage of female online shoppers is increasing.

Systems of Linear Equations

75. $y = -4.75x + 104.75$
$\underline{y = 4.75x - 4.75}$
$2y = 100$
$y = 50$

$50 = -4.75x + 104.75$
$-54.75 = -4.75x$
$x = 11.53$

$(11.5, 50)$

77. $x + y = a$
$\underline{x - y = b}$
$2x = a + b$
$x = \dfrac{a+b}{2}$

$\dfrac{a+b}{2} + y = a$
$a + b + 2y = 2a$
$b + 2y = a$
$2y = a - b$
$y = \dfrac{a-b}{2}$

$\left(\dfrac{a+b}{2}, \dfrac{a-b}{2}\right)$

79. $x - 2y = k$
$-2x + 4y = 10$

$2x - 4y = 2k$ (multiply by 2)
$\underline{-2x + 4y = 10}$
$0 = 2k + 10$
$-10 = 2k$
$k = -5$

81. $kx + 14y = 2$
$2x - 7y = 3$

$kx + 14y = 2$
$\underline{4x - 14y = 6}$ (multiply by 2)
$kx + 4x = 8$
$(k + 4)x = 8$

Let $k = -4$
$(-4 + 4)x = 80$
$0 = 8$

## Section 5.3 The Substitution Method

1. (a) Multiply the second equation by $-3$ and add the equations.

   (b) Solve the second equation for $y$.

3. $8x + (3 - 2x) = 9$
$6x + 3 = 9$
$6x = 6$
$x = 1$

$y = 3 - 2(1)$
$= 1$
$(1, 1)$

5. $(3y-6)-y=8$
$2y-6=8$
$2y=14$
$y=7$

$x=3(7)-6$
$x=15$

$(15,7)$

7. $3(5)-4y=3$
$15-4y=3$
$-4y=-12$
$y=3$

$(5,3)$

9. $3x+2(-2x-7)=-10$
$3x-4x-14=-10$
$-x-14=-10$
$-x=4$
$x=-4$

$y=-2(-4)-7$
$=8-7$
$=1$

$(-4,1)$

11. $\frac{1}{2}(4+6y)-4y=6$
$2+3y-4y=6$
$2-y=6$
$-y=4$
$y=-4$

$x=4+6(-4)$
$=4-24$
$=-20$

$(-20,-4)$

13. $2x+y=4$
$y=-2x+4$

$2x-(-2x+4)=0$
$4x-4=0$
$4x=4$
$x=1$

$y=-2(1)+4$
$=-2+4$
$=2$

$(1,2)$

15. $x + 2y = 8$
$x = -2y + 8$

$4(-2y + 8) - 5y = -20$
$-8y + 32 - 5y = -20$
$-13y + 32 = -20$
$-13y = -52$
$y = 4$

$x = -2(4) + 8$
$= -8 + 8$
$= 0$

$(0, 4)$

17. $y + 5 = 0$
$y = -5$

$2x + 3(-5) = 1$
$2x - 15 = 1$
$2x = 16$
$x = 8$

$(8, -5)$

19. $2x - 6y = 10$
$x - 3y = 5$
$x = 3y + 5$

$7 = 2(3y + 5) - 5y$
$7 = 6y + 10 - 5y$
$7 = y + 10$
$y = -3$

$x = 3(-3) + 5$
$= -9 + 5$
$= -4$

$(-4, -3)$

21. $5x - 4y = 4$
$-4y = -5x + 4$
$y = \dfrac{5}{4}x - 1$

$2x - 3 = 3\left(\dfrac{5}{4}x - 1\right)$
$2x - 3 = \dfrac{15}{4}x - 3$
$2x = \dfrac{15}{4}x$
$0 = \dfrac{2}{4}x$
$x = 0$

$y = \dfrac{5}{4}(0) - 1$
$= -1$

$(0, -1)$

Systems of Linear Equations

23. $\dfrac{1}{6}x + \dfrac{1}{4}y = \dfrac{1}{3}$

$x + \dfrac{6}{4}y = 2$

$x = \dfrac{-3}{2}y + 2$

$\dfrac{1}{3}\left(\dfrac{-3}{2}y + 2\right) - \dfrac{1}{2}y = \dfrac{-4}{3}$

$-\dfrac{1}{2}y + \dfrac{2}{3} - \dfrac{1}{2}y = \dfrac{-4}{3}$

$-y + \dfrac{2}{3} = \dfrac{-4}{3}$

$-y = \dfrac{-6}{3}$

$y = 2$

$x = \dfrac{-3}{2}(2) + 2$

$= -3 + 2$

$= -1$

$(-1, 2)$

25. The graphs coincide, and the system has infinitely many solutions.

27. $x - 2(2x) = -9$

$x - 4x = -9$

$-3x = -9$

$x = 3$

$y = 2(3)$

$= 6$

$(3, 6)$

29. $x = 4y + 2$

$3(4y + 2) - 12y = 0$

$12y + 6 - 12y = 0$

$6 = 0$

No solution

31. $x = 2y + 3$

$6y - 3(2y + 3) = -9$

$6y - 6y - 9 = -9$

$-9 = -9$

Infinitely many solutions

33. $x + 4y = 2$

$x = 2 - 4y$

$(2 - 4y) + 3y = -1$

$2 - 4y + 3y = -1$

$2 - y = -1$

$-y = -3$

$y = 3$

$x = 2 - 4(3)$

$= 2 - 12$

$= -10$

$(-10, 3)$

35. $2y = 6x$

$y = 3x$

$3x - (3x) = 0$

$0 = 0$

Infinitely many solutions

Systems of Linear Equations

37. $x - 5y = 0$
$x = 5y$

$2(5y) + 3y = 0$
$10y + 3y = 0$
$13y = 0$
$y = 0$

$x = 5(0)$
$= 0$

$(0, 0)$

39. $x = 2 - 5y$
$3(2 - 5y) = 5 - 15y$
$6 - 15y = 5 - 15y$
$6 = 5$

No solution

41. $x - y = 16$
$x + 3y = -24$
$x = y + 16$

$(y + 16) + 3y = -24$
$4y + 16 = -24$
$4y = -40$
$y = -10$

$x = -10 + 16$
$= 6$

So, $x = 6$ and $y = -10$.

43. $w = \dfrac{1}{2}l - 1$
$2w + 2l = 64$

$2\left(\dfrac{1}{2}l - 1\right) + 2l = 64$
$l - 2 + 2l = 64$
$3l - 2 = 64$
$3l = 66$
$l = 22$ feet

$w = \dfrac{1}{2}(22) - 1$
$= 11 - 1$
$= 10$ feet

The dimensions are 10 feet by 22 feet.

45. $(0, 2.7), (9, 1.5)$
$m = \dfrac{2.7 - 1.5}{0 - 9}$
$= \dfrac{1.2}{-9}$
$= -0.133$

Energy: $y = -0.133x + 2.7$

$(0, 1.4), (9, 2.0)$
$m = \dfrac{2.0 - 1.4}{9 - 0}$
$= \dfrac{0.6}{9}$
$= 0.067$

Environment: $y = 0.067x + 1.4$

$(0, 2.4), (9, 4.6)$

$m = \dfrac{4.6 - 2.4}{9 - 0}$

$= \dfrac{2.2}{9}$

$= 0.244$

General Science: $y = 0.244x + 2.4$

47. $y = -0.13x + 2.7$
$y = 0.07x + 1.4$

$-0.13x + 2.7 = 0.07x + 1.4$
$2.7 = 0.20x + 1.4$
$1.3 = 0.20x$
$x \approx 6.5$

$y = 0.07(6.5) + 1.4$
$\approx 1.9$

The solution indicates that in 1997, the amounts spent in the two categories were the same.

49. (a) $3y - (k_1 y + 12) = 21$
$3y - k_1 y - 12 = 21$
$y(3 - k_1) - 12 = 21$
$y(3 - k_1) = 33$
Let $k_1 = 3$
$y(3 - 3) = 33$
$0 = 33$
No solution

(b) $3\left(\dfrac{2}{3}x + 1\right) - 2x = k_2$
$2x + 3 - 2x = k_2$
$3 = k_2$
Let $k_2 = 3$
$3 = 3$
Infinitely many solutions

51. $x - 2 = 5$
$x = 7$

$7 + 2y = 13$
$2y = 6$
$y = 3$

$2(7) - 3 + z = 9$
$14 - 3 + z = 9$
$11 + z = 9$
$z = -2$

$(7, 3, -2)$

53. $x + 3 = 0$
$x = -3$

$6y - 7x = 15$
$6y = 7x + 15$
$y = \dfrac{7}{6}x + \dfrac{15}{6}$
$y = 6$

Vertex A: $(-3, 6)$

Vertex B:
$6 = \dfrac{7}{6}x + \dfrac{15}{6}$
$36 = 7x + 15$
$21 = 7x$
$x = 3$

$y = \dfrac{7}{6}(3) + \dfrac{5}{2}$
$= \dfrac{7}{2} + \dfrac{5}{2}$
$= 6$

$(3, 6)$

Vertex C:
$$y = \frac{7}{6}(-3) + \frac{15}{6}$$
$$= -\frac{7}{2} + \frac{5}{2}$$
$$= -1$$

$(-3, -1)$

AC is 7 units long, AB is 6 units long, and BC is $\sqrt{(6-(-1))^2 + (3-(-3))^2} = \sqrt{49+36} = \sqrt{85}$ units long. The triangle is not isosceles.

## Section 5.4 Applications

1. The product of the number of items and the unit value is the total value of the items.

3. $n + q = 43$
   $5n + 25q = 495$

   $n + q = 43$
   $\underline{-n - 5q = 99}$ (multiply by $-\frac{1}{5}$)
   $-4q = -56$
   $q = 14$

   $n + 14 = 43$
   $n = 29$

   There are 29 nickels and 14 quarters.

5. $5f + 20t = 275$
   $f = 3t - 1$

   $5(3t - 1) + 20t = 275$
   $15t - 5 + 20t = 275$
   $35t - 5 = 275$
   $35t = 280$
   $t = 8$

   $f = 3(8) - 1$
   $= 24 - 1$
   $= 23$

   There are 8 twenties and 23 fives.

7.  $b$ = amount of basil
    $s$ = amount of oregano

    $6.20b + 3.70s = 5.20\,(20)$
    $b + s = 20$

    $6.20\,(20 - s) + 3.70s = 104$
    $124 - 6.2s + 3.7s = 104$
    $-2.5s + 124 = 104$
    $-2.5s = -20$
    $s = 8$
    $b + 8 = 20$
    $b = 12$

    There are 12 ounces of basil and 8 ounces of oregano.

9.  $a$ = speed of airplane
    $w$ = speed of wind

    Total distance: 800 miles
    $d$ = rate · time

    $800 = (a + w)4$
    $800 = (a - w)5$

    $800 = 4a + 4w$
    $800 = 5a - 5w$

    $200 = a + w$
    $160 = a - w$
    $360 = 2a$
    $180 = a$

    $200 = 180 + w$
    $w = 20$

    The airplane's speed is 180 mph and the wind speed is 20 mph.

11. $b$ = boat speed
    $c$ = current speed

    $14 = 2(b + c)$
    $6 = 2(b - c)$

    $14 = 2b + 2c$
    $6 = 2b - 2c$
    $20 = 4b$
    $b = 5$

    $14 = 2(5) + 2c$
    $14 = 10 + 2c$
    $4 = 2c$
    $c = 2$

    The canoe traveled at 5 mph and the current moved at 2 mph.

13. $f$ = time on freeway
    $s$ = time on secondary roads

    $f + s = 5.5$
    $306 = 62f + 48s$

    $306 = 62(5.5 - s) + 48s$
    $306 = 341 - 62s + 48s$
    $306 = -14s + 341$
    $-35 = -14s$
    $s = 2.5$
    $f + 2.5 = 5.5$
    $f = 3$

    $62f = 62(3) = 186$ miles on freeway
    $48s = 48(2.5) = 120$ miles on secondary roads

15. The volume of the solution is needed. To determine the amount of acid, multiply the concentration (0, 2) by the volume of the solution.

## Systems of Linear Equations

17. $s$ = amount in savings
$c$ = amount in CD
$s + c = 8250$
$s(.04) \cdot 1 + c(0.055) \cdot 1 = 414$
$0.04s + 0.055c = 414$
$0.04(8250 - c) + 0.055x = 414$
$330 - 00.4c + 0.055c = 414$
$0.015c + 330 = 414$
$0.015c = 84$
$c = 5600$

$s + 5600 = 8250$
$s = 2650$

$2650 is invested in the savings account and $5600 is invested in the CD.

19. $c$ = clothing sales
$x$ = other sales

$c + x = 9950$
$0.20c + 0.15 = 1677.50$

$0.20(9950 - x) + 0.15x = 1677.50$
$1990 - 0.2x + 0.15x = 1677.50$
$-0.05x + 1990 = 1677.50$
$-0.05x = -322.5$
$x = 6450$

$c + 6450 = 9950$
$c = 3500$

Clothing sales total $3500 and other sales item sales total $6450.

21. $x$ = amount of 75 % solution
$y$ = amount of 45 % solution

$x + y = 15$
$0.75x + 0.45y = 0.65\,(15)$

$0.75(15 - y) + 0.45y = 9.75$
$11.25 - 0.75y + 0.45y = 9.75$
$-0.30y + 11.25 = 9.75$
$-0.3y = -1.5$
$y = 5$

$x + 5 = 15$
$x = 10$

There are 10 liters of the 75% solution and 5 liters of the 45% solution.

23. $c$ = car age
$t$ = truck age

$c = 2t$
$c - 2 = 4(t - 2)$

$(2t) - 2 = 4(t - 2)$
$2t - 2 = 4t - 8$
$-2 = 2t - 8$
$6 = 2t$
$t = 3$

$c = 2(3)$
$= 6$

The truck is 3 years old and the car is 6 years old.

163

25. $w = 4c$
$w + 20 = 3(c + 20)$

$(4c) + 20 = 3(c + 20)$
$4c + 20 = 3c + 60$
$c + 20 = 60$
$c = 40$

$w = 4(40)$
$= 160$

There are 40 acres in corn and 160 acres in wheat.

27. $h = 3a$
$h + 50 = 5(a - 50)$

$3a + 50 = 5(a - 50)$
$3a + 50 = 5a - 250$
$-2a + 50 = -250$
$-2a = -300$
$a = 150$
$h = 3(150)$
$= 450$

There are 450 homes and 150 apartments.

29. $c = 3t$
$c + t = 48$

$3t + t = 48$
$4t = 48$
$t = 12$

$c = 3(12)$
$c = 36$

There are 12 trucks and 36 cars.

31. $a + p = 15$
$1.10a + 1.40p = 18.75$

$1.10(15 - p) + 1.40p = 18.75$
$16.5 - 1.1p + 1.4p = 18.75$
$0.3p + 16.5 = 18.75$
$0.3p = 2.25$
$p = 7.5$

$a + 7.5 = 15$
$a = 7.5$

There are $7\frac{1}{2}$ pounds each of the apples and pears.

33. $x$ = speed on icy roads
$y$ = speed on wet roads

$0.75x + 1.5y = 87$
$x = \frac{1}{2}y - 4$

$0.75\left(\frac{1}{2}y - 4\right) + 1.5y = 87$
$0.375y - 3 + 1.5y = 87$
$1.875y - 3 = 87$
$1.875y = 90$
$y = 48$

$x = \frac{1}{2}(48) - 4$
$= 24 - 4$
$= 20$

The speed on icy roads was 20 mph and the speed on wet roads was 48 mph.

Systems of Linear Equations

35. $n = d - 4$

$\dfrac{n+2}{d+3} = \dfrac{1}{2}$

$\dfrac{(d-4)+2}{d+3} = \dfrac{1}{2}$

$\dfrac{d-2}{d+3} = \dfrac{1}{2}$

$2(d-2) = 1(d+3)$
$2d - 4 = d + 3$
$d - 4 = 3$
$d = 7$

$n = 7 - 4$
$= 3$

The numerator is 3 and the denominator is 7, so the fraction is $\dfrac{3}{7}$.

37. $2w + 2l = 30$
$l = 2w$

$2w + 2(2w) = 30$
$2w + 4w = 30$
$6w = 30$
$w = 5$

$l = 2(5)$
$= 10$

The length is 10 feet and the width is 5 feet.

39. $5c + 8b = 5056$
$c + b = 842$

$5(842 - b) + 8b = 5056$
$4210 - 5b + 8b = 5056$
$3b + 4210 = 5056$
$3b = 846$
$b = 282$

$c + 282 = 842$
$c = 560$

There were 282 bottles and 560 cans.

41. $x$ = amount of 25% solution
$y$ = amount of 40% solution

$x + y = 60$
$25x + 40y = 30(60)$

$25(60 - y) + 40y = 1800$
$1500 - 25y + 40y = 1800$
$15y + 1500 = 1800$
$15y = 300$
$y = 20$

$x + 20 = 60$
$x = 40$

They should combine 40 ml of the 25% solution and 20 ml of the 40% solution.

43. $w + l = 45$

$l = \dfrac{1}{2}w + 3$

$w + \left(\dfrac{1}{2}w + 3\right) = 45$

$\dfrac{3}{2}w + 3 = 45$

$\dfrac{3}{2}w = 42$

$w = \dfrac{2}{3}(42)$

$\phantom{w} = 28$

$l = \dfrac{1}{2}(28) + 3$

$\phantom{l} = 14 + 3$

$\phantom{l} = 17$

The winning team scored 28 points and the losing team scored 17 points.

45. $s$ = successful free throws
$m$ = missed free throws

$s = m$

$\dfrac{s + 20}{s + m + 20} = 0.75$

$\dfrac{s + 20}{s + s + 20} = \dfrac{3}{4}$

$4(s + 20) = 3(2s + 20)$

$4s + 80 = 6s + 60$

$80 = 2s + 60$

$20 = 2s$

$s = 10$

$m = 10$

The team has attempted 20 free throws and has made 10 of the shots.

47. $2l + 2w = 40$
$l = w + 4$

$2(w + 4) + 2w = 40$

$2w + 8 + 2w = 40$

$4w + 8 = 40$

$4w = 32$

$w = 8$

$l = 8 + 4$

$\phantom{l} = 12$

The length is 12 inches and the width is 8 inches.

49. $s$ = speed of southbound biker
$n$ = speed of northbound biker

$s = n + 5$

$2s + 2n = 50$

$2(n + s) + 2n = 50$

$2n + 10 + 2n = 50$

$4n + 10 = 50$

$4n = 40$

$n = 10$

$s = 10 + 5$

$\phantom{s} = 15$

The southbound biker traveled at 15 mph and the northbound biker traveled at 10 mph.

Systems of Linear Equations

51. $u$ = amount of upstairs carpet
$d$ = amount of downstairs carpet

$u + d = 72$
$18u + 14d = 1176$

$18(72 - d) + 14d = 1176$
$1296 - 18d + 14d = 1176$
$1296 - 4d = 1176$
$-4d = -120$
$d = 30$

$u + 30 = 72$
$u = 42$

They used 42 square yards upstairs and 30 square yards downstairs.

53. $r = 2d$
$r - 800 = 3(d - 800)$

$2d - 800 = 3d - 2400$
$-800 = d - 2400$
$1600 = d$
$r = 2(1600)$
$= 3200$

There are 1600 Democrats and 3200 Republicans.

55. $x$ = 45% alloy
$y$ = 20% alloy

$x + y = 30$
$45x + 20y = 25(30)$

$45(30 - y) + 20y = 750$
$1350 - 45y + 20y = 750$
$-25y + 1350 = 750$
$-25y = -600$
$y = 24$

$x + 24 = 30$
$x = 6$

There were 6 pounds of the 45% alloy and 24 pounds of the 20% alloy.

57. $f + a = 860$
$\dfrac{f}{a} = \dfrac{3}{2}$

$2f = 3a$
$2f = 3(860 - f)$
$2f = 2580 - 3f$
$5f = 2580$
$f = 516$

$516 + a = 860$
$a = 344$

There were 516 voters for the bond and 344 against the bond.

59. $5t + 2c = 5.50$
$3t + 1c = 3.15$

$5t + 2c = 5.50$
$\underline{-6t - 2c = -6.30}$ (multiply by $-2$)
$-t = -0.80$
$t = 0.80$

$3(0.80) + 1c = 3.15$
$2.40 + c = 3.15$
$c = 0.75$

The tacos cost $0.80 each and the colas cost $0.75 each.

61. (a) $(0, 183), (10, 154)$
$m = \dfrac{183 - 154}{0 - 10}$
$= \dfrac{29}{-10}$
$= -2.9$

Small: $y = -2.9x + 183$

$(0, 102), (10, 101)$
$m = \dfrac{102 - 101}{0 - 10}$
$= \dfrac{1}{-10}$
$= -0.1$

Large: $y = -0.1x + 102$

(b) $y = -2.9x + 183$
$y = -0.1x + 102$

$2.9x + y = 183$
$0.1x + y = 102$

$2.9x + y = 183$
$\underline{-0.1x - y = -102}$ (multiply by $-1$)
$2.8x = 81$
$x = 28.9$

$1987 + 29 = 2016$

In 2016, the farms in both categories will be equal.

63. $(0, 47.5), (5, 94.4)$
$m = \dfrac{94.4 - 47.5}{5 - 0}$
$= 9.38$
Yankees: $y = 9.38x + 47.5$

$(0, 41.1), (5, 49.4)$
$m = \dfrac{49.4 - 41.1}{5 - 0}$
$= 1.66$
Tigers: $y = 1.66x + 41.1$

$(0, 36.3), (5, 80.4)$
$m = \dfrac{80.4 - 36.3}{5 - 0}$
$= 8.82$
Red Sox: $y = 8.82x + 36.3$

## Systems of Linear Equations

65. $y = 1.66x + 41.1$
$y = 8.82x + 36.3$

$1.66x + 41.1 = 8.82x + 36.3$
$41.1 = 7.16x + 36.3$
$4.8 = 7.16x$
$x = 0.7$

According to the models, Boston and Detroit payrolls were the same 0.7 years after 1994(1995).

67. $w = l - 2$
$2w + 2l = 8(3)$

$2(l - 2) + 2l = 24$
$2l - 4 + 2l = 24$
$4l - 4 = 24$
$4l = 28$
$l = 7$

$w = 7 - 2$
$= 5$

The width of the ranch is 5 miles and the length is 7 miles.

69. $d$ = dog's speed
$c$ = cat's speed

$d = \frac{4}{5}c$

Distance traveled by cat: $2c$
Distance traveled by dog: $2d$

Distance from dog to tree at start:
$26 + 2d$

Distance from cat to tree at start:
$26 + 2d - 20$

The speed of the cat, $c$, is

$\frac{26 + 2d - 20}{2}$
$= \frac{2d + 6}{2}$
$= d + 3$
$c = d + 3$
$d = \frac{4}{5}c$

$c = \frac{4}{5}c + 3$
$\frac{1}{5}c = 3$
$c = 15$ ft/sec

$d = \frac{4}{5}(15)$
$= 12$ ft/sec

The cat moved at 15 ft/sec and the dog moved at 12 ft/sec.

## Section 5.5 Systems of Linear Inequalities

1. No, the pair must satisfy both inequalities.

3. (i) $-5 < 2(3) - 5$
   $-5 < 6 - 5$
   $-5 < 1$

   $2(3) + 3(-5) < 6$
   $6 - 15 < 6$
   $-9 < 6$

   $(3, -5)$ is a solution.

   (ii) $0 < 2(0) - 5$
   $0 < -5$
   $(0, 0)$ is not a solution.

5. (i) $2(5) - 7 \leq 3$
   $10 - 7 \leq 3$
   $3 \leq 3$

   $5 + 2(7) \geq 19$
   $5 + 14 \geq 19$
   $19 \geq 19$

   $(5, 7)$ is a solution.

   (ii) $2(5) - 8 \leq 3$
   $10 - 8 \leq 3$
   $2 \leq 3$

   $5 + 2(8) \geq 19$
   $5 + 16 \geq 19$
   $21 \geq 19$

   $(5, 8)$ is solution.

7.

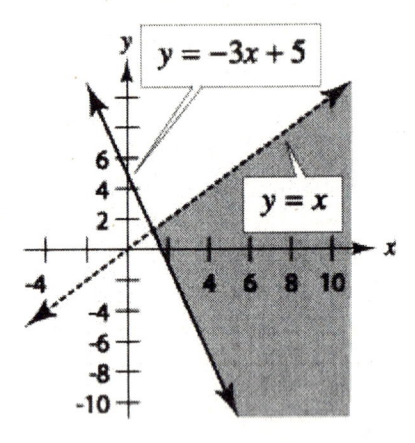

[Graph showing $y = -3x + 5$ (solid line) and $y = x$ (dashed line), with shaded region.]

9.

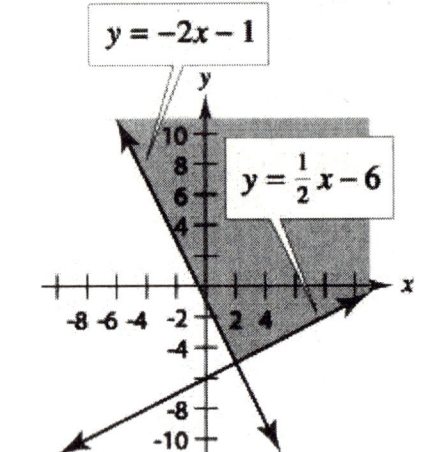

[Graph showing $y = -2x - 1$ and $y = \frac{1}{2}x - 6$, with shaded region.]

11.

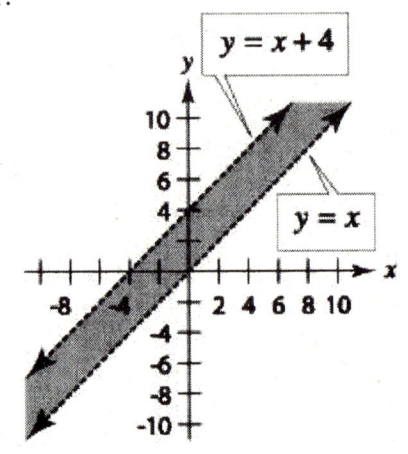

[Graph showing $y = x + 4$ and $y = x$, with shaded region between them.]

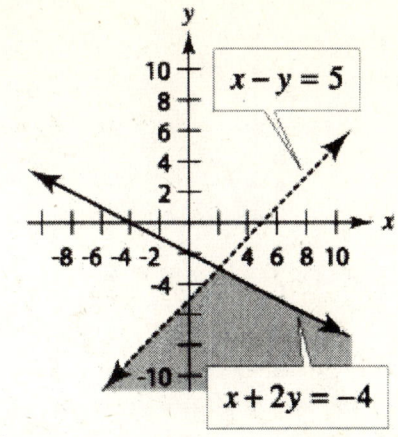

13.

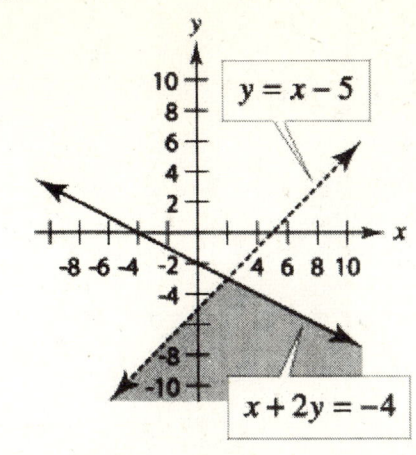

$y = x - 5$
$x + 2y = -4$

15.

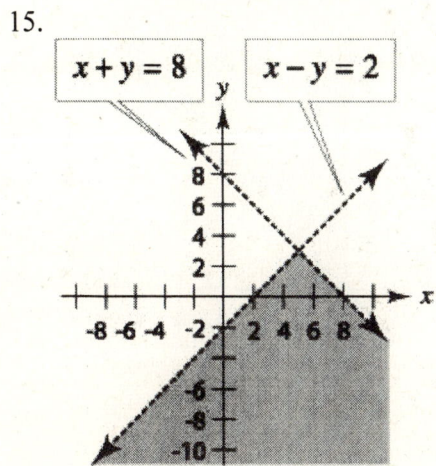

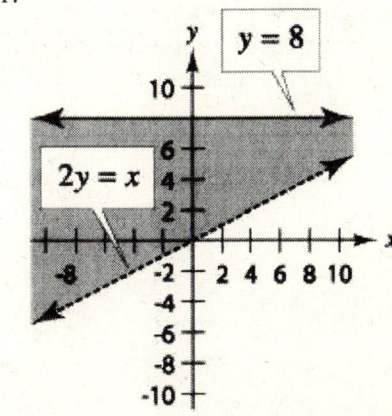

$x + y = 8$
$x - y = 2$

17.

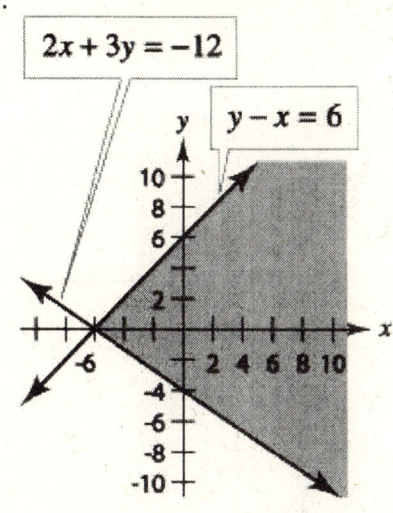

$2x + 3y = -12$
$y - x = 6$

19.

$x - y = 5$
$x + 2y = -4$

21.

$y = 8$
$2y = x$

23.

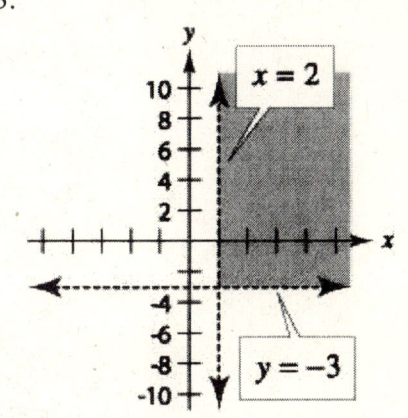

$x = 2$
$y = -3$

25.

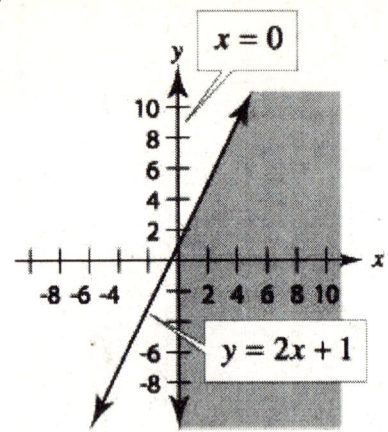

27. The boundary lines are parallel, and the half–planes are in opposite directions; that is, they do not intersect.

29. ∅

31.

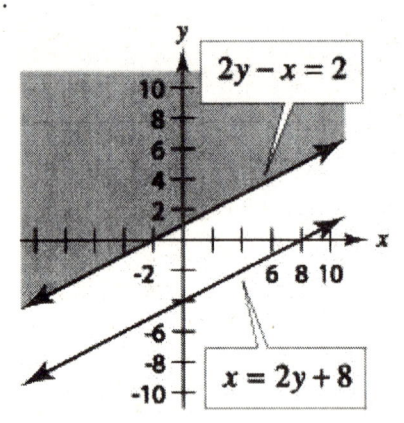

33.

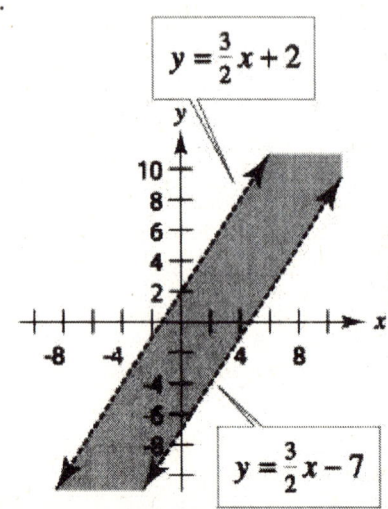

35.

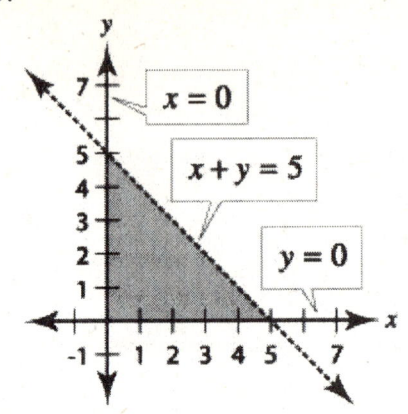

37.

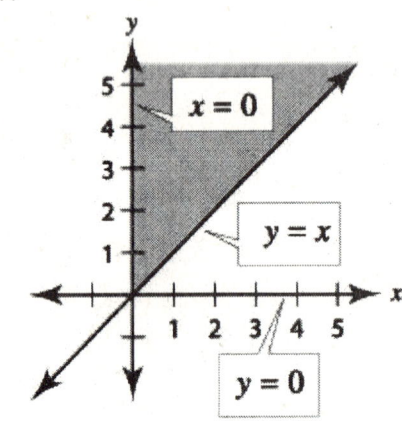

39. $y \leq -x + 5, x \geq 0, y \geq 0$

41.

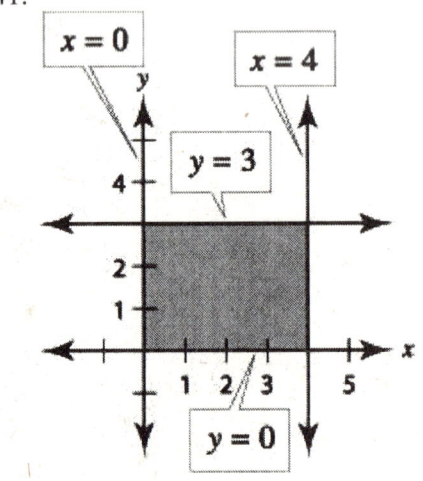

43. Points of the line $y = x - 5$.

45. ∅

47.

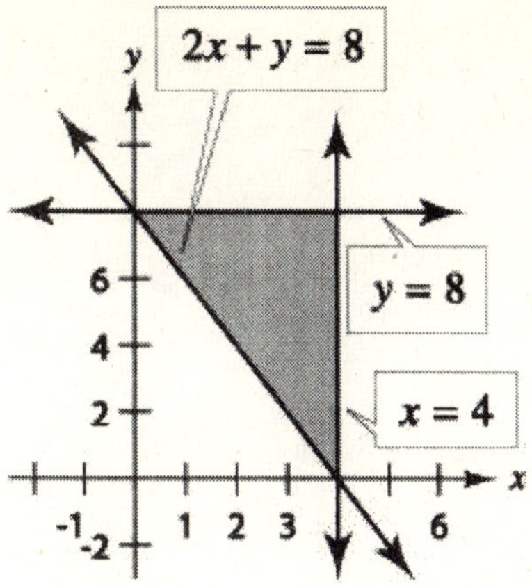
2x + y = 8
y = 8
x = 4
y
x

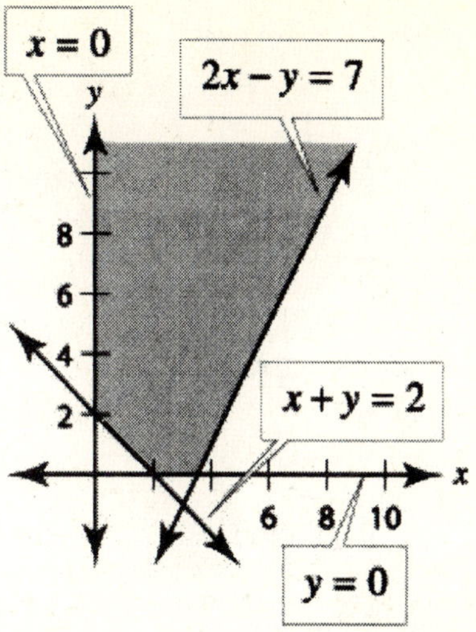
x = 0
2x − y = 7
x + y = 2
y = 0
y
x

49.

## Chapter 5 Review Exercises

1. A solution is a pair of numbers that satisfies each equation.

3. (i) $3 = 7(-1) - 4$
$3 = -7 - 4$
$3 = -11$
No

(ii) $3 = 7(1) - 4$
$3 = 7 - 4$
$3 = 3$

$5(1) - 2(3) = -1$
$5 - 6 = -1$
$-1 = -1$
Yes

(iii) $-3 = 7(1) - 4$
$-3 = 7 - 4$
$-3 = -3$

$5(1) - 2(-3) = -1$
$5 + 6 = -1$
$11 = -1$
No

5. $(0, -5)$

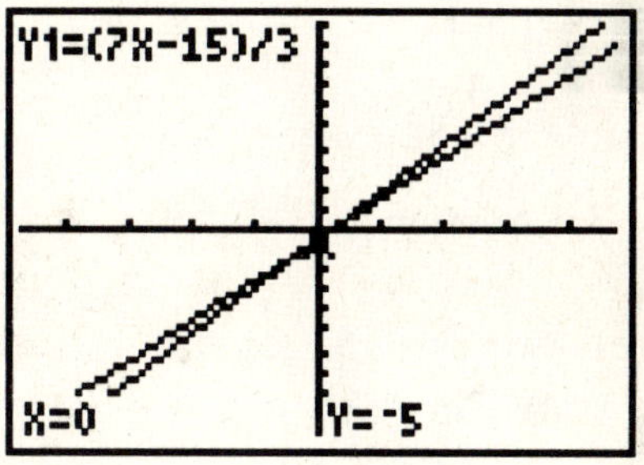
Y1=(7X-15)/3
X=0
Y=-5

7. (a) $2y - x = -10$
$2y = x - 10$
$y = \frac{1}{2}x - 5$

$\frac{1}{2}x - y = -3$
$-y = \frac{1}{2}x - 3$
$y = \frac{1}{2}x + 3$

0 solutions

(b) $2x - 3y = -3$
$-3y = -2x - 3$
$y = \frac{2}{3}x + 1$

$2y - 3x = -2$
$2y = 3x - 2$
$y = \frac{3}{2}x - 1$

1 solution

(c) $4x = 5y - 15$
$4x + 15 = 5y$
$y = \frac{4}{5}x + 3$

$y - 3 = \frac{4}{5}x$
$y = \frac{4}{5}x + 3$

Infinitely many solutions

9. $(8, -15)$

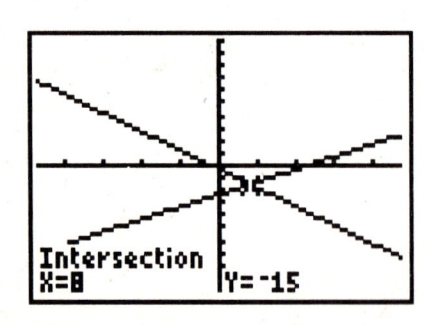

11. $2x + 3y = -11$
$\underline{5x - 3y = 25}$
$7x = 14$
$x = 2$

$2(2) + 3y = -11$
$4 + 3y = -11$
$3y = -15$
$y = -5$

$(2, -5)$ is the solution.

13. $6x + y = 19$
$3x + y = 7$

$6x + y = 19$
$\underline{-3x - y = -7}$ (multiply by $-1$)
$3x = 12$
$x = 4$

$3(4) + y = 7$
$12 + y = 7$
$y = -5$

$(4, -5)$ is the solution.

15. $10x + 6y = 22$ (multiply by 2)
$\underline{-10x + 35y = -145}$ (multiply by 5)
$41y = -123$
$y = -3$

$10x + 6(-3) = 22$
$10x - 18 = 22$
$10x = 40$
$x = 4$

$(4, -3)$ is a solution

17. $\frac{4}{7}x - 1 = y$
$4x - 7 = 7y$
$4x - 7y = 7$

$4x - 7y = 7$
$4x - 7y = -7$
$4x - 7y = 7$
$\underline{-4x + 7y = 7}$ (multiply by $-1$)
$0 = 14$

No solution

Systems of Linear Equations

19. The sum $x + y = 3$ does not give any information about the number of solutions to the original system.

21. It is not easy to solve for a variable in either equation.

23. $-x + 1 = 2x + 19$
$-3x + 1 = 19$
$-3x = 18$
$x = -6$

$y = -(-6) + 1$
$= 6 + 1$
$= 7$

$(-6, 7)$ is the solution.

25. $\frac{2}{3}(3) - 4y = 10$
$2 - 4y = 10$
$-4y = 8$
$y = -2$
$(3, -2)$ is the solution.

27. $y - 3 = \frac{6}{7}x$

$y = \frac{6}{7}x + 3$

$7\left(\frac{6}{7}x + 3\right) = 6x + 21$

$6x + 21 = 6x + 21$
$0 = 0$

Infinitely many solutions; dependent equations.

29. No, the resulting equation has two variables.

31. $x - y = 25$
$x + 2y = -2$

$-x + y = -25$ ( multiply by $-1$)
$\underline{x + 2y = -2}$
$3y = -27$
$y = -9$

$x - (-9) = 25$
$x = 16$

The numbers are $-9$ and 16.

33. $y = 7x + 2$
$x + y = 90$

$x + (7x + 2) = 90$
$8x + 2 = 90$
$8x = 88$
$x = 11$

$y = 7(11) + 2$
$= 77 + 2$
$= 79$

The angles are 11° and 79°.

35. $x$ = number of old cards
$y$ = number of new cards

$12x + 3y = 300$
$y = x + 25$

$12x + 3(x + 25) = 300$
$12x + 3x + 75 = 300$
$15x = 125$
$x = 15$

$y = 15 + 25$
$= 40$

There are 15 old cards and 40 new cards.

37. $26\dfrac{\text{miles}}{\text{hour}} \cdot \dfrac{1\,\text{hour}}{60\,\text{min}}$

$= \dfrac{26\,\text{miles}}{60\,\text{min}}$

$= \dfrac{13\,\text{miles}}{30\,\text{min}}$

$60\dfrac{\text{miles}}{\text{hour}} \cdot \dfrac{1\,\text{hour}}{60\,\text{min}}$

$= \dfrac{60\,\text{miles}}{60\,\text{min}}$

$= 1$ mile per min

$b =$ time on bus (in minutes)
$c =$ time in car (in minutes)

$b + c = 70$

$\dfrac{13}{30}b + 1 \cdot c = 53$

$\dfrac{13}{30}b + (70 - b) = 53$

$\dfrac{-17}{30}b + 70 = 53$

$\dfrac{-17}{30}b = -17$

$b = 30$

$30 + c = 70$

$c = 40$

Bus: 30 minutes   Car: 40 minutes

39. $h =$ amount of home sales (in thousands)
$l =$ amount of land sales (in thousands)

$h + l = 1000$
$0.06h + 0.10l = 66$

$0.06(1000 - l) + 0.10l = 66$
$60 - 0.06l + 0.10l = 66$
$0.04l + 60 = 66$
$0.04l = 6$
$l = 150$

$h + 150 = 1000$
$h = 850$

The agent sold $850,000 of homes and $150,000 of vacant land. The amount of the commission on home sales was 0.06 (850,000) = $51,000.

41. $f = 1.5n$

$\dfrac{3}{4}(f + 2000) = n + 2000$

$\dfrac{3}{4}(1.5n + 2000) = n + 2000$

$1.125n + 1500 = n + 2000$

$0.125n + 1500 = 2000$

$0.125n = 500$

$n = 4000$

$f = 1.5(4000)$
$= 6000$

Adding 2000 to both $f$ and $n$. We get 6000 nonfiction and 8000 fiction in the new library.

43. It is easier to produce the graphs of boundary lines and to determine which half-plane to shade

Systems of Linear Equations

45. (i) $3 \le -2(-4) - 5$
$3 \le 8 - 5$
$3 \le 3$

$-4 - 3 \le 3$
$-7 \le 3$

$(-4, 3)$ is a solution.

(ii) $-1 < -2(-5) - 5$
$-1 < 10 - 5$
$-1 < 5$

$-(-5) - (-1) \le 3$
$5 + 1 \le 3$
$6 \le 3$

$(-5, -1)$ is not a solution.

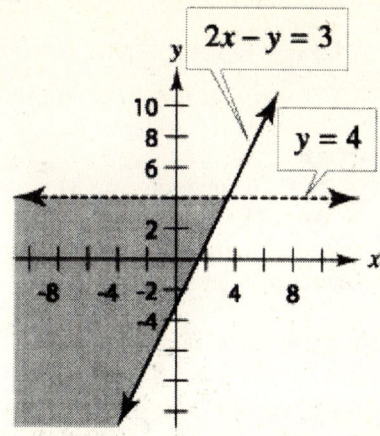

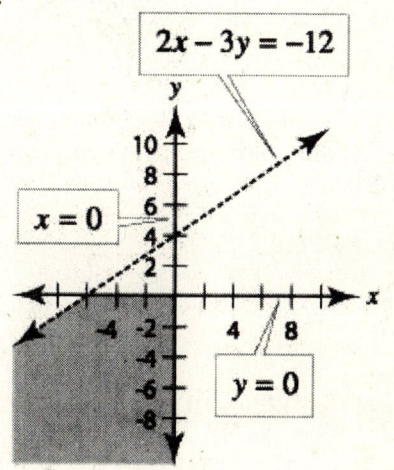

47.

[Graph showing lines $2x - y = 3$ and $y = 4$ with shaded region]

49.

[Graph showing lines $2x - 3y = -12$, $x = 0$, and $y = 0$ with shaded region]

## Chapter 5 Test

1. $a(2) + (-3) = -9$
$2a - 3 = -9$
$2a = -6$
$a = -3$

$2 - b(-3) = 8$
$2 + 3b = 8$
$3b = 6$
$b = 2$

3. (a) The lines coincide.

(b) The lines are parallel.

(c) The lines intersect at one point.

5. $3x - 2y = -4$
$-2x + 3y = -9$

$6x - 4y = -8$ (multiply by 2)
$-6x + 9y = -27$ (multiply by 3)
$5y = -35$
$y = -7$

$3x - 2(-7) = -4$
$3x + 14 = -4$
$3x = -18$
$x = -6$

$(-6, -7)$ is the solution.

7. $-4x + 3y = -15$

$\dfrac{-4}{3}x + y = 2$

$-4x + 3y = -15$
$\underline{4x - 3y = -6}$ (multiply by $-3$)
$0 = -21$

No solution.

9. Solve one equation for one variable.

11. $2ax - a(2x+1) = 2$

$2ax - 2ax - a = 2$

$-a = 2$

$a = -2$

If $a = -2$, the system has infinitely many solutions. Otherwise, the system has no solution.

13. $x + y = 17$

$\dfrac{1}{2}x = y + 10$

$\dfrac{1}{2}(17 - y) = y + 10$

$17 - y = 2(y + 10)$

$17 - y = 2y + 20$

$17 = 3y + 20$

$-3 = 3y$

$y = -1$

$x + (-1) = 17$

$x = 18$

15. $5n + 10(50) + 25q = 870$

$q = n - 2$

$5n + 500 + 25(n - 2) = 870$

$5n + 500 + 25n - 50 = 870$

$30n + 450 = 870$

$30n = 420$

$n = 14$

$q = 14 - 2$

$= 12$

There are 12 quarters and 14 nickels.

17. $x$ = amount of 6% loan
$y$ = amount of 10% loan

$x + y = 25{,}000$
$0.06x + 0.10y = 2260$

$0.06(25{,}000 - y) + 0.10y = 2260$
$1500 - 0.06y + 0.10y = 2260$
$0.04y + 1500 = 2260$
$0.04y = 760$
$y = 19{,}000$

$x + 19{,}000 = 25{,}000$
$x = 6000$

She borrowed $6,000 from her father for the down payment.

19.

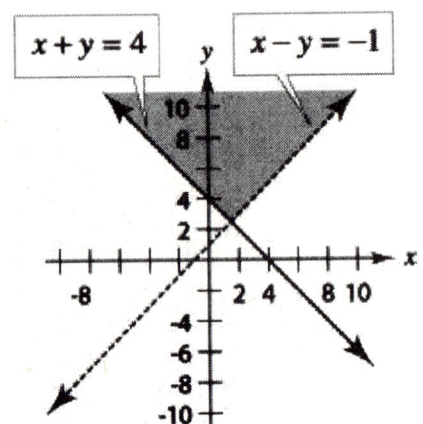

## Cumulative Test, Chapters 4 – 5.

1.

   | X | Y1 |
   |---|---|
   | 5 | -2 |
   | 6 | -1 |
   | 10 | 3 |

   X=

3. $4x - 3y = -12$

   $4(0) - 3y = -12$
   $-3y = -12$
   $y = 4$
   $(0, 4)$ is the $y$-intercept.

   $4x - 3(0) = -12$
   $4x = -12$
   $x = -3$
   $(-3, 0)$ is the $x$-intercept.

5. $m = \dfrac{5 - (-4)}{-3 - 7}$

   $= \dfrac{9}{-10}$

   $= -0.9$

7. $5x - 3y = 9$
   $-3y = -5x + 9$
   $y = \dfrac{5}{3}x - 3$

   Slope: $\dfrac{5}{3}$; $y$-intercept: $(0, -3)$

9. (i) $3x - 4y = 20$
   $-4y = -3x + 20$
   $y = \dfrac{3}{4}x - 5$

   (ii) $5y - 3x = 25$
   $5y = 3x + 25$
   $y = \dfrac{3}{5}x + 5$

   (iii) $y = \dfrac{3}{4}x + 4$

   Graph (ii) is not parallel to (i) and (iii).

11. $m = \dfrac{11 - 2}{5 - (-6)}$

    $= \dfrac{9}{11}$

    $y = \dfrac{9}{11}x + b$

    $11 = \dfrac{9}{11}(5) + b$

    $\dfrac{121}{11} = \dfrac{45}{11} + b$

    $\dfrac{76}{11} = b$

    $y = \dfrac{9}{11}x + \dfrac{76}{11}$

    $11y = 9x + 76$
    $-9x + 11y = 76$
    $9x - 11y = -76$

13. The inequality must be solved for $y$.

15. $3x - 2y = -8$
    $-2y = -3x - 8$
    $y = 1.5x + 4$

    Since the lines have the same slope, there are no solutions.

17. $25x + 10y = -85$ (multiply by 5)
    $\underline{6x - 10y = -8}$ (multiply by 2)
    $31x = -93$
    $x = -3$

   $6(-3) - 10y = -8$
   $-18 - 10y = -8$
   $-10y = 10$
   $y = -1$

   $(-3, -1)$ is the solution.

19. $3y + x = 23$
    $x = -3y + 23$

   $4(-3y + 23) - 2y = -34$
   $-12y + 92 - 2y = -34$
   $-14y + 92 = -34$
   $= -14y = -126$
   $y = 9$

   $x = -3(9) + 23$
   $= -27 + 23$
   $= -4$

   $(-4, 9)$ is the solution.

21. $w$ = time walking (in hours)
    $b$ = time on bus (in hours)

   $w + b = 1$
   $3w + 20b = 11.5$
   $3w + 20(1 - w) = 11.5$
   $3w + 20 - 20w = 11.5$
   $-17w + 20 = 11.5$
   $-17w = -8.5$
   $w = 0.5$

   $0.5 + b = 1$
   $b = 0.5$

   They walk for $\frac{1}{2}$ hour at 3 mph, so they walk $1\frac{1}{2}$ miles.

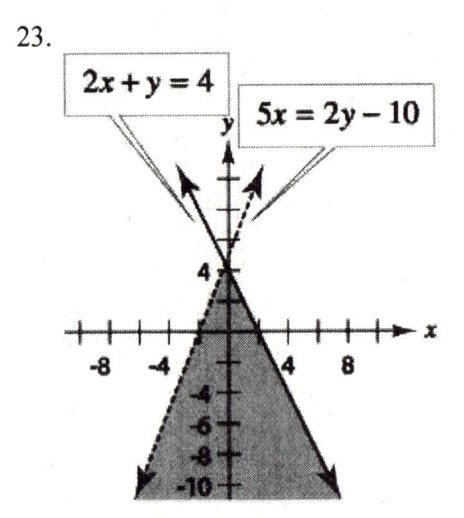

23.

# Chapter 6

# Exponents and Polynomials

## Section 6.1 Properties of Exponents

1. Because the bases are the same, we multiply $x^2 \cdot x^3$ by adding the exponents. Because they are not like terms, $x^2 + x^3$ cannot be simplified.

3. $-4^2 = -1 \cdot 4 \cdot 4 = -16$
   $(-4)^2 = (-4)(-4) = 16$
   $4^2 = 4 \cdot 4 = 16$

5. $7^2 = 7 \cdot 7 = 49$
   $-7^2 = -1 \cdot 7 \cdot 7 = -49$
   $(-7)^2 = (-7)(-7) = 49$

7. $7^3 \cdot 7$
   $= 7^3 \cdot 7^1$
   $= 7^{3+1}$
   $= 7^4$

9. $y^5 \cdot y^3$
   $= y^{5+3}$
   $= y^8$

11. $x^4 \cdot x^5 \cdot x$
    $= x^4 \cdot x^5 \cdot x^1$
    $= x^{4+5+1}$
    $= x^{10}$

13. $(3x)^5 \cdot (3x)^9$
    $= (3x)^{5+9}$
    $= (3x)^{14}$

15. $-4x^3 (x^6)$
    $= -4x^{3+6}$
    $= -4x^9$

17. $-4x^5 (-2x^7)$
    $= (-4)(-2) x^{5+7}$
    $= 8x^{12}$

19. $(2x^2 y^5)(-5x^3 y^2)$
    $= (2)(-5) x^{2+3} y^{5+2}$
    $= -10x^5 y^7$

21. $(3x^3)(-2x^4)(5x)$
    $= (3)(-2)(5) x^{3+4+1}$
    $= -30x^8$

23. $(xy^2)(x^3 y)(x^4 y^2)$
    $= x^{1+3+4} y^{2+1+2}$
    $= x^8 y^5$

25. (a) $-5x^4 + 7x^4$
    $= 2x^4$

(b) $-5x^4 \cdot 7x^4$
$= (-5)(7)x^{4+4}$
$= -35x^8$

27. (a) $8y^5 + 8y^4$
can't add

(b) $8y^5 \cdot 8y^4$
$= (8)(8)y^{5+4}$
$= 64y^9$

29. (a) $n + 3n + (-6n)$
$= 4n - 6n$
$= -2n$

(b) $n(3n)(-6n)$
$= (3)(-6)n^{1+1+1}$
$= -18n^3$

31. $(5^2)^4$
$= 5^{2 \cdot 4}$
$= 5^8$

33. $(a^3)^5$
$= a^{3 \cdot 5}$
$= a^{15}$

35. $(-x^4)^7$
$= (-1)^7 (x^4)^7$
$= -1 \cdot x^{4 \cdot 7}$
$= -1 \cdot x^{28}$
$= -x^{28}$

37. $(xy)^4$
$= x^4 y^4$

39. $(4y)^3$
$= 4^3 y^3$
$= 64y^3$

41. $\left(\dfrac{a}{b}\right)^5$
$= \dfrac{a^5}{b^5}$

43. $\left(\dfrac{-2}{x}\right)^5$
$= \dfrac{(-2)^5}{x^5}$
$= \dfrac{(-1)^5 (2^5)}{x^5}$
$= \dfrac{-1 \cdot 32}{x^5}$
$= \dfrac{-32}{x^5}$

45. $(3x^4)^2$
$= 3^2 x^{4 \cdot 2}$
$= 9x^8$

47. $(-2y^3)^5$
$= (-2)^5 y^{3 \cdot 5}$
$= -32y^{15}$

49. $\left(\dfrac{a^4}{b^3}\right)^5$
$= \dfrac{a^{4 \cdot 5}}{b^{3 \cdot 5}}$
$= \dfrac{a^{20}}{b^{15}}$

51. $\left(\dfrac{3n}{5m}\right)^3$
$= \dfrac{3^3 n^3}{5^3 m^3}$
$= \dfrac{27n^3}{125m^3}$

Exponents and Polynomials

53. $\left(\dfrac{-8x^3}{y}\right)^2$

$= \dfrac{(-8)^2 x^{3 \cdot 2}}{y^2}$

$= \dfrac{64x^6}{y^2}$

55. $(n^2)^4 (n^3)^2$

$= n^{2 \cdot 4} \cdot n^{3 \cdot 2}$

$= n^8 \cdot n^6$

$= n^{8+6}$

$= n^{14}$

57. $3x^2 (x^2)^4$

$= 3x^2 (x^{2 \cdot 4})$

$= 3x^2 \cdot x^8$

$= 3x^{2+8}$

$= 3x^{10}$

59. $(2x^2)^3 (-3x^5)^2$

$= 2^3 (x^2)^3 (-3)^2 (x^5)^2$

$= 8x^{2 \cdot 3} \cdot 9x^{5 \cdot 2}$

$= 8x^6 \cdot 9x^{10}$

$= 8 \cdot 9 x^{6+10}$

$= 72x^{16}$

61. $4^2 \cdot 4^4$

$= 4^{2+4}$

$= 4^6$

$\neq 16^6$

False

63. $(t^4)^3$

$= t^{4 \cdot 3}$

$= t^{12}$

$(t^6)^2$

$= t^{6 \cdot 2}$

$= t^{12}$

True

65. $(y^2)^3$

$= y^{2 \cdot 3}$

$= y^6$

$\neq y^8$

False

67. Use the Quotient Rule for Exponents to simplify $\dfrac{x^5}{x^3} = x^2$ and then use the Power to a Power Rule to simplify $(x^2)^2 = x^4$. Use the Quotient to Power Rule and the Power to Power Rule to write $\dfrac{x^{10}}{x^6}$ and then use the Quotient Rule for Exponents to obtain $x^4$.

69. $\dfrac{x^8}{x^5}$

$= x^{8-5}$

$= x^3$

71. $\dfrac{t^5}{t}$

$= t^{5-1}$

$= t^4$

183

73. $\dfrac{x^6 y^3}{x^2 y^2}$
$= x^{6-2} y^{3-2}$
$= x^4 y$

75. $\dfrac{12 c^5 d^2}{9 c^3 d}$
$= \dfrac{\cancel{3} \cdot 4}{\cancel{3} \cdot 3} c^{5-3} d^{2-1}$
$= \dfrac{4}{3} c^2 d$

77. $\left(\dfrac{a^4}{a^2}\right)^6$
$= \left(a^2\right)^6$
$= a^{12}$

79. $\dfrac{\left(a^2\right)^3}{a^5}$
$= \dfrac{a^6}{a^5}$
$= a$

81. $\dfrac{\left(2y^2\right)^3}{y^3}$
$= \dfrac{8 y^6}{y^3}$
$= 8 y^3$

83. $\left(\dfrac{t^2}{3t}\right)^3$
$= \left(\dfrac{t}{3}\right)^3$
$= \dfrac{t^3}{3^3}$
$= \dfrac{t^3}{27}$

85. 1
87. 1

89. 3

91. $\dfrac{x^0 y^3}{x^2 y}$
$= \dfrac{y^2}{x^2}$

93. $\dfrac{6 x^4}{2 x \left(3 x^0\right)}$
$= \dfrac{\cancel{6} x^4}{\cancel{6} x}$
$= x^3$

95. $\left(t^2\right)^0 (3t)^2$
$= 1 \cdot 3^2 t^2$
$= 9 t^2$

97. $A = (3d)(2d^2)$
$= 6 d^3$

99. $V = (2x)(2x)(2x)$
$= (2x)^3$
$= 8 x^3$

101. (a) $\dfrac{1}{2}\left(\dfrac{x}{2}\right)^2$
$= \dfrac{1}{2} \cdot \dfrac{x^2}{4}$
$= \dfrac{x^2}{8}$
$= \dfrac{1}{8} x^2$

(b) $\dfrac{(3x)^2}{30x}$

$=\dfrac{9x^2}{30x}$

$=\dfrac{\cancel{3}\cdot 3x^2}{\cancel{3}\cdot 10x}$

$=\dfrac{3x}{10}$

$=\dfrac{3}{10}x$

103. $\left(x^n\right)^2 = x^{10}$

$x^{2n} = x^{10}$

$2n = 10$

$n = 5$

105. $\dfrac{x^7}{x^n} = x^4$

$x^{7-n} = x^4$

$7 - n = 4$

$7 = 4 + n$

$n = 3$

107. $3^n \cdot 3^{n+1}$

$= 3^{n+n+1}$

$= 3^{2n+1}$

109. $\left(y^{2n}\right)^n$

$= y^{2n \cdot n}$

$= y^{2n^2}$

## Section 6.2 Introduction to Polynomials

1. Because a number is a monomial, 1 is a monomial. Because a polynomial is a monomial or sum of monomials, 1 is also a polynomial.

3. No, because $x^2$ is in the denominator

5. Yes

7. Yes

9. No, because $y$ is in the denominator.

11. binomial
    terms: $-y, -7$
    coefficients: $-1, -7$

13. monomial
    term: $-a^2 b$
    coefficient: $-1$

15. trinomial
    terms: $6, -t, -t^2$
    coefficients: $6, -1, -1$

17. polynomial
    terms: $2x^2 y, 5yz, -7z, 1$
    coefficients: $2, 5, -7, 1$

19. coefficient: $-1$
    degree: 1

21. coefficient: 1
    degree: 5

23. coefficient: 5
    degree: 1

25. coefficient: 2
    degree: $1 + 2 + 3 = 6$

27. 1

29. 3

31. 6 since $4 + 2 = 6$

33. 3 since $1 + 1 + 1 = 3$

35. (a) $-y^2 + 16$

    (b) $16 - y^2$

37. (a) $-x^2 + 3x + 4$

    (b) $4 + 3x - x^2$

39. (a) $-3t^3 + t^2 - t + 2$

    (b) $2 - t + t^2 - 3t^3$

41. (a) $-3x^5 + 8x^4 - 6x^2 + x$

    (b) $x - 6x^2 + 8x^4 - 3x^5$

43. $G(3)$ is the value of the polynomial when the variable is replaced with 3.

45. $5 + 2(0) = 5$
$5 + 2(-3)$
$5 - 6$
$= -1$

47. $(1)^2 + 3(1) + 2$
$= 1 + 3 + 2$
$= 6$

$(-4)^2 + 3(-4) + 2$
$= 16 - 12 + 2$
$= 6$

49. $1^4 - 3(1)^3 + 2(1)^2 + (1) - 5$
$= 1 - 3 + 2 + 1 - 5$
$= -4$

$(-1)^4 - 3(-1)^3 + 2(-1)^2 + (-1) - 5$
$= 1 - 3(-1) + 2(1) - 1 - 5$
$= 1 + 3 + 2 - 1 - 5$
$= 0$

51. $5(3) + 4$

53. $4 - (2)^2$

55. $f(2)$

57. $-(-2)^2 + 2(-2) + 5$

59. $P(3) = 6 - 3$
$= 3$

$P(-2) = 6 - (-2)$
$= 8$

61. $F(-3) = 2(-3) + 9$
$= 3$

$F(4) = 2(4) + 9$
$= 17$

63. $P(-2) = -(-2)^2 + (-2) + 3$
$= -4 - 2 + 3$
$= -3$

$P(3) = -(3)^2 + (3) + 3$
$= -9 + 6$
$= -3$

Exponents and Polynomials

65. (a) $F(-4) = 2(-4) + 6$
$= -8 + 6$
$= -2$

(b) $P(-3) = -(-3) + 6$
$= 3 + 6$
$= 9$

(c) $F(1) - P(-2) = 2(1) + 6 - [-(-2) + 6]$
$= 2 + 6 - (2 + 6)$
$= 0$

(d) $P(5) + F(0) = -(5) + 6 + [2(0) + 6]$
$= -5 + 6 + 6$
$= 7$

67. (a) $\frac{1}{3}(6)^2$
$= \frac{1}{3}(36)$
$= 12$

(b) $(-3)^3$
$= -27$

(c) $\frac{1}{3}(0)^2 + 1^3$
$= 1$

(d) $\frac{1}{3}(3)^2 - (-1)^3$
$= \frac{1}{3}(9) - (-1)$
$= 3 + 1$
$= 4$

69. $Q(-3) = 0$ ; $Q(0) = 9$

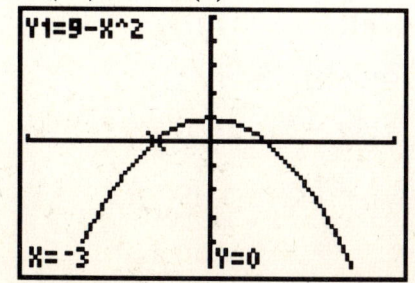

71. $P(-1) = 2$ ; $P(1) = 2$

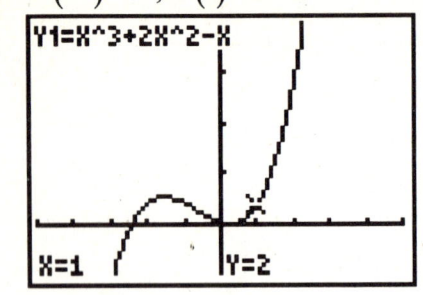

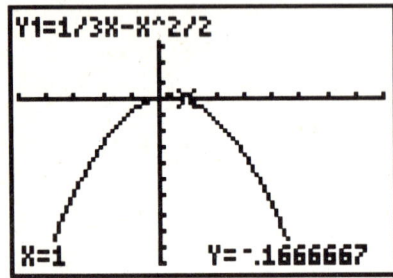

    Y1=X^3+2X^2-X
    X=1   Y=2

73. $P(6) = -16$ ; $P(1) = -\dfrac{1}{6}$

    Y1=1/3X-X^2/2
    X=1   Y=-.1666667

75. $0 \cdot 3 + 3 - 3$
    $= 0$

77. $(-5)^2 - (-6)^2$
    $= 25 - 36$
    $= -11$

79. $(2)^2 + 3(2)(-3) + (-3)^2$
    $= 4 - 18 + 9$
    $= -5$

81. $2 \cdot (-4)(5) + (2)(-4) - 5$
    $= -40 - 8 - 5$
    $= -53$

83. $x \cdot x \cdot (x - y)$
    $= x^2 (x - y)$
    $= x^3 - x^2 y$

85. $B(8) = -2(8)^2 - 10(8) + 1500$
    $= -2(64) - 80 + 1500$
    $= -128 + 1420$
    $= 1292$ bacteria

87. (a) 2

    (b) $S(12) = 0.9(12)^2 + 6.7(12) + 54.1$
    $= 129.6 + 80.4 + 54.1$
    $= 264.1$

    In 2002, credit card spending between Thanksgiving and Christmas is expected to be $264.1 billion.

89.

    | X  | Y1    |
    |----|-------|
    | 21 | 36.96 |
    | 30 | 33.9  |
    | 55 | 25.4  |
    | 70 | 20.3  |

    X=

    $P_1(21) = 37$
    $P_1(30) = 34$
    $P_1(55) = 25$
    $P_1(70) = 20$

91. $P_2(x)$ appears to fit the data better.

93. $F(t + 2) = 4(t + 2) + 5$
    $= 4t + 8 + 5$
    $= 4t + 13$

95. $(2a)(3a)^2 + 3(2a)$
    $2a(9a^2) + 6a$
    $= 18a^3 + 6a$

97. $-3x^{3n} + 5x^{3n-1} + x^n$

## Section 6.3 Addition and Subtraction

1. To add or subtract polynomials, we remove parentheses and combine like terms. That is, we simplify.

3. $(3x+7)+(5x-8)$
   $=3x+5x+7-8$
   $=8x-1$

5. $(x^2+3x-7)+(-x+2)$
   $=x^2+3x-x-7+2$
   $=x^2+2x-5$

7. $(4x^2+3x+5)+(2x^2-x-2)$
   $=4x^2+2x^2+3x-x+5-2$
   $=6x^2+2x+3$

9. $(2y^3+5y+1)+(y^2-7y)$
   $=2y^3+y^2+5y-7y+1$
   $=2y^3+y^2-2y+1$

11. $(x^2-5xy+y^2)+(3x^2+4xy-2y^2)$
    $=x^2+3x^2-5xy+4xy+y^2-2y^2$
    $=4x^2-xy-y^2$

13. $\phantom{+}4x-7$
    $\underline{+3x+8}$
    $\phantom{+}7x-1$

15. $x^2+4x-6$
    $\underline{+\phantom{x^2+}3x+6}$
    $x^2+7x+0$

17. $-w^2-4w$
    $\underline{+w^2+3w+2}$
    $\phantom{-w^2+}-w+2$

19. $\phantom{+}-x^2+2x+4$
    $\underline{+-x^2\phantom{+2x}+5}$
    $\phantom{+}-2x^2+2x+9$

21. $4x^3\phantom{+x^2}+x-8$
    $\underline{+\phantom{4x^3+}x^2-6x\phantom{-8}}$
    $4x^3+x^2-5x-8$

23. $\phantom{+x^2-}2x+3$
    $\phantom{+x^2-5}-5x+4$
    $\underline{+\phantom{}x^2-3x-1}$
    $\phantom{+}x^2-6x+6$

25. $-3w^2\phantom{+2w}+1$
    $\phantom{-}3w^2+2w-1$
    $\underline{+\phantom{3w^2+}-2w+6}$
    $\phantom{-3w^2+2w+}6$

27. $\phantom{+-4x^3-}3x^2\phantom{+2x}+5$
    $\phantom{+-4}x^3\phantom{-3x^2}+2x-1$
    $\underline{+-4x^3-3x^2\phantom{+2x}+4}$
    $\phantom{+}-3x^3\phantom{-3x^2}+2x+8$

29. $y^2+6y+2-2$
    $=y^2+6y$

31. $-b^2+6b+2b+3$
    $=-b^2+8b+3$

33. $6x-3x+1-2$
    $=3x-1$

35. $2x^2+6-7x-5x^2-x-(-4)$
    $=2x^2-5x^2-7x-x+6+4$
    $=-3x^2-8x+10$

37. $5y^2 + y + 3 - y^2 - (-3)$
$= 5y^2 - y^2 + y + 3 + 3$
$= 4y^2 + y + 6$

39. $x^2 + 16 - x^3 - (-8x)$
$= -x^3 + x^2 + 8x + 16$

41. $y^2 - 3xy - 2x^2 - y^2 - xy - x^2$
$= y^2 - y^2 - 3xy - xy - 2x^2 - x^2$
$= -4xy - 3x^2$

43. $xy^2 + x^2y + 3 - xy^2 - (-x^2y) - (-2x)$
$= xy^2 - xy^2 + x^2y + x^2y + 2x + 3$
$= 2x^2y + 2x + 3$

45. $\phantom{-(}6 - 5x$
$\underline{-(5 + 7x)}$
$\phantom{-(}1 - 12x$

47. $\phantom{-(}x^2 - 3x - 1$
$\phantom{-(}\underline{-(2x - 3)}$
$\phantom{-(}x^2 - 5x + 2$

49. $\phantom{-(}-t^2 - 9t$
$\phantom{-(}\underline{-(t^2 - 2t + 6)}$
$\phantom{-(}-2t^2 - 7t - 6$

51. $\phantom{-(}-2x^2 + 6x + 7$
$\phantom{-(}\underline{-(-4x^2 \phantom{+6x} + 1)}$
$\phantom{-(}2x^2 + 6x + 6$

53. $\phantom{-(}6x^3 + 2x^2 - 5$
$\phantom{-(}\underline{-(\phantom{6x^3 +} 2x^2 + 4)}$
$\phantom{-(}6x^3 \phantom{+ 2x^2} - 9$

55. $\phantom{-(}6x + 2$
$\phantom{-(}\underline{-(3x - 5)}$
$\phantom{-(}3x + 7$

57. $\phantom{-(}3x^2 - 4x + 8$
$\phantom{-(}\underline{-(x^2 - 7x - 5)}$
$\phantom{-(}2x^2 + 3x + 13$

59. $6x^2 - (4 - 5x) + (8x^2 + 15)$
$= 6x^2 + 8x^2 - (-5x) - 4 + 15$
$= 14x^2 + 5x + 11$

61. $w^3 + 2w - 1 - w^2 - w - (-2) - 3w^2 - (-w^3)$
$= w^3 + w^3 - w^2 - 3w^2 + 2w - w - 1 + 2$
$= 2w^3 - 4w^2 + w + 1$

63.
$3x^2 - 5x - 1 + 4x^2 + 11 - 2x^2 - (-4x) - (-7)$
$= 3x^2 + 4x^2 - 2x^2 - 5x + 4x - 1 + 11 + 7$
$= 5x^2 - x + 17$

65. $(2t - 5) + (5t^2 + 4) - (t^2 - 3t + 7)$
$= 2t - 5 + 5t^2 + 4 - t^2 - (-3t) - 7$
$= 5t^2 - t^2 + 2t + 3t - 5 + 4 - 7$
$= 4t^2 + 5t - 8$

67. $2z^3 + 5z + 6 - (z^3 + 2z^2 - 5 + 4z^2 + 3z + 2)$
$= 2z^3 + 5z + 6 - z^3 - 2z^2 + 5 - 4z^2 - 3z - 2$
$= 2z^3 - z^3 - 2z^2 - 4z^2 + 5z - 3z + 6 + 5 - 2$
$= z^3 - 6z^2 - 2z + 9$

69. Adding $-x^2$ to both sides of the equation eliminates the second-degree term. Then the remaining variable can be isolated.

71. $x^2 + 2x + 5 = x^2 + x + 2$
$-x^2 + x^2 + 2x + 5 = -x^2 + x^2 + x + 2$
$2x + 5 = x + 2$
$2x + 5 - 5 = x + 2 - 5$
$2x = x - 3$
$2x - x = x - 3 - x$
$x = -3$

**Exponents and Polynomials**

73. $x^3 + x - 4 + 2 - x^3 = -2$
    $x - 2 = -2$
    $x - 2 + 2 = -2 + 2$
    $x = 0$

75. $(3x + x^2) - (x^2 + x - 1) = 7 - x$
    $3x + x^2 - x^2 - x + 1 = 7 - x$
    $2x + 1 = 7 - x$
    $2x = 7 - x - 1$
    $2x = -x + 6$
    $2x + x = x - x + 6$
    $3x = 6$
    $x = \dfrac{6}{3}$
    $x = 2$

77. $0 - (2x - 1) = -2x + 1$

79. $(4 - x) - 3$
    $= 4 - x - 3$
    $= 1 - x$

81. $ax^2 + bx - 4 - 3x^2 - 2x - c = -2x^2 - 5$
    $ax^2 - 3x^2 + bx - 2x - c - 4 = 2x^2 - 5$
    $(a - 3)x^2 + (b - 2)x - c - 4 = 2x^2 - 5$

    $a - 3 = 2$
    $a = 5$

    $b - 2 = 0$
    $b = 2$

    $-c - 4 = -5$
    $-c = -5 + 4$
    $-c = -1$
    $c = 1$

83.
$(2y + 3) + (2y + 3) + (y^2 - y - 1) + (y^2 - y - 1)$
$= 2y + 2y + 3 + 3 + y^2 + y^2 - y - y - 1 - 1$
$= 4y + 6 + 2y^2 - 2y - 2$
$= 2y^2 + 4y - 2y + 6 - 2$
$= 2y^2 + 2y + 4$

85. $180 - (x^3 + 3x + x^3 - x - 50)$
    $= 180 - (2x^3 + 2x - 50)$
    $= 180 - 2x^3 - 2x + 50$
    $= -2x^3 - 2x + 230$

87. $P(x) = R(x) - C(x)$
    $= 2x^3 + 3x - 1 - (x^3 + x^2 + 4)$
    $= 2x^3 + 3x - 1 - x^3 - x^2 - 4$
    $= 2x^3 - x^3 - x^2 + 3x - 1 - 4$
    $= x^3 - x^2 + 3x - 5$

89. $-16t^2 + 80t = -16t^2 + 60t + 40$
    $80t = 60t + 40$
    $20t = 40$
    $t = 2$ seconds

91. (a) $(0, 5.5), (28, 12.7)$

$$m = \frac{12.7 - 5.5}{28 - 0}$$
$$= 0.257$$

$$F = 0.257x + 5.5$$

(b) $(0, 1.2), (28, 3.9)$

$$m = \frac{3.9 - 1.2}{28 - 0}$$
$$= 0.096$$

$$m = 0.096x + 1.2$$

93. $(8x + 3) - [(2x - 1) + (x + 3)]$
$= 8x + 3 - (2x - 1 + x + 3)$
$= 8x + 3 - (3x + 2)$
$= 8x + 3 - 3x - 2$
$= 5x + 1$

95. $7y^2 + [(5y^2 + 3) - (y^2 - 6)]$
$= 7y^2 + (5y^2 + 3 - y^2 + 6)$
$= 7y^2 + 5y^2 - y^2 + 3 + 6$
$= 11y^2 + 9$

97. $P(2a) - P(a)$
$= 1 - 3(2a) - (1 - 3a)$
$= 1 - 6a - 1 + 3a$
$= -3a$

## Section 6.4 Multiplication

1. The product of binomials may have two terms, such as $(x + 3)(x - 3) = x^2 - 9$, or it may have four terms, such as $(x + 3)(a - 2) = ax - 2x + 3a - 6$.

3. $-15x^2$

5. $-10y^{2+1}$
$= -10y^3$

7. $3x^{2+1}y^{1+4}$
$= 3x^3y^5$

9. $3x^2 \cdot 4x - 3x^2 \cdot 7$
$= 12x^3 - 21x^2$

11. $-5x \cdot 3x^2 + (-5x)x - (-5x)2$
$= -15x^3 - 5x^2 + 10x$

13. $-4x^2(x^4 - 3x^2 + 2x)$
$= -4x^2 \cdot x^4 - (-4x^2) \cdot 3x^2 + (-4x^2) \cdot 2x$
$= -4x^6 + 12x^4 - 8x^3$

15. $6x(3x^2 + 4x - 5)$
$= 6x \cdot 3x^2 + 6x \cdot 4x - 6x(5)$
$= 18x^3 + 24x^2 - 30x$

17. $xy^2(x^2 + 3xy + y^2)$
$= xy^2 \cdot x^2 + xy^2 \cdot 3xy + xy^2 \cdot y^2$
$= x^3y^2 + 3x^2y^3 + xy^4$

19. $x^3(xy^2 + y + 2x)$
$= x^3 \cdot xy^2 + x^3 \cdot y + x^3 \cdot 2x$
$= x^4y^2 + x^3y + 2x^4$

Exponents and Polynomials

21. $(x+7)(x-3)$
$= x \cdot x - 3 \cdot x + 7 \cdot x + 7(-3)$
$= x^2 - 3x + 7x - 21$
$= x^2 + 4x - 21$

23. $(y+6)(y+1)$
$= y \cdot y + 1 \cdot y + 6 \cdot y + 6 \cdot 1$
$= y^2 + y + 6y + 6$
$= y^2 + 7y + 6$

25. $(2x+1)(x-3)$
$= 2x \cdot x - 3 \cdot 2x + 1 \cdot x + 1(-3)$
$= 2x^2 - 6x + x - 3$
$= 2x^2 - 5x - 3$

27. $(3a-5)(2a+1)$
$= 3a \cdot 2a + 1 \cdot 3a - 5 \cdot 2a - 5 \cdot 1$
$= 6a^2 + 3a - 10a - 5$
$= 6a^2 - 7a - 5$

29. $(x+4)(x+4)$
$= x \cdot x + 4 \cdot x + 4 \cdot x + 4 \cdot 4$
$= x^2 + 4x + 4x + 16$
$= x^2 + 8x + 16$

31. $(x+2)(x^2+2x-5)$
$= x \cdot x^2 + x \cdot 2x - 5 \cdot x + 2 \cdot x^2 + 2 \cdot 2x + 2(-5)$
$= x^3 + 2x^2 - 5x + 2x^2 + 4x - 10$
$= x^3 + 4x^2 - x - 10$

33. $(x-3)(x^2+3x+9)$
$= x \cdot x^2 + x \cdot 3x + x \cdot 9 - 3 \cdot x^2 - 3 \cdot 3x - 3 \cdot 9$
$= x^3 + 3x^2 + 9x - 3x^2 - 9x - 27$
$= x^3 - 27$

35. $(y-4)(3y^2-y+2)$
$= y \cdot 3y^2 + y \cdot y + 2 \cdot y - 4 \cdot 3y^2 + 4(-y) - 4 \cdot 2$
$= 3y^3 - y^2 + 2y - 12y^2 + 4y - 8$
$= 3y^3 - 13y^2 + 6y - 8$

37. $(2x-1)(x^2-x-3)$
$= 2x \cdot x^2 - x \cdot 2x - 3 \cdot 2x - 1 \cdot x^2 - 1(-x) - 1(-3)$
$= 2x^3 - 2x^2 - 6x - x^2 + x + 3$
$= 2x^3 - 3x^2 - 5x + 3$

39. $(x^2+x-4)(x^2+2x-1)$
$= x^2 \cdot x^2 + x^2 \cdot 2x - 1 \cdot x^2 + x \cdot x^2 + x \cdot 2x - 1 \cdot x - 4 \cdot x^2 - 4 \cdot 2x - 4(-1)$
$= x^4 + 2x^3 - x^2 + x^3 + 2x^2 - x - 4x^2 - 8x + 4$
$= x^4 + 3x^3 - 3x^2 - 9x + 4$

41. 
$$\begin{array}{r} 3t - 7 \\ \times \quad t + 5 \\ \hline 15t - 35 \\ 3t^2 - 7t \quad\quad \\ \hline 3t^2 + 8t - 35 \end{array}$$

43.
$$\begin{array}{r} x^2 - 4x + 7 \\ \times \quad\quad x - 2 \\ \hline -2x^2 + 8x - 14 \\ x^3 - 4x^2 + 7x \quad\quad \\ \hline x^3 - 6x^2 + 15x - 14 \end{array}$$

45. The result is the same. The Associative Property of Multiplication allows us to group any two factors.

47. $2x(x+3)(x-6)$
$= 2x(x^2 - 6x + 3x - 18)$
$= 2x(x^2 - 3x - 18)$
$= 2x^3 - 6x^2 - 36x^2$

49. $(x+1)(x-2)(x+3)$
$= (x+1)(x^2 + 3x - 2x - 6)$
$= (x+1)(x^2 + x - 6)$
$= x^3 + x^2 - 6x + x^2 + x - 6$
$= x^3 + 2x^2 - 5x - 6$

51. $(4x+3)(x-1)(2x+1)$
$= (4x+3)(2x^2 + x - 2x - 1)$
$= (4x+3)(2x^2 - x - 1)$
$= 8x^3 - 4x^2 - 4x + 6x^2 - 3x - 3$
$= 8x^3 + 2x^2 - 7x - 3$

53. $2y(y+1)^2$
$= 2y(y+1)(y+1)$
$= 2y(y^2 + y + y + 1)$
$= 2y(y^2 + 2y + 1)$
$= 2y^3 + 4y^2 + 2y$

55. $-3x^3(x-1)(x+2)$
$= -3x^3(x^2 + 2x - x - 2)$
$= -3x^3(x^2 + x - 2)$
$= -3x^5 - 3x^4 + 6x^3$

57. $(x+1)^3$
$= (x+1)(x+1)(x+1)$
$= (x+1)(x^2 + x + x + 1)$
$= (x+1)(x^2 + 2x + 1)$
$= x^3 + 2x^2 + x + x^2 + 2x + 1$
$= x^3 + 3x^2 + 3x + 1$

59. $3x + (x+1)(x-4)$
$= 3x + x^2 - 4x + x - 4$
$= x^2 - 4$

61. $7x - (x+2)(x-3)$
$= 7x - (x^2 - 3x + 2x - 6)$
$= 7x - (x^2 - x - 6)$
$= 7x - x^2 + x + 6$
$= -x^2 + 8x + 6$

63. $(x+2)(x-4) + (x^2 + 2x)$
$= x^2 - 4x + 2x - 8 + x^2 + 2x$
$= 2x^2 - 8$

65. $(7x-2) - (x+6)(x-7)$
$= 7x - 2 - (x^2 - 7x + 6x - 42)$
$= 7x - 2 - (x^2 - x - 42)$
$= 7x - 2 - x^2 + x + 42$
$= -x^2 + 8x + 40$

67. $2x + 5 - x(x-3)$
$= 2x + 5 - (x^2 - 3x)$
$= 2x + 5 - x^2 + 3x$
$= -x^2 + 5x + 5$

69. $2x^2 + 5x - 6 + (2x-1)(x+2)$
$= 2x^2 + 5x - 6 + 2x^2 + 4x - x - 2$
$= 4x^2 + 8x - 8$

71. $(x+3)(2x-1)x^2$
$= (2x^2 - x + 6x - 3)x^2$
$= (2x^2 + 5x - 3)x^2$
$= 2x^4 + 5x^3 - 3x^2$

73. $x(x+3) = x^2 + 15$
$x^2 + 3x = x^2 + 15$
$3x = 15$
$x = 5$

75. $(x+2)(x-3) = x(x-2)$
$x^2 - 3x + 2x - 6 = x^2 - 2x$
$x^2 - x - 6 = x^2 - 2x$
$-x - 6 = -2x$
$x - 6 = 0$
$x = 6$

77. $(2x+1)(x^2+3x-1)$
$= 2x^3 + 6x^2 - 2x + x^2 + 3x - 1$
$= 2x^3 + 6x^3 + x - 1$

79. Area of pool $= x(x+7)$
$= x^2 + 7x$

Area of pool and walk
$= (x+6+6)(x+7+6+6)$
$= (x+12)(x+19)$
$= x^2 + 19x + 12x + 228$
$= x^2 + 31x + 228$

Note: We add 6 twice to the length and the width since the walkway is on both sides and both ends of the pool.

Area of walk =
$x^2 + 31x + 228 - (x^2 + 7x)$
$= x^2 + 31x + 228 - x^2 - 7x$
$= 24x + 228$

81. $l = 3 + w$
$(l+5)(w+5) = lw + 160$
$(3+w+5)(w+5) = (3+w)w + 160$
$(w+8)(w+5) = w^2 + 3w + 160$
$w^2 + 5w + 8w + 40 = w^2 + 3w + 160$
$w^2 + 13w + 40 = w^2 + 3w + 160$
$13w + 40 = 3w + 160$
$10w + 40 = 160$
$10w = 120$
$w = 12$

$l = 3 + 12$
$= 15$

The dimensions are 12 inches by 15 inches.

83. $w = l - 4$
area of each side flower bed $= l \cdot 4$
area of end flower bed $= w \cdot 4$
area of each square corner flower bed
$= 4 \cdot 4 = 16$

$4l + 4l + 4w + 16 + 16 = 184$
$8l + 4w + 32 = 184$
$8l + 4w = 152$
$8l + 4(l-4) = 152$
$8l + 4l - 16 = 152$
$12l = 168$
$l = 14$

$w = 14 - 4$
$= 10$

The patio is 10 feet by 14 feet.

Exponents and Polynomials

85. $l = 2w + 1$

$(l + 2)(w + 2) = lw + 12$

$lw + 2w + 2l + 4 = lw + 12$

$(2w + 1)w + 2w + 2(2w + 1) + 4 = (2w + 1)w + 12$

$2w^2 + w + 2w + 4w + 2 + 4 = 2w^2 + w + 12$

$2w^2 + 7w + 6 = 2w^2 + w + 12$

$7w + 6 = w + 12$

$6w + 6 = 12$

$6w = 6$

$w = 1$ miles

$l = 2(1) + 1$
$= 3$ miles

The dimensions are 1 mile by 3 miles.

87. (a) Total expenditures on home video games

(b) $T(x) = (1.2x + 12.3)(2.2x + 207)$
$= 2.64x^2 + 248.4x + 27.06x + 2546.1$
$= 2.64x^2 + 275.46x + 2546.1$

89. $n(7) = 1.9(7) + 50.8$
$= 13.3 + 50.8$
$= 64.1$

$r(7) = 1.4(7) + 16.7$
$= 9.8 + 16.7$
$= 26.5$

$n(7) \cdot r(7) = (64.1)(26.5)$
$= 1698.65$

91. Both numbers approximate the total monthly cable TV revenue (in millions of dollars) in 1997. The table value is $1700.02 million.

93. $x^2(3x^{2n} + 2x^n - 5)$
$= 3x^{2n+2} + 2x^{n+2} - 5x^2$

95. $(x^2 + 5x - 1)(x^3 - 2x^2 + 4x - 5)$
$= x^5 - 2x^4 + 4x^3 - 5x^2 + 5x^4 - 10x^3 + 20x^2 - 25x - x^3 + 2x^2 - 4x + 5$
$= x^5 + 3x^4 - 7x^3 + 17x^2 - 29x + 5$

97. $3x^3(ax^2 + bx - 4) = -6x^5 + 3x^4 - 12x^3$
$3ax^5 + 3bx^4 - 12x^3 = -6x^5 + 3x^4 - 12x^3$

$3a = -6$
$a = -2$

$3b = 3$
$b = 1$

## Section 6.5 Special Products

1. Because (ii) is the product of binomials, the FOIL method applies. Because (i) is the product of a binomial and a monomial, FOIL does not apply.

3. $(x + 5)(x + 2)$
$= x^2 + 2x + 5x + 10$
$= x^2 + 7x + 10$

5. $(y - 7)(y + 3)$
$= y^2 + 3y - 7y - 21$
$= y^2 - 4y - 21$

7. $(8 + x)(2 - x)$
$= 16 - 8x + 2x - x^2$
$= 16 - 6x - x^2$

9. $(3x + 2)(2x - 3)$
$= 6x^2 - 9x + 4x - 6$
$= 6x^2 - 5x - 6$

11. $(4x - 1)(x - 2)$
$= 4x^2 - 8x - x + 2$
$= 4x^2 - 9x + 2$

13. $(x + 5)(2x - 7)$
$= 2x^2 - 7x + 10x - 35$
$= 2x^2 + 3x - 35$

15. $(2a - 5)(4a - 3)$
$= 8a^2 - 6a - 20a + 15$
$= 8a^2 - 26a + 15$

17. $(x + y)(x - 3y)$
$= x^2 - 3xy + xy - 3y^2$
$= x^2 - 2xy - 3y^2$

19. $(5x - 4y)(4x + 5y)$
$= 20x^2 + 25xy - 16xy - 20y^2$
$= 20x^2 + 9xy - 20y^2$

21. $(x^2 - 5)(x^2 - 3)$
$= x^4 - 3x^2 + 5x^2 - 15$
$= x^4 + 2x^2 - 15$

Exponents and Polynomials

23. $(a^2+4)(a^2+2)$
$= a^4 + 2a^2 + 4a^2 + 8$
$= a^4 + 6a^2 + 8$

25. $(ab+3)(ab+2)$
$= a^2b^2 + 2ab + 3ab + 6$
$= a^2b^2 + 5ab + 6$

27. In (i) the terms are identical, $2x$ and $2x$. In (ii) the terms are opposites, $2x$ and $-2x$.

29. $x^2 - 7^2$
$= x^2 - 49$

31. $3^2 - y^2$
$= 9 - y^2$

33. $(3x)^2 - 5^2$
$= 9x^2 - 25$

35. $6^2 - (5y)^2$
$= 36 - 25y^2$

37. $(3x)^2 - (7y)^2$
$= 9x^2 - 49y^2$

39. $(2a)^2 - b^2$
$= 4a^2 - b^2$

41. $(x^2)^2 - (3)^2$
$= x^4 - 9$

43. $x^2 + 2x + 4$

45. $9^2 - 2(9x) + x^2$
$= 81 - 18x + x^2$

47. $(2x)^2 + 2(2x) + 1^2$
$= 4x^2 + 4x + 1$

49. $4^2 - 2(28x) + (7x)^2$
$= 16 - 56x + 49x^2$

51. $a^2 - 2(3ab) + (3b)^2$
$= a^2 - 6ab + 9b^2$

53. $(x^3)^2 + 2(2x^3) + 2^2$
$= x^6 + 4x^3 + 4$

55. $y^2 - 2y - 3y + 6$
$= y^2 - 5y + 6$

57. $c^2 - (5)^2$
$= c^2 - 25$

59. $3^2 + 2(3b) + b^2$
$= 9 + 6b + b^2$

61. $4^2 - (7b)^2$
$= 16 - 49b^2$

63. $6y^2 + 10y + 9y + 15$
$= 6y^2 + 19y + 15$

65. $(3x)^2 + 2(15x) + 5^2$
$= 9x^2 + 30x + 25$

67. $ab + 2a^2 - 6b^2 - 12ab$
$= 2a^2 - 11ab - 6b^2$

69. $y^2 - (6x)^2$
$= y^2 - 36x^2$

71. $x^2 + 3x + 7^2 - x^2$
$= x^2 + 3x + 49 - x^2$
$= 3x + 49$

73. $x^2 - (6^2 - 2(6x) + x^2)$
$= x^2 - (36 - 12x + x^2)$
$= x^2 - 36 + 12x - x^2$
$= 12x - 36$

75. $y((3y)^2 + 2(6y) + 2^2)$
$= y(9y^2 + 12y + 4)$
$= 9y^3 + 12y^2 + 4y$

77. $(x+2)(x-2)(x^2+4)$
$= (x^2 - 2^2)(x^2 + 4)$
$= (x^2 - 4)(x^2 + 4)$
$= (x^2)^2 - 4^2$
$= x^4 - 16$

79. $(x-5)^2 + 1 + 2x - x^2$
$= x^2 - 2(5x) + 5^2 + 1 + 2x - x^2$
$= x^2 - 10x + 25 + 1 + 2x - x^2$
$= -8x + 26$

81. $(6x^2 - 3x) - (2x - 5)(2x + 5)$
$= 6x^2 - 3x - ((2x)^2 - 5^2)$
$= 6x^2 - 3x - (4x^2 - 25)$
$= 6x^2 - 3x - 4x^2 + 25$
$= 2x^2 - 3x + 25$

83. $x^2 - 2x + x - 2 = x^2 + 2x + 3x + 6$
$x^2 - x - 2 = x^2 + 5x + 6$
$-x - 2 = 5x + 6$
$-6x - 2 = 6$
$-6x = 8$
$x = \dfrac{-8}{6}$
$x = \dfrac{-4}{3}$

85. $x^2 - 1 = x^2 + 2x$
$-1 = 2x$
$x = -\dfrac{1}{2}$

87. $4x^2 - 9 = 4x^2 - 4x + 3x - 3$
$4x^2 - 9 = 4x^2 - x - 3$
$-9 = -x - 3$
$-6 = -x$
$x = 6$

89. $x^2 + (x+1)^2 = 2(x+1)^2 - 7$
$x^2 + x^2 + 2x + 1 = 2(x^2 + 2x + 1) - 7$
$2x^2 + 2x + 1 = 2x^2 + 4x + 2 - 7$
$2x + 1 = 4x - 5$
$-2x + 1 = -5$
$-2x = -6$
$x = 3$

The integers are 3 and 4.

91. $x^2 = (x+1)^2$
$x^2 = x^2 + 2x + 1$
$0 = 2x + 1$
$-2x = 1$
$x = -\dfrac{1}{2}$

93. $(3x+7)(3x-7)$
$= (3x)^2 - 7^2$
$= 9x^2 - 49$

**Exponents and Polynomials**

95. Area of picture and frame $= (x+2+2)^2$
$= (x+4)^2$
$= x^2 + 2(4x) + 4^2$
$= x^2 + 8x + 16$

Area of picture $= x^2$

Area of frame $= x^2 + 8x + 16 - x^2$
$= 8x + 16$

97. $l = 6 + w$
$l(w+3) = lw + 33$
$(6+w)(w+3) = (6+w)w + 33$
$6w + 18 + w^2 + 3w = 6w + w^2 + 33$
$w^2 + 9w + 18 = w^2 + 6w + 33$
$9w + 18 = 6w + 33$
$3w + 18 = 33$
$3w = 15$
$w = 5$

$l = 6 + 5$
$= 11$

The dimensions are 5 feet by 11 feet.

99. Area of room $= (x+2+2)^2$
$= (x+4)^2$
$= x^2 + 8x + 16$

Area of carpeted area $= x^2$

Area of border $= x^2 + 8x + 16 - x^2$
$= 8x + 16$
$8x + 16 + = 60$
$8x = 44$
$x = 5.5$

The dimensions of the carpet are 5.5 feet by 5.5 feet.

101.
$\frac{1}{2}(x-1+x+3)x + 122 = \frac{1}{2}(x-1+x+3)(x+2)$

$\frac{1}{2}(2x+2)x + 122 = \frac{1}{2}(2x+2)(x+2)$
$(x+1)x + 122 = (x+1)(x+2)$
$x^2 + x + 122 = x^2 + 2x + x + 2$
$x + 122 = 3x + 2$
$-2x + 122 = 2$
$-2x = -120$
$x = 60$

$x + 3 = 60 + 3$
$= 63$ feet

103. $p(x) = (-0.35x + 15.65)(0.009x + 0.075)$
$= -0.00315x^2 + 0.1146x + 1.17375$
$= -0.003x^2 + 0.115x + 1.174$

105. $0.299(6.9)$
$= 2.06$¢

The per-mile tax can be approximated by evaluating $p(25)$.

107. $(x^2+9)(x^2-3^2)$
$=(x^2+9)(x-9)$
$=(x^2)^2-(9)^2$
$=x^4-81$

109. $(2x+1)^2-(2x+1)(2x-1)$
$=(2x)^2+2(2x)+1^2-\left((2x)^2-1^2\right)$
$=4x^2+4x+1-(4x^2-1)$
$=4x^2+4x+1-4x^2+1$
$=4x^2+2$

111. $P(t+2)=(t+2)^2+3(t+2)-1$
$=t^2+4t+4+3t+6-1$
$=t^2+7t+9$

113. $\dfrac{10}{2}=5$
$5^2=25$
$x^2+10x+25=(x+5)^2$

## Section 6.6 Division

1. A is the dividend, B is the divisor, and C is the quotient.

3. $\dfrac{14}{21}\cdot x^{4-3}$
$=\dfrac{2}{3}x$

5. $\dfrac{3}{12}\cdot x^{6-6}y^{3-2}$
$=\dfrac{1}{4}y$

7. $\dfrac{10x}{2}-\dfrac{16}{2}$
$=5x-8$

9. $\dfrac{2y^6}{2y^2}+\dfrac{6y^4}{2y^2}-\dfrac{2y^2}{2y^2}$
$=y^4+3y^2-1$

11. $\dfrac{3x^4-9x^3+6x^2}{6x^2}$
$=\dfrac{3x^4}{6x^2}-\dfrac{9x^3}{6x^2}+\dfrac{6x^2}{6x^2}$
$=\dfrac{1}{2}x^2-\dfrac{3}{2}x+1$

13. $\dfrac{6a^2b^3-8a^3b^2}{2ab}$
$=\dfrac{6a^2b^3}{2ab}-\dfrac{8a^3b^2}{2ab}$
$=3ab^2-4a^2b$

15. $\dfrac{2x^4}{4x^2}+\dfrac{12x^3}{4x^2}-\dfrac{4x^2}{4x^2}$
$=\dfrac{1}{2}x^2+3x-1$

17. $\dfrac{6}{2x}-\dfrac{4x}{2x}$
$=\dfrac{3}{x}-2$

Exponents and Polynomials

19. $\dfrac{6x^3}{6x} + \dfrac{6x^2}{6x} - \dfrac{2x}{6x}$

$= x^2 + x - \dfrac{1}{3}$

21. $\dfrac{6x^3}{2x^2} - \dfrac{2x^2}{2x^2} + \dfrac{2x}{2x^2}$

$= 3x - 1 + \dfrac{1}{x}$

23. First, divide $x^3$ by $x$ and place the result $x^2$ above the $3x^2$. Second, multiply $x^2$ by $x-4$ and place the result $x^3 - 4x^2$ below $x^3 + 3x^2$.

25.
$$\require{enclose}\begin{array}{r} x+3 \phantom{xxxxx} \\ x-7 \enclose{longdiv}{x^2 - 4x - 21} \\ \underline{x^2 - 7x} \phantom{xxxxx} \\ 3x - 21 \\ \underline{3x - 21} \\ 0 \end{array}$$
$Q = x + 3, R = 0$

27.
$$\begin{array}{r} -x + 3 \phantom{xxxx} \\ x+4 \enclose{longdiv}{-x^2 - x + 12} \\ \underline{-x^2 - 4x} \phantom{xxxx} \\ 3x + 12 \\ \underline{3x + 12} \\ 0 \end{array}$$
$Q = -x + 3, R = 0$

29.
$$\begin{array}{r} 2x + 5 \phantom{xxxx} \\ x+4 \enclose{longdiv}{2x^2 + 13x + 20} \\ \underline{2x^2 + 8x} \phantom{xxxxx} \\ 5x + 20 \\ \underline{5x + 20} \\ 0 \end{array}$$
$Q = 2x + 5, R = 0$

31.
$$\begin{array}{r} x + 3 \phantom{xxxx} \\ x-5 \enclose{longdiv}{x^2 - 2x - 5} \\ \underline{x^2 - 5x} \phantom{xxxx} \\ 3x - 5 \\ \underline{3x - 15} \\ 10 \end{array}$$
$Q = x + 3, R = 10$

33.
$$\begin{array}{r} 2x + 5 \phantom{xxxx} \\ x-7 \enclose{longdiv}{2x^2 - 9x - 15} \\ \underline{2x^2 - 14x} \phantom{xxxx} \\ 5x - 15 \\ \underline{5x - 35} \\ 20 \end{array}$$
$Q = 2x + 5, R = 20$

35.
$$\begin{array}{r} 3x - 5 \phantom{xxxx} \\ x+2 \enclose{longdiv}{3x^2 + x - 12} \\ \underline{3x^2 + 6x} \phantom{xxxx} \\ -5x - 12 \\ \underline{-5x - 10} \\ -2 \end{array}$$
$Q = 3x - 5, R = -2$

**37.**
$$\begin{array}{r} 2x - 1 \\ x+3\overline{\smash{)}2x^2+5x+1} \\ \underline{2x^2+6x} \\ -x+1 \\ \underline{-x-3} \\ 4 \end{array}$$
$Q = 2x - 1, R = 4$

**39.**
$$\begin{array}{r} 2x - 3 \\ 3x+1\overline{\smash{)}6x^2-7x-3} \\ \underline{6x^2+2x} \\ -9x-3 \\ \underline{-9x-3} \\ 0 \end{array}$$
$Q = 2x - 3, R = 0$

**41.**
$$\begin{array}{r} -3x - 1 \\ 2x-7\overline{\smash{)}-6x^2+19x+7} \\ \underline{-6x^2+21x} \\ -2x+7 \\ \underline{-2x+7} \\ 0 \end{array}$$
$Q = -3x - 1, R = 0$

**43.**
$$\begin{array}{r} x + 2 \\ 3x-5\overline{\smash{)}3x^2+x-2} \\ \underline{3x^2-5x} \\ 6x-2 \\ \underline{6x-10} \\ 8 \end{array}$$
$Q = x + 2, R = 8$

**45.**
$$\begin{array}{r} x - 2 \\ 4x+3\overline{\smash{)}4x^2-5x-9} \\ \underline{4x^2+3x} \\ -8x-9 \\ \underline{-8x-6} \\ -3 \end{array}$$
$Q = x - 2, R = -3$

**47.**
$$\begin{array}{r} 2x + 1 \\ 3x+5\overline{\smash{)}6x^2+13x+2} \\ \underline{6x^2+10x} \\ 3x+2 \\ \underline{3x+5} \\ -3 \end{array}$$
$Q = 2x + 1, R = -3$

**49.** Include the missing terms with 0 coefficients: $x^3 + 0x^2 + 0x - 5$.

**51.**
$$\begin{array}{r} 2x + 5 \\ 2x-5\overline{\smash{)}4x^2-25} \\ \underline{4x^2-10x} \\ 10x-25 \\ \underline{10x-25} \\ 0 \end{array}$$
$Q = 2x + 5, R = 0$

**53.**
$$\begin{array}{r} x + 5 \\ x-5\overline{\smash{)}x^2-5} \\ \underline{x^2-5x} \\ 5x-5 \\ \underline{5x-25} \\ 20 \end{array}$$
$Q = x + 5, R = 20$

**Exponents and Polynomials**

**55.**
$$\begin{array}{r} 3x + 4 \\ 2x-1 \overline{\smash{\big)}\ 6x^2 + 5x} \\ \underline{6x^2 - 3x} \\ 8x \\ \underline{8x - 4} \\ 4 \end{array}$$
$Q = 3x + 4, R = 4$

**57.**
$$\begin{array}{r} x^2 + 3x - 1 \\ x+1 \overline{\smash{\big)}\ x^3 + 4x^2 + 2x - 1} \\ \underline{x^3 + x^2} \\ 3x^2 + 2x \\ \underline{3x^2 + 3x} \\ -x - 1 \\ \underline{-x - 1} \\ 0 \end{array}$$
$Q = x^2 + 3x - 1, R = 0$

**59.**
$$\begin{array}{r} 2x^2 + x - 2 \\ x-6 \overline{\smash{\big)}\ 2x^3 - 11x^2 - 8x + 12} \\ \underline{2x^3 - 12x^2} \\ x^2 - 8x \\ \underline{x^2 - 6x} \\ -2x + 12 \\ \underline{-2x + 12} \\ 0 \end{array}$$
$Q = 2x^2 + x - 2, R = 0$

**61.**
$$\begin{array}{r} 2x^2 + x + 3 \\ x-4 \overline{\smash{\big)}\ 2x^3 - 7x^2 - x - 4} \\ \underline{2x^3 - 8x^2} \\ x^2 - x \\ \underline{x^2 - 4x} \\ 3x - 4 \\ \underline{3x - 12} \\ 8 \end{array}$$
$Q = 2x^2 + x + 3, R = 8$

**63.**
$$\begin{array}{r} 2x^2 - x - 1 \\ 3x+1 \overline{\smash{\big)}\ 6x^3 - x^2 - 4x - 1} \\ \underline{6x^3 + 2x^2} \\ -3x^2 - 4x \\ \underline{-3x^2 - x} \\ -3x - 1 \\ \underline{-3x - 1} \\ 0 \end{array}$$
$Q = 2x^2 - x - 1, R = 0$

**65.**
$$\begin{array}{r} x^3 + x^2 + x + 1 \\ x-1 \overline{\smash{\big)}\ x^4 \qquad\qquad - 1} \\ \underline{x^4 - x^3} \\ x^3 \\ \underline{x^3 - x^2} \\ x^2 \\ \underline{x^2 - x} \\ x - 1 \\ \underline{x - 1} \\ 0 \end{array}$$
$Q = x^3 + x^2 + x + 1, R = 0$

67.
$$\begin{array}{r}x^3-3x^2+9x\phantom{+}-7\phantom{0}\\x+3\overline{)x^4\phantom{+0x^3+0x^2}+20x-10}\\\underline{x^4+3x^3}\phantom{+20x-10}\\-3x^3\phantom{+20x-10}\\\underline{-3x^3-9x^2}\phantom{+20x-10}\\9x^2+20x\phantom{-10}\\\underline{9x^2+27x}\phantom{-10}\\-7x-10\\\underline{-7x-21}\\11\end{array}$$

$Q = x^3 - 3x^2 + 9x - 7 \quad R = 11$

69.
$$\begin{array}{r}3x+4\phantom{0}\\x-1\overline{)3x^2+x-4}\\\underline{3x^2-3x}\phantom{-4}\\4x-4\\\underline{4x-4}\\0\end{array}$$

71.
$$\begin{array}{r}x+3\phantom{0}\\x-7\overline{)x^2-4x-11}\\\underline{x^2-7x}\phantom{-11}\\3x-11\\\underline{3x-21}\\10\end{array}$$

$= x+3+\dfrac{10}{x-7}$

73. $\dfrac{(2x+1)(x+2)}{x+3}$

$= \dfrac{2x^2+4x+x+2}{x+3}$

$= \dfrac{2x^2+5x+2}{x+3}$

$$\begin{array}{r}2x-1\phantom{0}\\x+3\overline{)2x^2+5x+2}\\\underline{2x^2+6x}\phantom{+2}\\-x+2\\\underline{-x-3}\\5\end{array}$$

$= 2x-1+\dfrac{5}{x+3}$

75. $\dfrac{(x^3+2x^2+x-5x^2-x+5)}{x-2}$

$= \dfrac{x^3-3x^2+5}{x-2}$

$$\begin{array}{r}x^2-x-2\phantom{0}\\x-2\overline{)x^3-3x^2\phantom{+0x}+5}\\\underline{x^3-2x^2}\phantom{+0x+5}\\-x^2\phantom{+0x+5}\\\underline{-x^2+2x}\phantom{+5}\\-2x+5\\\underline{-2x+4}\\1\end{array}$$

$= x^2-x-2+\dfrac{1}{x-2}$

77. $\dfrac{3x^3 + x^2 - 1}{2x - 2 + 1 - x}$

$= \dfrac{3x^3 + x^2 - 1}{x - 1}$

$= x - 1 \overline{\smash{\big)}\begin{array}{l} 3x^2 + 4x + 4 \\ 3x^3 + x^2 \phantom{+00} - 1 \end{array}}$

$\phantom{= x - 1)}\underline{3x^3 - 3x^2}$
$\phantom{= x - 1)00000}4x^2$
$\phantom{= x - 1)0000}\underline{4x^2 - 4x}$
$\phantom{= x - 1)000000000}4x - 1$
$\phantom{= x - 1)000000000}\underline{4x - 4}$
$\phantom{= x - 1)0000000000000}3$

$= 3x^2 + 4x + 4 + \dfrac{3}{x - 1}$

79. $\dfrac{x^4 - (x^3 + 28x - 40)}{x + 4}$

$x + 4 \overline{\smash{\big)}\begin{array}{l} x^3 - 5x^2 + 20x - 108 \\ x^4 - x^3 \phantom{000} - 28x - 40 \end{array}}$

$\phantom{x + 4)}\underline{x^4 + 4x^3}$
$\phantom{x + 4)0000}-5x^3$
$\phantom{x + 4)000}\underline{-5x^3 - 20x^2}$
$\phantom{x + 4)000000000}20x^2 - 28x$
$\phantom{x + 4)000000000}\underline{20x^2 + 80x}$
$\phantom{x + 4)00000000000000}108x - 40$
$\phantom{x + 4)00000000000000}\underline{108x - 432}$
$\phantom{x + 4)000000000000000000}392$

$= x^3 - 5x^2 + 20x - 108x + \dfrac{392}{x + 4}$

81.
$t - 2 \overline{\smash{\big)}\begin{array}{l} 3t + 1 \\ 3t^2 - 5t - 2 \end{array}}$

$\phantom{t - 2)}\underline{3t^2 - 6t}$
$\phantom{t - 2)00000}t - 2$
$\phantom{t - 2)00000}\underline{t - 2}$
$\phantom{t - 2)0000000}0$

The length is $3t + 1$.

83.
$2x + 3 \overline{\smash{\big)}\begin{array}{l} x^2 + 4 \\ 2x^3 + 3x^2 + 8x + 12 \end{array}}$

$\phantom{2x + 3)}\underline{2x^3 + 3x^2}$
$\phantom{2x + 3)00000}0 + 8x + 12$
$\phantom{2x + 3)000000000}\underline{8x + 12}$
$\phantom{2x + 3)0000000000000}0$

The price per share is $x^2 + 4$.

85.
$n + 1 \overline{\smash{\big)}\begin{array}{l} n + 2 \\ n^2 + 3n + 2 \end{array}}$

$\phantom{n + 1)}\underline{n^2 + n}$
$\phantom{n + 1)0000}2n + 2$
$\phantom{n + 1)0000}\underline{2n + 2}$
$\phantom{n + 1)000000}0$

$(n + 1)$ and $(n + 2)$ are consecutive integers.

87.
$4x + 160 \overline{\smash{\big)}\begin{array}{l} 27x - 1039 \\ 108x^2 + 164x + 27{,}150 \end{array}}$

$\phantom{4x + 160)}\underline{108x^2 + 4320x}$
$\phantom{4x + 160)00000}-4156x + 27{,}150$
$\phantom{4x + 160)00000}\underline{-4156x - 166{,}240}$
$\phantom{4x + 160)0000000000000}193{,}390$

$Q = 27x - 1039, R = 193{,}390$

89. $Q(12) = 27(12) - 1039 + \dfrac{193,390}{4(12) + 160}$

$= 215$

The estimated number of volunteers per country at the 2000 Olympics is 215.

91.
$$x^2 - 1 \overline{\smash{\big)}\, x^2 + 3}$$
$$\underline{x^2 - 1}$$
$$4$$

$Q = 1, R = 4$

93. $(x+3)(2x-5)$

$= 2x^2 - 5x + 6x - 15$

$= 2x^2 + x - 15$

95.
$$x + 5 \overline{\smash{\big)}\, x^2 + 3x + k}$$ with quotient $x - 2$
$$\underline{x^2 + 5x}$$
$$-2x + k$$
$$\underline{-2x - 10}$$
$$k + 10$$

let $k = -10$

97. $2x(x+1) = 2x^2 + 2x$

$$2x^2 + 2x \overline{\smash{\big)}\, 2x^3 + 8x^2 + 6x}$$ with quotient $x + 3$
$$\underline{2x^3 + 2x^2}$$
$$6x^2 + 6x$$
$$\underline{6x^2 + 6x}$$
$$0$$

The side is $x - 3$.

## Section 6.7 Negative Exponents

1. If the exponent is the denominator is larger than the exponent in the numerator, applying the Quotient Rule for Exponents results in a negative exponent.

3. $\dfrac{1}{6^2}$

$= \dfrac{1}{36}$

5. $\dfrac{1}{(-2)^3}$

$\dfrac{1}{-8}$

7. $\left(\dfrac{3}{4}\right)^2$

$= \dfrac{9}{16}$

9. $\dfrac{1}{3} + \dfrac{1}{5}$

$\dfrac{1.5 + 3.1}{3.5}$

$= \dfrac{5 + 3}{15}$

$= \dfrac{8}{15}$

11. $5^{-1}$

$= \dfrac{1}{5}$

**Exponents and Polynomials**

13. $8 \cdot \dfrac{1}{4}$
    $= \dfrac{8}{4}$
    $= 2$

15. $-3^{-2}$
    $= \dfrac{1}{-3^2}$
    $= -\dfrac{1}{9}$

17. $\left(3^{-2}\right)^{-2}$
    $= 3^4$
    $= 81$

19. $3^{-2} \cdot 3$
    $= \dfrac{1}{3^2} \cdot 3$
    $= \dfrac{3}{9}$
    $= \dfrac{1}{3}$

21. $\dfrac{1}{4^3} \cdot \left(6^{-2}\right)^{-1}$
    $= \dfrac{1}{64} \cdot 6^2$
    $= \dfrac{36}{64}$
    $= \dfrac{9}{16}$

23. $x^{-1}$
    $= \dfrac{1}{x}$

25. $\left(\dfrac{3}{10}\right)^{-1}$
    $= \dfrac{10}{3}$

27. $\dfrac{1}{(-12)^9}$

29. $x^5$

31. $y^{-5}$

33. $\dfrac{1}{a^{-6}}$

35. $-7x^{-4}$

37. $\dfrac{5}{x^4}$

39. $\dfrac{1}{3}x^5$

41. $\dfrac{1}{7t}$

43. $\dfrac{-2y}{x^2}$

45. $\dfrac{7}{a} \cdot \dfrac{1}{b^6}$
    $= \dfrac{7}{ab^6}$

47. $\dfrac{x}{3}$

49. $\dfrac{1}{x^2 y^3}$

51. $\dfrac{1}{a^3 b^2}$

53. $\dfrac{4}{5} \dfrac{b}{a^3}$
    $= \dfrac{4b}{5a^3}$

55. Apply the Quotient to a Power Rule and then the Power to a Power Rule to obtain
$$\frac{(x^{-2})^{-1}}{(y^{-3})^{-1}} = \frac{x^2}{y^3}.$$

57. $a^{-5} \cdot a^{-2}$
$= a^{-7}$
$= \dfrac{1}{a^7}$

59. $y^{3-12}$
$= y^{-9}$
$= \dfrac{1}{y^9}$

61. $4y(-2y^{-5})$
$-8y^{1-5}$
$= -8y^{-4}$
$= \dfrac{-8}{y^4}$

63. $7 \cdot 5 \cdot a^{-7+9}$
$= 35a^2$

65. $y^{(-4)(-2)}$
$= y^8$

67. $\dfrac{1}{7^6}$

69. $x^{4-(-2)}$
$= x^6$

71. $a^{-5-4}$
$= a^{-9}$
$= \dfrac{1}{a^9}$

73. $\dfrac{5}{10} x^{-7-(-2)}$
$= \dfrac{1}{2} x^{-5}$
$= \dfrac{1}{2x^5}$

75. $\dfrac{10}{15} x^{-2-3}$
$= \dfrac{2}{3} x^{-5}$
$= \dfrac{2}{3x^5}$

77. $\left(\dfrac{x^2}{x^{-3}}\right)^2$
$= \dfrac{x^4}{x^{-6}}$
$= x^4 \cdot x^6$
$= x^{10}$

79. $\dfrac{1}{3a^{-2}}$
$= \dfrac{a^2}{3}$

81. $x^{3(-2)} y^{-2(-2)}$
$= x^{-6} y^4$
$= \dfrac{y^4}{x^6}$

83. $a^{-3(-2)} b^{-2} a^{4(-1)} b^{(-3)(-1)}$
$= a^6 b^{-2} a^{-4} b^3$
$= a^{6-4} b^{-2+3}$
$= a^2 b$

Exponents and Polynomials

85. $\dfrac{a^{-4(-3)}}{b^{2(-3)}}$

$= \dfrac{a^{12}}{b^{-6}}$

$= a^{12} b^{6}$

87. $\dfrac{z^{-2}}{x^{2} y^{-3}}$

$= \dfrac{y^{3}}{x^{2} z^{2}}$

89. $\dfrac{2 x^{-4}}{3^{-2} x^{5(-2)}}$

$= \dfrac{3^{2} \cdot 2 x^{-4}}{x^{-10}}$

$= 9 \cdot 2 x^{-4-(-10)}$

$= 9 \cdot 2 x^{6}$

$= 18 x^{6}$

91. $x^{-5-(-3)} y^{2-(-1)}$

$= x^{-2} y^{3}$

$= \dfrac{y^{3}}{x^{2}}$

93. $y^{-6} \cdot y^{4} = y^{-2}$

95. $x^{n-(-1)} y^{-2-m} = x^{4} y^{-1}$

$n + 1 = 4$
$n = 3$

$-2 - m = -1$
$-m = 1$
$m = -1$

97. (a) $\dfrac{260}{x}$

(b) $\dfrac{260}{15}$
17.3 ¢ per minute

99. $H_{1}(10) = 44$: the result is close to 43, the actual number of hours.

101. $-x^{3}\left(105 x^{-3} - 33.1 x^{-2} + 2.5 x^{-1} - 0.068\right)$

$= -105 + 33.1 x - 2.5 x^{2} + 0.068 x^{3}$

$= H_{2}(x)$

103. $\dfrac{32}{12} x^{3} y^{-3-2}$

$= \dfrac{8}{3} x^{3} y^{-5}$

$= \dfrac{8 x^{3}}{3 y^{5}}$

105. $\dfrac{24 x^{3} y^{-2}}{2^{3} x^{-3} y^{-9}}$

$= \dfrac{24 x^{3} y^{-2}}{8 x^{-3} y^{-9}}$

$= 3 x^{3-(-3)} y^{-2-(-9)}$

$= 3 x^{6} y^{7}$

Copyright © Houghton Mifflin Company. All rights reserved

107. $\dfrac{\frac{1}{4}}{3+\frac{1}{2}}$

$= \dfrac{\frac{1}{4}}{\frac{6}{2}+\frac{1}{2}}$

$= \dfrac{\frac{1}{4}}{\frac{7}{2}}$

$= \dfrac{1}{4} \cdot \dfrac{2}{7}$

$= \dfrac{2}{28}$

$= \dfrac{1}{14}$

## Section 6.8 Scientific Notation

1. Each time we multiply by 10, the decimal point is moved one place to the right.

3. 2900

5. 0.04

7. 639,000

9. 0.0067

11. 70,000

13. 0.00000756

15. 4,285,000,000

17. 0.00000000025534

19. $5.6 \cdot 1000$
    $= 5.6 \cdot 10^3$

21. $3.07 \cdot 100,000$
    $= 3.07 \cdot 10^5$

23. $5.2 \cdot 10,000,000$
    $= 5.2 \cdot 10^7$

25. $2 \cdot 1,000,000,000$
    $= 2 \cdot 10^9$

27. $3 \cdot \dfrac{1}{100}$
    $= 3 \cdot 10^{-2}$

Exponents and Polynomials

29. $4.74 \cdot \dfrac{1}{100,000}$
$= 4.74 \cdot 10^{-5}$

31. $8 \cdot \dfrac{1}{100,000,000}$
$= 8 \cdot 10^{-8}$

33. $3.4569 \cdot \dfrac{1}{10}$
$= 3.4569 \cdot 10^{-1}$

35. $9.0 \cdot 10^{19}$

37. $9.3 \cdot 10^{-10}$

39. $4.0 \cdot 10^{-4}$

41. $3.1 \cdot 10^{10}$

43. $4.7 \cdot 10^2 \cdot 10^3$
$= 4.7 \cdot 10^3$

45. $2.9 \cdot 10^{-2} \cdot 10^{-2}$
$= 2.9 \cdot 10^{-4}$

47. $7 \cdot 10^{-2} \cdot 10^5$
$= 7 \cdot 10^3$

49. $5 \cdot 100 + 2 \cdot 10 + 4$
$= 500 + 20 + 4$
$= 524$
$= 8000 + 30 + 2$
$= 8032$

51. $4 + 7 \cdot \dfrac{1}{10} + 3 \cdot \dfrac{1}{100}$
$= 4 + 0.7 + 0.03$
$= 4.73$

53. $3 \cdot 2 \cdot 10^5 \cdot 10^{-2}$
$= 6 \cdot 10^3$

55. $4 \cdot 6 \cdot 10^4 \cdot 10^8$
$= 24 \cdot 10^{12}$
$= 2.4 \cdot 10^{13}$

57. $3 \cdot 10^{9-5}$
$= 3 \cdot 10^4$

59. $\dfrac{1}{2} \cdot 10^{-3-8}$
$= 0.5 \cdot 10^{-11}$
$= 5 \cdot 10^{-12}$

61. $3^3 \cdot 10^{5 \cdot 3}$
$= 27 \cdot 10^{15}$
$= 2.7 \cdot 10^{16}$

63. $5^3 \cdot 10^{-4 \cdot 3}$
$= 125 \cdot 10^{-12}$
$= 1.25 \cdot 10^{-10}$

64. $2^4 \cdot 10^{6 \cdot 4}$
$= 16 \cdot 10^{24}$
$= 1.6 \cdot 10^{25}$

66. $43.2915 \cdot 10^{-9+3}$
$= 43.2915 \cdot 10^{-6}$
$= 4.3 \cdot 10^{-5}$

68. $2.48 \cdot 10^{15-(-4)}$
$= 2.5 \cdot 10^{19}$

70. $(5.34)^{-2} \cdot 10^{5 \cdot (-2)}$
$= 0.035 \cdot 10^{-10}$
$= 3.5 \cdot 10^{-12}$

71. $1.98 \cdot 10^{11}$

73. $1 \cdot 10^{-12}$

75. $7 \cdot 10^{-4}$

77. $3.68 \cdot 10^9$

79. $1 \cdot 10^{20}$

81. $9.3 \cdot 10^{-5}$

Copyright © Houghton Mifflin Company. All rights reserved

83. $\dfrac{28}{1} = \dfrac{x}{500,000}$
$x = 14,000,000$
$= 1.4 \cdot 10^7$

85. $\dfrac{29 \cdot 11}{280 \cdot 10^6}$
$= 1.14 \cdot 10^{-6}$

87. $\dfrac{1}{4000}$
$= 2.5 \cdot 10^{-4}$ inches

89. $186,000 \cdot 60 \cdot 60$
$= 669,600,000$
$= 6.69 \cdot 10^8$ miles

91. $1.491 \cdot 10^9 \cdot 2.25$
$= \$3.35 \cdot 10^9$

93. $289 \cdot 18.8 \cdot 10^6 \cdot 8$
$= 43465.6 \cdot 10^6$
$= 4.35 \cdot 10^4 \cdot 10^6$
$= 4.35 \cdot 10^{10}$ ounces

95. $1985 : \$5.25 \cdot 10^5$
$2000 : \$2.2 \cdot 10^6$

97. 10 minutes = 20   30-second ads
4 hours $\cdot 20 \dfrac{\text{ads}}{\text{hour}} = 80$ ads
$80 \cdot 2.2 \cdot 10^6$
$= 176 \cdot 10^6$
$= \$1.76 \cdot 10^8$

99. $0.5 \dfrac{\text{inches}}{\cancel{\text{min}}} \times \dfrac{60 \cancel{\text{min}}}{1 \cancel{\text{hour}}} \times \dfrac{24 \cancel{\text{hour}}}{1 \cancel{\text{day}}} \times \dfrac{365 \cancel{\text{day}}}{1 \text{ year}}$
$= 262,800 \dfrac{\text{inches}}{\text{years}}$
$262,800 \dfrac{\cancel{\text{inches}}}{\text{year}} \cdot \dfrac{1 \cancel{\text{foot}}}{12 \cancel{\text{inches}}} \cdot \dfrac{1 \text{ mile}}{5280 \cancel{\text{feet}}}$
$= 4.1477 \dfrac{\text{miles}}{\text{year}}$
$\dfrac{25,000}{4.1477} = 6027$ years

## Chapter 6 Review Exercises

1. (a) Base: $-4$
$(-4)^2$
$= (-4)(-4)$
$= 16$

   (b) Base: 4
$-4^2$
$= -(4)(4)$
$= -16$

3. $y^{3 \cdot 4}$
$= y^{12}$

5. $\dfrac{9}{6} x^{6 \cdot 2}$
$= \dfrac{3}{2} x^4$

7. $y^2 \cdot y^{3 \cdot 2}$
$= y^2 \cdot y^6$
$= y^{2+6}$
$= y^8$

Exponents and Polynomials

9. $\dfrac{2^3 x^{2\cdot 3}}{4\cdot 3 x^{1+4}}$

$= \dfrac{8x^6}{12x^5}$

$= \dfrac{2}{3} x^{6-5}$

$= \dfrac{2}{3} x$

11. (i) and (iii) are polynomials. (ii) is not because of the $x^2$ term in the denominator.

13. In (a), the degree is 3, the value of the exponent. In (b), the degree is 5, the sum of the exponents. In (c), the degree is 0, because 9 can be written as $9x^0$, and the exponent on the variable is 0.

15. $2x^5 + 4x^3 - x^2 + 6$

17. $P(-2) = 6(-2) + 5 + 3(-2)^2$
$= -12 + 5 + 3 \cdot 4$
$= -12 + 5 + 12$
$= 5$

19. $(1)^2(-1) - 3(1)(-1)(2) - 2(-1)^2(2)$
$-1 + 6 - 4$
$= 1$

21. $3t^4 + t^3 + 2t^2 - 3t - 1$

23. $9x^2 y - 9xy + 15xy^2$

25. $\phantom{-}3x + 2$
$\underline{-(x^2 - 2x - 4)}$
$-x^2 + 5x + 6$

27.
$5x^2 - x^2 - 2x^2 - 6x - 2x + 3x + 7 - (-5) + 4$
$= 2x^2 - 5x + 16$

29. $13x + 18 - (x^2 - 2x + 8 + 6x + 2)$
$= 13x + 18 - (x^2 + 4x + 10)$
$= -x^2 + 9x + 8$

31. $-3(2) x^{2+1} y^{3+2}$
$= -6x^3 y^5$

33. $\phantom{00}8x - 3$
$\phantom{00}5x + 2$
$\overline{\phantom{0}16x - 6}$
$\underline{40x^2 - 15x}$
$40x^2 + x - 6$

35. $3x(x^2 + 2x + x + 2)$
$= 3x^3 + 6x^2 + 3x^2 + 6x$
$= 3x^3 + 9x^2 + 6x$

37. $3x^2 + x - 2 - (x^2 + 4x - 3x - 12)$
$= 3x^2 + x - 2 - (x^2 + x - 12)$
$= 3x^2 + x - 2 - x^2 - x + 12$
$= 2x^2 + 10$

39. $3x - x^2 - x = 2 - 2x + x - x^2$
$-x^2 + 2x = -x^2 - x + 2$
$2x = -x + 2$
$3x = 2$
$x = \dfrac{2}{3}$

41. $x^2 - x + 3x - 3$
$= x^2 + 2x - 3$

43. $4^2 - a^2$
$= 16 - a^2$

45. $(2a)^2 + 2(6ab) + (3b)^2$
$= 4a^2 + 12ab + 9b^2$

47. $1^2 - (3x)^2 - \left[1 - 2(2x) + (2x)^2\right]$

$= 1 - 9x^2 - \left(1 - 4x + 4x^2\right)$

$= 1 - 9x^2 - 1 + 4x - 4x^2$

$= -13x^2 + 4x$

49. $x^2 + 2x + 1 - \left(x^2 - 1\right) = x$

$x^2 + 2x + 1 - x^2 + 1 = x$

$2x + 2 = x$

$2 = -x$

$x = -2$

51. We stop dividing when the remainder is 0, or when the degree of the remainder is less than the degree of the divisor.

53. $\dfrac{21a^5 - 14a^4 + 28a^2}{7a^2}$

$= \dfrac{21a^5}{7a^2} - \dfrac{14a^4}{7a^2} + \dfrac{28a^2}{7a^2}$

$= 3a^{5-2} - 2a^{4-2} + 4a^{2-2}$

$= 3a^3 - 2a^2 + 4$

55.
$$\begin{array}{r} x^2 + 2x \phantom{aaaa} \\ x-2\overline{)\,x^3 \phantom{aaaa} - 4x\phantom{a}} \\ \underline{x^3 - 2x^2} \phantom{aaaaa} \\ 2x^2 - 4x \\ \underline{2x^2 - 4x} \\ 0 \end{array}$$

57.
$$\begin{array}{r} x^2 + 2x \phantom{aaaaa} \\ x-2\overline{)\,x^3 \phantom{aaaa} - 4x + 2} \\ \underline{x^3 - 2x^2} \phantom{aaaaaa} \\ 2x^2 - 4x \phantom{aa} \\ \underline{2x^2 - 4x} \phantom{aa} \\ 0 + 2 \end{array}$$

$Q = x^2 + 2x, R = 2$

59.
$$\begin{array}{r} x^2 - x + 3 \phantom{aaaa} \\ x+3\overline{)\,x^3 + 2x^2 \phantom{aaaa} + 10} \\ \underline{x^3 + 3x^2} \phantom{aaaaaaa} \\ -x^2 \phantom{aaaaaaa} \\ \underline{-x^2 - 3x} \phantom{aaaa} \\ 3x + 10 \\ \underline{3x + 9} \\ 1 \end{array}$$

$= x^2 - x + 3 + \dfrac{1}{x+3}$

61. $(-3)^{-2}$

$= \dfrac{1}{(-3)^2}$

$= \dfrac{1}{9}$

63. $\dfrac{-5}{x^2}$

65. $\dfrac{1}{x^2} \cdot y^3$

$= \dfrac{y^3}{x^2}$

67. $(-2)^{-3} x^{-2(-3)}$

$= \dfrac{1}{(-2)^3} x^6$

$= -\dfrac{1}{8} x^6$

69. $x^{-2-4} y^{1-5}$

$= x^{-6} y^{-4}$

$= \dfrac{1}{x^6 y^4}$

71. If $n < 0$, then $-n$ is a positive number. Therefore, $b^{-n}$ already has a positive exponent.

Exponents and Polynomials

73. $0.00028$

75. $6.7 \cdot 10^{-5}$

77. $1.9 \cdot 10^{-7}$

79. $4.8 \cdot 10^{21}$

81. $16,000 \cdot 250,000,000$
$= 1.6 \cdot 10^4 \cdot 2.5 \cdot 10^8$
$= 4 \cdot 10^{12}$ dollars

## Chapter 6 Test

1. $-8x^{3+2}y^{1+5}$
$= -8x^5 y^6$

3. $\left(\dfrac{20x^3 y^2}{4x^2 y}\right)^2$
$= \left(5x^{3-2} y^{2-1}\right)^2$
$= (5xy)^2$
$= 5^2 x^2 y^2$
$= 25x^2 y^2$

5. $p(-2) = 6 + 2(-2) - (-2)^2$
$= 6 - 4 - 4$
$= -2$

7. $3x^3 + x^3 + x^2 - 2x + 3x - 4 - 1$
$= 4x^3 + x^2 + x - 5$

9. $x^2 + 3x - 8 - 2 - x^2 = x$
$3x - 10 = x$
$2x - 10 = x$
$\phantom{2x-1}2x = 10$
$\phantom{2x-10}x = 5$

11. $-6x^{2+3} y^{3+6}$
$= -6x^5 y^9$

13. $(2x+1)(x^2 + 4x - 1x - 4)$
$= (2x+1)(x^2 + 3x - 4)$
$= 2x^3 + 6x^2 - 8x + x^2 + 3x - 4$
$= 2x^3 + 7x^2 - 5x - 4$

15. $(2x-5)^2$
$= (2x)^2 - 2(2x)(5) + (-5)^2$
$= 4x^2 - 20x + 25$

17. $\dfrac{8x^7}{4x^2} - \dfrac{12x^5}{4x^2} + \dfrac{20x^4}{4x^2}$
$= 2x^{7-2} - 3x^{5-2} + 5x^{4-2}$
$Q = 2x^5 - 3x^3 + 5x^2,\ R = 0$

19.
$$\begin{array}{r}
x^3 + x^2 + 2x + 2 \phantom{)} \\
x-1\overline{)x^4 \phantom{+x^3} + x^2 \phantom{+2x} + 1} \\
\underline{x^4 - x^3} \phantom{+x^2+2x+1} \\
x^3 + x^2 \phantom{+2x+1} \\
\underline{x^3 - x^2} \phantom{+2x+1} \\
2x^2 \phantom{+2x+1} \\
\underline{2x^2 - 2x} \phantom{+1} \\
2x + 1 \\
\underline{2x - 2} \\
3
\end{array}$$
$Q = x^3 + x^2 + 2x + 2,\ R = 3$

217

21. $a^{-3(-4)}b^{2(-4)}$
    $= a^{12}b^{-8}$
    $= \dfrac{a^{12}}{b^8}$

23. The number $n$ is a real number such that $1 \le n < 10$, and $p$ is an integer.

25. $\dfrac{13}{87,261}$
    $= 0.000148978$
    $= 1.5 \cdot 10^{-4}$

# Chapter 7

# Factoring

## Section 7.1 Common Factors and Grouping

1. To factor a number completely, write the number as a product of prime numbers.

3. $2 \cdot 3 = 6$
$5 \cdot 3 = 15$
$GCF = 3$

5. $1 \cdot 11 = 11$
$1 \cdot 3 \cdot 3 \cdot 2 = 18$
$GCF = 1$

7. $3 \cdot 5 = 15$
$3 \cdot 4 = 12$
$3 \cdot 7 = 21$
$GCF = 3$

9. $3 \cdot 3 \cdot y = 9y$
$4 \cdot 3 \cdot y \cdot y = 12y^2$
$GCF = 3y$

11. $x \cdot x \cdot y \cdot y \cdot y = x^2 y^3$
$x \cdot y \cdot y \cdot y \cdot y \cdot y = xy^5$
$GCF = xy^3$

13. $2 \cdot 3 \cdot x \cdot x \cdot x = 6x^3$
$3 \cdot x \cdot x = 3x^2$
$4 \cdot 3 \cdot x = 12x$
$GCF = 3x$

15. (a) All are correct factorizations of the polynomial.

(b) Part (ii) and (iii) are complete factorizations.

17. 8

19. $4 \cdot 3x + 4$
$= 4(3x + 1)$

21. $5x^2 + 5 \cdot 2x + 5 \cdot 3$
$= 5(x^2 + 2x + 3)$

23. $7 \cdot x + 7 \cdot 3$
$= 7(x + 3)$

25. $8 \cdot 4x + 8 \cdot 3y - 8 \cdot 7$
$= 8(4x + 3y - 7)$

27. $6 \cdot 3 \cdot a \cdot a - 6 \cdot 4 \cdot a$
$= 6a(3a - 4)$

29. $2 \cdot 2 \cdot 2x^2 - 5$
Prime

31. $3 \cdot x \cdot x \cdot x - 3 \cdot 3 \cdot x \cdot x + 3 \cdot x$
$= 3x(x^2 - 3x + 1)$

33. $5 \cdot 2x^2 \cdot x \cdot y - 5x^2 \cdot y \cdot y^2$
$= 5x^2 y(2x - y^2)$

219

35. $8 \cdot 3 \cdot a^3 \cdot a \cdot b^4 + 8 \cdot 2 \cdot a^3 \cdot b^4$
    $= 8a^3 b^4 (3a + 2)$

37. $x \cdot x \cdot y + x \cdot y \cdot y - 5 \cdot x \cdot y$
    $= xy(x + y - 5)$

39. $8 \cdot 3y + 8$
    $= 8(3y + 1)$

41. $7 \cdot 3a + 7 \cdot 5b - 7$
    $= 7(3a + 5b - 1)$

43. $4 \cdot 3 \cdot x \cdot x + 4 \cdot x \cdot 1$
    $= 4x(3x + 1)$

45. $a^2 \cdot a \cdot b^3 \cdot b - a^2 \cdot b^3 \cdot b + a^2 \cdot b^3 \cdot 1$
    $= a^2 b^3 (ab - b + 1)$

47. $-6 \cdot 4y - 6 \cdot 3z$
    $= -6(4y + 3z)$

49. $-3y \cdot 2x - 3y(-1)$
    $= -3y(2x - 1)$

51. $3 \cdot x - 3 \cdot 5$
    $= 3(x - 5)$
    $= -3(-x + 5)$

53. $4 \cdot 3 - 4x$
    $= 4(3 - x)$
    $= -4(-3 + x)$

55. $-2 \cdot x^2 + (-2)(-x) + (-2)(-5)$
    $= -2(x^2 - x - 5)$
    $= 2(-x^2 + x + 5)$

57. The expression is the sum of the terms $x(a+b)$ and $y(a+b)$, not a product.

59. $(x - 8)(x + 5)$

61. $(y + 3)(3y - 2)$

63. $(x - 4)(x + 1)$

65. $(x - 3)(x^2 + 2)$

67. $a(x + 5) + 2(x + 5)$
    $= (x + 5)(a + 2)$

69. $2x(y - 1) + 3(y - 1)$
    $= (y - 1)(2x + 3)$

71. $y(x + 3) - 1(x + 3)$
    $= (x + 3)(y - 1)$

73. $3x - 3ax + y + ay$
    $3x(1 - a) + y(1 + a)$
    Prime

75. $t^2(t + 1) + 3(t + 1)$
    $= (t + 1)(t^2 + 3)$

77. $a(2y - 1) - 1(2y - 1)$
    $= (2y - 1)(a - 1)$

79. $z(x + y) + 1(x + y)$
    $= (x + y)(z + 1)$

81. $a(x + 1) - 1(x + 1)$
    $= (x + 1)(a - 1)$

83. $xz - 3x - 6y + 2yz$
    $= x(z - 3) + 2y(-3 + z)$
    $= x(z - 3) + 2y(z - 3)$
    $= (z - 3)(x + 2y)$

85. $3a(a + b) - 6(a + b)$
    $= (a + b)(3a - 6)$
    $= 3(a + b)(a - 2)$

Factoring

87. $3t^3(t-2)+15t(t-2)$
$=(t-2)(3t^3+15t)$
$=(t-2)3t(t^2+5)$
$=3t(t-2)(t^2+5)$

89. NF

91. NF

93. CF

95. F

97. $x^2+15x$
$=x\cdot x+15\cdot x$
$=x(x+15)$

99. $3x^2(x+2)$
$=3x\cdot x\cdot(x+2)$
The height is $x+2$.

101. $C(x)=2(x+12)$

103. The common factor is the annual increase in the percentage of people who use computers in their jobs.

105. $x(x-y+z)+3(x-y+z)$
$=(x-y+z)(x+3)$

107. $2\cdot\dfrac{1}{3}x-\dfrac{1}{3}\cdot 5$
$=\dfrac{1}{3}(2x-5)$

109. $\dfrac{1}{4}\cdot 3x-\dfrac{1}{4}(5)$
$=\dfrac{1}{4}(3x-5)$

## Section 7.2 Special Factoring

1. In (i), 10 is not a perfect square, and in (iii), the expression is a sum, not a difference.

3. $(x-5)(x+5)$

5. $(10-a)(10+a)$

7. $(8-7w)(8+7w)$

9. $(a-2b)(a+2b)$

11. $1-x^2$
$=(1+x)(1-x)$

13. $\left(x-\dfrac{1}{2}\right)\left(x+\dfrac{1}{2}\right)$

15. $(x^2-10)(x^2+10)$

17. $2(x^2-4)$
$=2(x+2)(x-2)$

19. $4(x^2-4)$
$=4(x-2)(x+2)$

21. $y^5(100-y^2)$
$=y^5(10-y)(10+y)$

23. $x(x^2-y^2)$
$=x(x-y)(x+y)$

25. $(x^2-9)(x^2+9)$
$=(x+3)(x-3)(x^2+9)$

27. $(x^4+1)(x^4-1)$
$=(x^4+1)(x^2+1)(x^2-1)$
$=(x^4+1)(x^2+1)(x+1)(x-1)$

29. The constant term is $-9$. The constant term of a perfect square trinomial must be positive.

31. $(x+3)^2$

33. $1-2(2y)+(2y)^2$
$=(1-2y)^2$

35. $x^2+2(4xy)+(4y)^2$
$=(x+4y)^2$

37. $(7x)^2-2(7x\cdot 3y)+(3y)^2$
$=(7x-3y)^2$

39. $(y^2+3)^2$

41. $(x^3)^2-2(2x^3)+2^2$
$=(x^3-2)^2$

43. $5(x^2-6x+9)$
$=5(x-3)^2$

45. $-x(49+28x+4x^2)$
$=-x(7+2x)^2$

47. Yes. First factor out of common factor $x$. The result $x^3+1$ is the sum of two cubes.

49. $(x+7)(x^2-7x+49)$

51. $(z+1)(x^2-z+1)$

53. $5(1+y^3)$
$=5(1+y)(1-y+y^2)$

55. $(11-t)(11^2+11t+t^2)$
$=(11-t)(121+11t+t^2)$

57. $4^3-y^3$
$=(4-y)(4^2+4y+y^2)$
$=(4-y)(16+4y+y^2)$

59. $x^3-(2y)^3$
$=(x-2y)(x^2+2xy+(2y)^2)$
$=(x-2y)(x^2+2xy+4y^2)$

61. $5^3+x^3$
$=(5+x)(25-5x+x^2)$

63. $(4y)^3+z^3$
$=(4y+z)(16y^2-4yz+z^2)$

65. $(2a)^3-b^3$
$=(2a-b)(4a^2+2ab+b^2)$

### Factoring

67. $9(x^3 - 8)$
    $= 9(x^3 - 2^3)$
    $= 9(x-2)(x^2 + 2x + 4)$

69. $27x^6(1 + x^3)$
    $= 27x^6(1+x)(1 - x + x^2)$

71. $x^2 - 18x + 81$
    $= (x-9)^2$

73. $\left(\dfrac{3}{5}t - 1\right)\left(\dfrac{3}{5}t + 1\right)$

75. $2mn(m^3 + 125n^3)$
    $= 2mn(m^3 + (5n)^3)$
    $= 2mn(m + 5n)(m^2 + 5mn + 25n^2)$

77. $4a^2(a^2 + 2a + 1)$
    $= 4a^2(a+1)^2$

79. Prime

81. $-2(1 - 36y^2)$
    $= -2(1 - 6y)(1 + 6y)$
    $= -2(6y - 1)(6y + 1)$

83. $(10x)^2 - (3y \cdot 10x) + (3y)^2$
    $= (10x - 3y)^2$

85. $3y(y^2 - 16x^2)$
    $= 3y(y - 4x)(y + 4x)$

87. $(2a)^3 - 3^3$
    $= (2a - 3)(4a^2 + 6a + 9)$

89. $(7x)^2 - (9y^3)^2$
    $= (7x - 9y^3)(7x + +9y^3)$

91. $(x^2)^3 - (3y)^3$
    $= (x^2 - 3y)(x^4 + 3x^2y + 9y^2)$

93. $x^2 - 4(3)^2$
    $= x^2 - 36$
    $= (x-6)(x+6)$

95. $5^3 - (3x)^3$
    $= 125 - 27x^3$
    $= (5 - 3x)(25 + 15x + 9x^2)$

97. The factor $25x + 585$ represents weekly salary.

99. $B(x) = 1872x + 28,548$
    Weekly: $B(x) = 52(36x + 549)$
    Monthly: $B(x) = 12(156x + 2379)$

101. $x^4 - 8x^2 + 16$
    $(x^2 - 4)^2$
    $= [(x+2)(x-2)]^2$
    $= (x+2)^2(x-2)^2$

103. $(x + 3 - y)(x + 3 + y)$

105. $((x - y) + 3)^2$
    $= (x - y + 3)^2$

## Section 7.3 Factoring Trinomials of the Form $x^2 + bx + c$

1. Both $m$ and $n$ are negative.

3. $(x+3)(x+1)$

5. $(y+3)(y-1)$ since $3(-1) = -3$ and $3 + (-1) = 2$.

7. $(a+3)(a+2)$ since $2 \cdot 3 = 6$ and $2 + 3 = 5$.

9. $(x+2)(x+3)$ since $2 \cdot 3 = 6$ and $2 + 3 = 5$.

11. $(x-6)(x-2)$ since $-6 \cdot (-2) = 12$ and $-6 - 2 = -8$.

13. $(y-5)(y-4)$ since $-5 \cdot (1) = 20$ and $-5 - 4 = -9$.

15. $(x+3)(x+4)$ since $3 \cdot 4 = 12$ and $3 + 4 = 7$.

17. The numbers $m$ and $n$ have opposite signs, and the one with the larger absolute value is negative.

19. $(x+2)(x-1)$

21. $(y-5)(y+2)$

23. $(y+7)(y-2)$

25. $(x-9)(x+1)$

27. $x^2 + 9x + 8$
    $(x+1)(x+8)$

29. $z^2 - 4z - 45$
    $= (z-9)(z+5)$

31. $x^2 + 9x + 14$
    $= (x+7)(x+2)$

33. The leading coefficient is $-1$. A possible first step is to factor out the $-1$. Writing it in ascending order makes it unnecessary to factor out a common factor.

## Factoring

35. $(x+2y)(x+3y)$

37. $(x-2y)(x+7y)$

39. $(2-xy)(1-xy)$

41. $(x^2+2)(x^2+7)$

43. $(x^3-5)(x^3-2)$

45. $-1(x^2+4x-5)$
$=-1(x+5)(x-1)$

47. $-1(t^2-4t-32)$
$=-1(t-8)(t+4)$

49. $-1(x^2-12x+35)$
$=-1(x-7)(x-5)$

51. $5(x^2+7x+12)$
$=5(x+3)(x+4)$

53. $4y^7(y^2+2y+1)$
$=4y^7(y+1)(y+1)$
$=4y^7(y+1)^2$

55. $x^2(x^2+3x+1)$

57. $-3x^2(x^2-5x-6)$
$=-3x^2(x-6)(x+1)$

59. $(x-10)(x-8)$

61. $x^2+11+28$
$=(x+7)(x+4)$

63. $(w+10)(w-4)$

65. $(c-7)(c+5)$

67. Not factorable

69. $-2(x^2-x-42)$
$=-2(x-7)(x+6)$

71. $x^2(x^2-8x-48)$
$=x^2(x-12)(x+4)$

73. $(x-y)(x-12y)$

75. $x(x^2-5x+4)$
$=x(x-1)(x-4)$

77. $-3(x^2-4x+12)$

79. $x^2-5x+4$
$=(x-1)(x-4)$

81. $-w^2-15w+34$
$=-1(w^2+15w-34)$
$=-1(w+17)(w-2)$

83. $(x+4y)(x-y)$

85. $x^5(x^2+2x-80)$
$=x^5(x+10)(x-8)$

87. $xy(x^2+7x-10)$

89. $(a+12)(a-3)$

91. $(a-4x)(a+3x)$

93. Not factorable

95. Profit $=2x^2+5x+8-(x^2-4x)$
$=2x^2+5x+8-x^2+4x$
$=x^2+9x+8$
$=(x+1)(x+8)$

97. $h(t) = -16(t^2 - 3t - 4)$
$= -16(t-4)(t+1)$

99. (a) $-0.6t + 77.4$
$= 0.6(-t + 129)$

(b) $-0.6(15) + 77.4 = 68.4$

$0.6(-15 + 129)$
$= 0.6(114)$
$= 68.4$

101. $(x-7)(x+3) - w(x+3)$
$= (x+3)(x-7-w)$

103. $\frac{1}{3}(x^2 - 2x - 3)$
$= \frac{1}{3}(x-3)(x+1)$

105. $x^2 + kx + 6$
$1 \cdot 6 = 6$ and $1+6=7$
$2 \cdot 3 = 6$ and $2+3=5$
$-1 \cdot (-6) = 6$ and $-1-6=-7$
$-2 \cdot (-3) = 6$ and $-2-3=-5$

$k$ can be $-7, -5, 5,$ or $7$.

## Section 7.4 Factoring Trinomials of the Form $ax^2 + bx + c$

1. Because 11 and 37 are prime, there are a few possible factors, and the trial and check method is a good choice for (ii). In (i), because 12 and 36 have several factors, there are many possible factors to try.

3. $(3x+2)(x+3)$

5. $(2x+3)(x-4)$

7. $(x-3)(6x-5)$

9. $(2x+1)(x+1)$ since $1 \cdot 2 = 1$ and $1+2=3$.

11.

| Factors of $-72$ | Sums |
|---|---|
| $1, -72$ and $-1, 72$ | $-71, 71$ |
| $2, -36$ and $-2, 36$ | $-34, 34$ |
| $3, -24$ and $-3, 24$ | $-21, 21$ |
| $4, -18$ and $-4, 18$ | $-14, 14$ |
| $6, -12$ and $-6, 12$ | $-6, 6$ |
| $8, -9$ and $-8, 9$ | $-1, 1$ |

$12x^2 + x - 6$
$= 12x^2 - 8x + 9x - 6$
$= 4x(3x-2) + 3(3x-2)$
$= (4x+3)(3x-2)$

# Factoring

**13.**

| Factors of $-60$ | Sums |
|---|---|
| $1, -60$ and $-1, 60$ | $-59, 59$ |
| $2, -30$ and $-2, 30$ | $-28, 28$ |
| $3, -20$ and $-3, 20$ | $-17, 17$ |
| $4, -15$ and $-4, 15$ | $-11, 11$ |
| $5, -12$ and $-5, 12$ | $-7, 7$ |
| $6, -10$ and $-6, 10$ | $-4, 4$ |

$12x^2 - 7x - 5$
$= 12x^2 + 5x - 12x - 5$
$= x(12x + 5) - (12x + 5)$
$= (12x + 5)(x - 1)$

**15.**

| Factors of 200 | Sums |
|---|---|
| $5, 40$ and $-5, -40$ | $45, -45$ |

$8y^2 - 45y + 25$
$= 8y^2 - 5y - 40y + 25$
$= y(8y - 5) - 5(8y - 5)$
$= (8y - 5)(y - 5)$

**17.**

| Factors of 60 | Sums |
|---|---|
| $1, 60$ and $-1, -60$ | $61, -61$ |
| $2, 30$ and $-2, -30$ | $32, -32$ |
| $3, 20$ and $-3, -20$ | $23, -23$ |
| $4, 15$ and $-4, -15$ | $19, -19$ |
| $5, 12$ and $-5, -12$ | $17, -17$ |
| $6, 10$ and $-6, -10$ | $16, -16$ |

$4x^2 + 16x + 15$
$= 4x^2 + 6x + 10x + 15$
$= 2x(2x + 3) + 5(2x + 3)$
$= (2x + 3)(2x + 5)$

**19.**

| Factors of $-110$ | Sums |
|---|---|
| $1, -110$ and $-1, 110$ | $-109, 109$ |
| $2, -55$ and $-2, 55$ | $-53, 53$ |
| $5, -22$ and $-5, 22$ | $-17, 17$ |
| $10, -11$ and $-10, 11$ | $-1, 1$ |

$11x^2 - x - 10$
$= 11x^2 + 10x - 11x - 10$
$= x(11x + 10) - (11x + 10)$
$= (11x + 10)(x - 1)$

**21.**

| Factor of 18 | Sums |
|---|---|
| $1, 18$ and $-1, -18$ | $19, -19$ |
| $2, 9$ and $-2, -9$ | $11, -11$ |
| $3, 6$ and $-3, -6$ | $9, -9$ |

$6x^2 + 11xy + 3y^2$
$= 6x^2 + 2xy + 9xy + 3y^2$
$= 2x(3x + y) + 3y(3x + y)$
$= (3x + y)(2x + 3y)$

**23.**

| Factor of $-126$ | Sum |
|---|---|
| $-1, 126$ and $1, -126$ | $125, -125$ |
| $-2, 63$ and $2, -63$ | $61, -61$ |
| $-3, 42$ and $3, -42$ | $39, -39$ |
| $-6, 21$ and $6, -21$ | $15, -15$ |

$14a^2 + 15ab - 9b^2$
$= 14a^2 - 6ab + 21ab - 9b^2$
$= 2a(7a - 3b) + 3b(7a - 3b)$
$= (7a - 3b)(2a + 3b)$

**25.**

| Factors of $-70$ | Sums |
|---|---|
| $1, -70$ and $-1, 70$ | $-69, 69$ |
| $2, -35$ and $-2, 35$ | $-33, 33$ |

$7x^2 - 33xz - 10z^2$
$= 7x^2 + 2xz - 35xz - 10z^2$
$= x(7x + 2z) - 5z(7x + 2z)$
$= (7x + 2z)(x - 5z)$

27. First factor out the common factor 2. Then use either the trial–and–check method or the grouping method to factor the trinomial.

29. $2x^2 + 7x + 5$
 $= (2x+5)(x+1)$

31. $16x^2 - 24x + 9$

| Factors of 144 | Sums |
|---|---|
| 12, 12 and −12, −12 | 24, −24 |

$16x^2 - 24x + 9$
$= 16x^2 - 12x - 12x + 9$
$= 4x(4x-3) - 3(4x-3)$
$= (4x-3)^2$

33. $5y^2 - 11y + 2$

| Factors of 10 | Sums |
|---|---|
| 1, 10 and −1, −10 | 11, −11 |

$5y^2 - 11y + 2$
$= 5y^2 - y - 10y + 2$
$= y(5y-1) - 2(5y-1)$
$= (5y-1)(y-2)$

35. (a) $14 - 3x - 2x^2$

| Factors of −28 | Sums |
|---|---|
| 4, −7 and −4, 7 | −3, 3 |

$14 - 3x - 2x^2$
$= 14 + 4x - 7x - 2x^2$
$= 2(7+2x) - x(7+2x)$
$= (2-x)(7+2x)$

(b) $-1(2x^2 + 3x - 14)$
$= -1(2x^2 - 4x + 7x - 14)$
$= -1[2x(x-2) + 7(x-2)]$
$= -1(2x+7)(x-2)$

37. (a) $42 + x - x^2$

| Factors of −42 | Sums |
|---|---|
| −6, 7 and 6, −7 | 1, −1 |

$42 + x - x^2$
$= 42 - 6x + 7x - x^2$
$= 6(7-x) + x(7-x)$
$= (7-x)(6+x)$

(b) $-1(x^2 - x - 42)$
$= -1(x^2 + 6x - 7x - 42)$
$= -1[x(x+6) - 7(x+6)]$
$= -1(x+6)(x-7)$

39. (a) $36 + 5x - x^2$

| Factors of −36 | Sums |
|---|---|
| 9, −4 and −9, 4 | 5, −5 |

$36 + 5x - x^2$
$= 36 + 9x - 4x - x^2$
$= 9(4+x) - x(4+x)$
$= (4+x)(9-x)$

(b) $-1(x^2 - 5x - 36)$
$= -1(x^2 - 9x + 4x - 36)$
$= -1[x(x-9) + 4(x-9)]$
$= -1(x-9)(x+4)$

# Factoring

41. $5(2x^2 - 7x + 6)$

| Factors of 12 | Sums |
|---|---|
| 3, 4 and −3, −4 | 7, −7 |

$5(2x^2 - 3x - 4x + 6)$
$= 5[x(2x-3) - 2(2x-3)]$
$= 5(2x-3)(x-2)$

43. $-7(12x^2 + x - 6)$

| Factors of −72 | Sums |
|---|---|
| 8, −9 and −8, 9 | −1, 1 |

$-7(12x^2 - 8x + 9x - 6)$
$= -7[4x(3x-2) + 3(3x-2)]$
$= -7(3x-2)(4x+3)$

45. $10(t^2 - t - 2)$
$= 10(t-2)(t+1)$

47. No. For the polynomial $x^2 + 4x + 3$, $ac = 3$ is prime, and the polynomial can be factored as $(x+3)(x+1)$.

49. $(3x+2)(x+1)$

51.

| Factors of 24 | Sums |
|---|---|
| 3, 8 and −3, −8 | 11, −11 |

$4x^2 - 11x + 6$
$= 4x^2 - 3x - 8x + 6$
$= x(4x-3) - 2(4x-3)$
$= (x-2)(4x-3)$

53. $(2x+3)(x-1)$

55.

| Factors of 36 | Sums |
|---|---|
| 1, 36 and −1, −36 | 37, −37 |
| 2, 18 and −2, −18 | 20, −20 |
| 3, 12 and −3, −12 | 15, −15 |
| 4, 9 and −4, −9 | 13, −13 |
| 6, 6 and −6, −6 | 12, −12 |

$6x^2 - 5x + 6$ is prime.

57. $2x(3x^2 + 2x - 8)$

| Factors of −24 | Sums |
|---|---|
| 1, −24 and −1, 24 | −23, 23 |
| 4, −6 and −4, 6 | −2, 2 |

$2x(3x^2 - 4x + 6x - 8)$
$= 2x[x(3x-4) + 2(3x-4)]$
$= 2x(3x-4)(x+2)$

59.

| Factors of 55 | Sums |
|---|---|
| 5, 11 and −5, −11 | 16, −16 |

$5t^2 - 5t - 11t + 11$
$= 5t(t-1) - 11(t-1)$
$= (t-1)(5t-11)$

61.

| Factors of −120 | Sums |
|---|---|
| 8, −15 and −8, 15 | −7, 7 |

$10b^2 + 8b - 15b - 12$
$= 2b(5b+4) - 3(5b+4)$
$= (5b+4)(2b-3)$

63.

| Factors of 15 | Sums |
|---|---|
| 3, 5 and −3, − 5 | 8, − 8 |

$5x^2 + 3x + 5x + 3$
$= x(5x+3) + 1(5x+3)$
$= (5x+3)(x+1)$

65. $(3x-1)(x+2)$

67.

| Factors of 36 | Sums |
|---|---|
| 2, 18 and −2, − 18 | 20, − 20 |

$12x^2 - 20xy + 3y^2$
$= 12x^2 - 2xy - 18xy + 3y^2$
$= 2x(6x-y) - 3y(6x-y)$
$= (6x-y)(2x-3y)$

69.

| Factors of −60 | Sums |
|---|---|
| − 10, 6 and 10, − 6 | − 4, 4 |

$4a^2 + 4a - 15$
$= 4a^2 + 10a - 6a - 15$
$= 2a(2a+5) - 3(2a+5)$
$= (2a+5)(2a-3)$

71.

| Factors of − 70 | Sums |
|---|---|
| 7, − 10 and − 7, 10 | − 3, 3 |

$10y^2 - 3y - 7$
$= 10y^2 + 7y - 10y - 7$
$= y(10y+7) - 1(10y+7)$
$= (10y+7)(y-1)$

73. $w^4(2w^2 - 3w - 20)$

| Factors of − 40 | Sums |
|---|---|
| 5, − 8 and − 5, 8 | − 3, 3 |

$w^4(2w^2 + 5w - 8w - 20)$
$= w^4[w(2w+5) - 4(2w+5)]$
$= w^4(2w+5)(w-4)$

75. $(4x+5)(x-1)$

77.

| Factors of 12 | Sums |
|---|---|
| 1, 12 and − 1, − 12 | 13, − 13 |
| 2, 6 and − 2, − 6 | 8, − 8 |
| 3, 4 and − 3, − 4 | 7, − 7 |

Prime

79.

| Factors of − 126 | Sums |
|---|---|
| 9, − 14 and − 9, 14 | − 5, 5 |

$7z^2 + 9z - 14z - 18$
$= z(7z+9) - 2(7z+9)$
$= (7z+9)(z-2)$

81.

| Factors of − 168 | Sums |
|---|---|
| 8, − 21 and − 8, 21 | − 13, 13 |

$28t^2 - 8t + 21t - 6$
$= 4t(7t-2) + 3(7t-2)$
$= (7t-2)(4t+3)$

83.

| Factors of 42 | Sums |
|---|---|
| − 14, − 3 and 14, 3 | − 17, 17 |

$2x^2 - 14x - 3x + 21$
$= 2x(x-7) - 3(x-7)$
$= (x-7)(2x-3)$

Factoring

85. $(3a-w)(a+3w)$

87. $(5x+6)(x+3)$

89. $(7x-3)(x+1)$

91. $(3a+2b)(a-b)$

93.
| Factors of $-360$ | Sums |
|---|---|
| $18, -20$ and $-18, 20$ | $-2, 2$ |

$8x^2 - 18x - 20x - 45$
$= 2x(4x-9) - 5(4x-9)$
$= (4x-9)(2x-5)$

95.
| Factors of 48 | Sums |
|---|---|
| 3, 16 and $-3, -16$ | $19, -19$ |

$2a^2 - 3a - 16a + 24$
$= a(2a-3) - 8(2a-3)$
$= (2a-3)(a-8)$

97. $6x + 5 + 2x^2 + 5x$
$= 2x^2 + 11x + 5$
$= (2x+1)(x+5)$

99.
$$\begin{array}{r} 2x+1 \\ x+4 \overline{\smash{)}2x^2+9x+4} \\ \underline{2x^2+8x} \\ x+4 \end{array}$$

The length of base is $2x+1$.

101. (a) $P(x) = 0.01x^2 - 0.78x + 66.31$
$= 0.01(x^2 - 78x + 6631)$

(b) The possible middle terms are $-6632x$ and $-368x$ neither which are equal to $-78x$.

(c) $P(52) = 0.01(52)^2 - 0.78(52) + 66.31$
$= 52.79$
About 53%

103. $\dfrac{1}{25}(50x^2 + 15x - 2)$

| Factors of $-100$ | Sums |
|---|---|
| $-20, 5$ and $20, -5$ | $-15, 15$ |

$\dfrac{1}{25}(50x^2 + 20x - 5x - 2)$
$= \dfrac{1}{25}[10x(5x+2) - 1(5x+2)]$
$= \dfrac{1}{25}(5x+2)(10x-1)$

105. $(2x^5 + 1)(x^5 - 4)$

107.
| Factors of 144 | Sums |
|---|---|
| 9, 16 and $-9, -16$ | $25, -25$ |

$36x^4 - 9x^2 - 16x^2 + 4$
$9x^2(4x^2 - 1) - 4(4x^2 - 1)$
$= (4x^2 - 1)(9x^2 - 1)$
$= (2x-1)(2x+1)(3x-1)(3x+1)$

## Section 7.5 General Strategy

1. $(1-x)(1+x)$

3. $(2+3x)^2$

5. $4y^2(y^2+3y+2)$
   $=4y^2(y+1)(y+2)$

7. $(2x+5)(4x-1)$

9. $(y+5)(y+9)$

11. $-4(y^2-2y-3)$
    $=-4(y-3)(y+1)$

13.
   | Factors of 30 | Sums |
   |---|---|
   | 5, 6 and $-5, -6$ | 11, $-11$ |

   $6x^2+5x+6x+5$
   $=x(6x+5)+1(6x+5)$
   $=(6x+5)(x+1)$

15. $x^2(12x^2+8xy-15y^2)$

   | Factors of $-180$ | Sums |
   |---|---|
   | $-10, 18$ and $10, -18$ | 8, $-8$ |

   $x^2(12x^2-10xy+18xy-15y^2)$
   $=x^2[2x(6x-5y)+3y(6x-5y)]$
   $=x^2(6x-5y)(2x+3y)$

17. $(8-y)(7-y)$

19. $a(x+7)+(x+7)$
    $=(x+7)(a+1)$

21. Prime, since a sum of two squares.

23. $x^3(x^2y+2xy^2+1)$

25. $(ab+5)^2$

27. $(b+6)(b-5)$

29. $(5x+6)(x-1)$

31.
   | Factors of $-112$ | Sums |
   |---|---|
   | 16, $-7$ and $-16, 7$ | 9, $-9$ |

   $8x^2-16x+7x-14$
   $=8x(x-2)+7(x-2)$
   $=(x-2)(8x+7)$

33. $(c+2)(a-2b)$

35. $(t-8)(t+3)$

37.
   | Factors of 16 | Sums |
   |---|---|
   | 1, 16 and $-1, -16$ | 17, $-17$ |
   | 2, 8 and $-2, -8$ | 10, $-10$ |
   | 4, 4 and $-4, -4$ | 8, $-8$ |

   Prime, since no sum equals $-9$.

39. $(8t^5-1)(8t^5+1)$

41. $(3-y)(9+3y+y^2)$

43. $t^2(t-5)-1(t-5)$
    $=(t-5)(t^2-1)$
    $=(t-5)(t+1)(t-1)$

45. $-3x^2(2x^2-3x+1)$
    $=-3x^2(2x-1)(x-1)$

Factoring

47. $(5x+1)(x-1) = 5x^2 - 4x - 1$
$(5x-1)(x+1) = 5x^2 + 4x - 1$
Prime

49. $3x(2x^2 - 3x - 5)$
$= 3x(2x-5)(x+1)$

51. $c(-2x + c - 5)$

53. $a^2(c^2 + d) - b(c^2 + d)$
$= (c^2 + d)(a^2 - b)$

55. $(x - 5b)(x - 4b)$

57. $2x(x^3 - 27)$
$= 2x(x-3)(x^2 + 3x + 9)$

59.
| Factors of 24 | Sums |
|---|---|
| 6, 4 and -6, -4 | 10, -10 |

$8b^2 + 6bc + 4bc + 3c^2$
$= 2b(4b + 3c) + c(4b + 3c)$
$= (4b + 3c)(2b + c)$

61. $3a(a^2 + 2a - 8)$
$= 3a(a-2)(a+4)$

63.
| Factors of -120 | Sums |
|---|---|
| -10, 12 and 10, -12 | 2, -2 |

$8y^2 - 10z + 12yz - 15z^2$
$= 2y(4y - 5z) + 3z(4y - 5z)$
$= (4y - 5z)(2y + 3z)$

65. $8x(x^2 - 4)$
$= 8x(x - 2)(x + 2)$

67. $x^3(x+1) + 2(x+1)$
$= (x+1)x^3 + 2$

69. $(x - 12y)(x + 2y)$

71. $x(x^2 + 12xy + 36y^2)$
$= x(x + 6y)^2$

73. $b^2(a^2 - 8ab + 24b^2)$

75. $2x^2(x^2 - 5x - 24)$
$= 2x^2(x - 8)(x + 3)$

77. $a(8a^3 - 1)$
$= a(2a - 1)(4a^2 + 4a + 1)$

79. $-6xyz^4(x^3z + 4y^4)$

81. $(a + 11b)(a - 2b)$

83.
| Factors of -588 | Sums |
|---|---|
| -12, 28 and 21, -28 | 7, -7 |

$12x^2 + 21xz - 28xz - 49z^2$
$= 3x(4x + 7z) - 7z(4x + 7z)$
$= (4x + 7z)(3x - 7z)$

85. $(4y^2 - 1)(4y^2 + 1)$
$= (2y - 1)(2y + 1)(4y^2 + 1)$

87. $(10x - 9)^2$

89. $yz^2(12y^2 - yz - z^2)$
$= yz^2(4y + z)(3y - z)$

91. $2x(x^2 - 3x + 18)$

93. $(4a)^3 + b^3$
$= (4a+b)(16a^2 - 4ab + b^2)$

## Section 7.6 Solving Equations by Factoring

1. For $a = 0$, the equation becomes $bx + c = 0$, which for $b \neq 0$ is a linear equation.

3. (a) 2    (the x-intercepts)

   (b) 2

   (c) 1    (the vertex)

   (d) 0

5. 10, 25

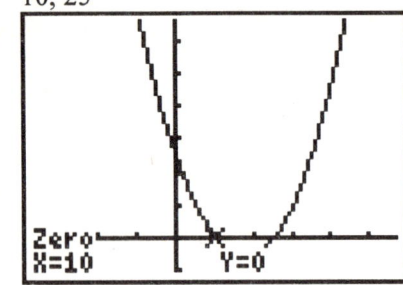

7. No Solution

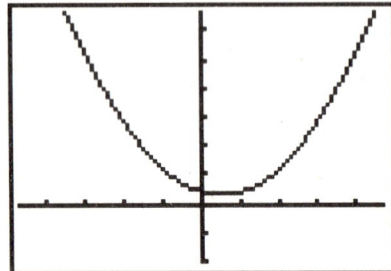

9. $-10, 30$

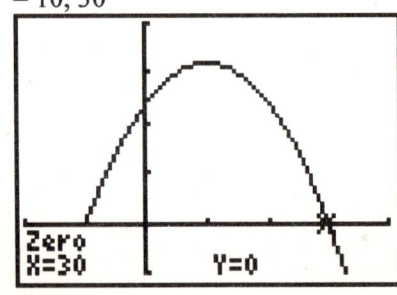

11. (a) 2

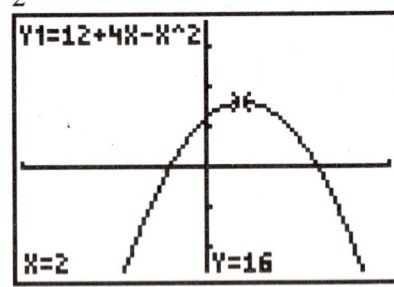

    (b) No solution

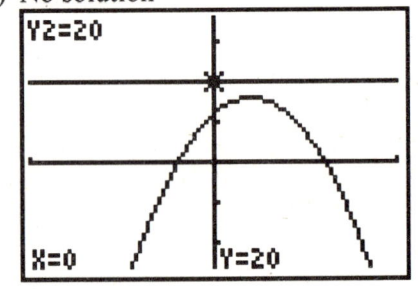

    (c) $-2, 6$

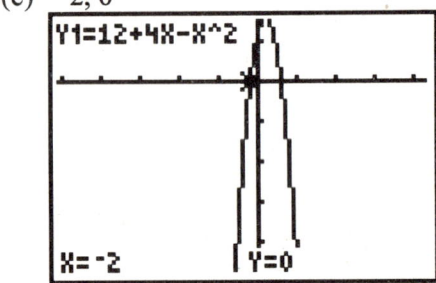

    (d) $-4, 8$

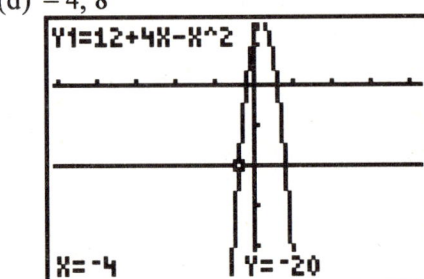

13. One side of the equation must be 0, and the other side must be written in factored form.

Factoring

15. $x + 5 = 0$
    $x = -5$

    $x - 7 = 0$
    $x = 7$

17. $y - 5 = 0$
    $y = 5$

    $3y + 7 = 0$
    $3y = -7$
    $y = -\dfrac{7}{3}$

19. $2x - 1$
    $2x = 1$
    $x = \dfrac{1}{2}$

    $x - 8 = 0$
    $x = 8$

21. $1 - t = 0$
    $t = 1$

    $3 + t = 0$
    $t = -3$

23. $5x = 0$
    $x = 0$

    $1 - 4x = 0$
    $-4x = -1$
    $x = \dfrac{1}{4}$

25. $x + 1 = 0$
    $x = -1$

    $x + 4 = 0$
    $x = -4$

    $x - 3 = 0$
    $x = 3$

27. $-6x = 0$
    $x = 0$

    $x + 9 = 0$
    $x = -9$

    $x - 2 = 0$
    $x = 2$

29. $(x + 2)(x + 4) = 0$
    $x = -2$
    $x = -4$

31. $(x + 7)(x - 3) = 0$
    $x = -7$
    $x = 3$

33. $x(x - 1) = 0$
    $x = 0$
    $x = 1$

35. $(y + 8)(y - 8) = 0$
    $y = -8$
    $y = 8$

37. $(x + 5)^2 = 0$
    $x = -5$

39. $6\left(\dfrac{1}{6}x^2 + x - \dfrac{9}{2}\right) = 6 \cdot 0$

$x^2 + 6x - 27 = 0$

$(x+9)(x-3) = 0$

$x = -9$

$x = 3$

41. $t^2 + 5t - 6 = 0$

$(t-1) = 0$

$t + 6$

$t = -6$

$t = 1$

43. $x^2 + x + 20 = 0$

$x^2 - x - 20 = 0$

$(x-5)(x+4) = 0$

$x = 5$

$x = -4$

45. $5x(x+4) = 0$

$x = 0$

$x = -4$

47. $(2x-5)^2 = 0$

$2x - 5 = 0$

$2x = 5$

$x = \dfrac{5}{2}$

49. $(7-2x)(7+2x) = 0$

$7 - 2x = 0$

$-2x = -7$

$x = \dfrac{7}{2}$

$7 + 2x = 0$

$2x = -7$

$x = -\dfrac{7}{2}$

51. $3x(2x-5) = 0$

$3x = 0$

$x = 0$

$2x - 5 = 0$

$2x = 5$

$x = \dfrac{5}{2}$

53. $(2x+1)(x-4) = 0$

$2x + 1 = 0$

$2x = -1$

$x = \dfrac{-1}{2}$

$x - 4 = 0$

$x = 4$

55. $(4x+1)(x+4) = 0$

$4x + 1 = 0$

$4x = -1$

$x = \dfrac{-1}{4}$

$x = -4$

Factoring

57. $4\left(2x^2 - \frac{1}{2}x - \frac{1}{4}\right) = 0$
$8x^2 - 2x - 1 = 0$
$(4x + 1)(2x - 1) = 0$

$4x + 1 = 0$
$4x = -1$
$x = -\frac{1}{4}$

$2x - 1 = 0$
$2x = 1$
$x = \frac{1}{2}$

59. $6x^2 - 17x + 10 = 0$
Factors of 60    Sums
$-12, -5$ and $12, 5$    $-17, 17$
$6x^2 - 12x - 5x + 10 = 0$
$6x(x - 2) - 5(x - 2) = 0$
$(6x - 5)(x - 2 = 0)$

$6x - 5 = 0$
$6x = 5$
$x = \frac{5}{6}$

$x = 2$

61. $9x^2 + 14x - 8 = 0$

Factors of $-72$    Sums
$-18, 4$ and $18, -4$    $-14, 14$

$9x^2 + 18x - 4x - 8 = 0$
$9x(x + 2) - 4(x + 2) = 0$
$(x + 2)(9x - 4) = 0$

$x = -2$

$9x - 4 = 0$
$9x = 4$
$x = \frac{4}{9}$

63. (a) There are 3 factors: 2, $x$ and $x - 1$.

(b) No, the factor 2 can never be 0.

65. $x^2 + 3x = 28$
$x^2 + 3x - 28 = 0$
$(x + 7)(x - 4) = 0$

$x = -7$
$x = 4$

67. $x^2 + 6x + 9 = 16$
$x^2 + 6x - 7 = 0$
$(x - 1)(x + 7) = 0$

$x = 1$
$x = -7$

69. $3y - y^2 = 2$
$y^2 - 3y + 2 = 0$
$(y - 1)(y - 2) = 0$

$y = 1$
$y = 2$

71. $x^2 + 5x + 4x + 20 = 2$
    $x^2 + 9x + 20 = 2$
    $x^2 + 9x + 18 = 0$
    $(x+6)(x+3) = 0$

    $x = -3$
    $x = -6$

73. $2x^2 - x = 45$
    $2x^2 - x - 45 = 0$
    $(2x+9)(x-5) = 0$

    $2x + 9 = 0$
    $2x = -9$
    $x = \dfrac{-9}{2}$

    $x = 5$

75. $2x^2 + 4x + x + 2 = -1$
    $2x^2 + 5x + 2 = -1$
    $2x^2 + 5x + 3 = 0$
    $(2x+3)(x+1) = 0$

    $2x + 3 = 0$
    $2x = -3$
    $x = \dfrac{-3}{2}$

    $x = -1$

77. $10y^2 - 10y = 3 + 3y$
    $10y^2 - 13y - 3 = 0$
    $10y^2 - 15y + 2y - 3 = 0$
    $5y(2y-3) + 1(2y-3) = 0$
    $(2y-3)(5y+1) = 0$

    $2y - 3 = 0$
    $2y = 3$
    $y = \dfrac{3}{2}$

    $5y + 1 = 0$
    $5y = -1$
    $y = \dfrac{-1}{5}$

79. $x(x^2 + 3x - 28) = 0$
    $x(x+7)(x-4) = 0$

    $x = 0$
    $x = -7$
    $x = 4$

81. $8x(x^2 - 1) = 0$
    $8x(x+1)(x-1) = 0$

    $x = 0$
    $x = -1$
    $x = 1$

83. $-x^3 - 7x^2 + 18x = 0$
    $-x(x^2 + 7x - 18) = 0$
    $-x(x+9)(x-2) = 0$

    $x = 0$
    $x = -9$
    $x = 2$

**Factoring**

85. $5x(9x^2 - 16) = 0$

$5x(3x - 4)(3x + 4) = 0$

$x = 0$

$3x - 4 = 0$
$3x = 4$
$x = \dfrac{4}{3}$

$3x + 4 = 0$
$3x = -4$
$x = \dfrac{-4}{3}$

87. $x^2(x + 2) - 4(x + 2) = 0$
$(x + 2)(x^2 - 4) = 0$
$(x + 2)(x + 2)(x - 2) = 0$
$(x + 2)^2(x - 2) = 0$

$x = 2$
$x = -2$

89. $9x^2(x - 3) - 1(x - 3) = 0$
$(x - 3)(9x^2 - 1) = 0$
$(x - 3)(3x - 1)(3x + 1) = 0$

$x = 3$

$3x - 1 = 0$
$3x = 1$
$x = \dfrac{1}{3}$

$3x + 1 = 0$
$3x = -1$
$x = \dfrac{-1}{3}$

91. $-1, 4, 2.4$

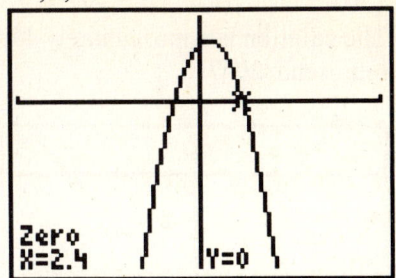

93. $1.4$

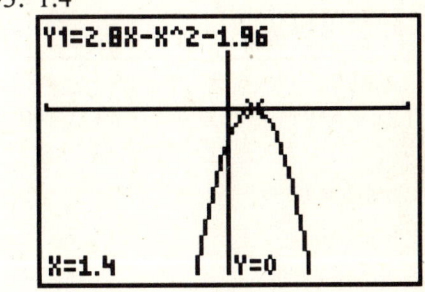

95. $2.1, -0.7$

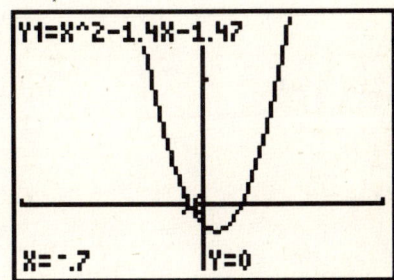

97. (a) $-0.0082(x^2 - 45x - 2800) = 0$

$x^2 - 45x - 2800 = 0$
$(x - 80)(x + 35) = 0$
$x = 80$
$x = -35$

(b) Because age cannot be negative, $-35$ is not meaningful.

(c) No one of age 80 received assistance.

99. $(36x + 1132)(0.11x + 3.80)$

101. $(36x+1132)(0.11x+3.80)=10,000$;
The solution is approximately 17, which represents 2007.

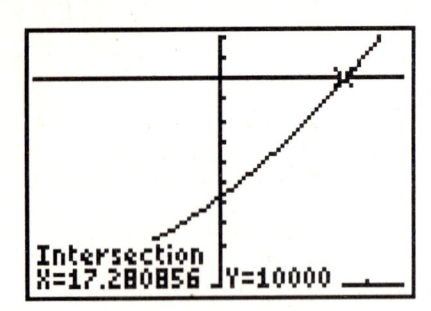

103. $(2x+5)(x-2)(3x+4)(x+3)=0$

$2x+5=0$
$2x=-5$
$x=\dfrac{-5}{2}$

$x=2$
$3x+4=0$
$3x=-4$
$x=\dfrac{-4}{3}$

$x=-3$

105. $(x^2+9)(x+9)^2(x-3)(x+3)=0$

$x^2+9=0$
No solution

$x=-9$
$x=3$
$x=-3$

107. $(x^2-9)(x^2-1)$
$=(x+3)(x-3)(x-1)(x+1)$

$x=-3$
$x=3$
$x=1$
$x=-1$

109. $(x+3b)^2=0$

$x+3b=0$
$x=-3b$

## Section 7.7 Applications

1. $x(x+2)=63$

$x^2+2x=63$
$x^2+2x-63=0$
$(x+9)(x-7)=0$

$x=-9$ or $x=7$
Since the integers are supposed to be positive, $x=7$ and $x+2=9$ are the integers.

3. $x^2+(x+1)^2=25$

$x^2+x^2+2x+1=25$
$2x^2+2x-24=0$
$2(x^2+x-12)=0$
$2(x+4)(x-3)=0$

$x=-4$
$x+1=-3$
The numbers are $-4$ and $-3$.

Factoring

5. $x(x+11) = -30$
$x^2 + 11x + 30 = 0$
$(x+6)(x+5) = 0$

$x = -6$
$x + 11 = 5$
or
$x = -5$
$x + 11 = 6$

7. $x(x+4) = 21$
$x^2 + 4x = 21$
$x^2 + 4x - 21 = 0$
$(x+7)(x-3) = 0$

$x = -7$
or
$x = 3$

9. $x^2 = 24 - 5x$
$x^2 + 5x - 24 = 0$
$(x+8)(x-3) = 0$

$x = -8$
or
$x = 3$

11. $11x + 7 = 6x^2$
$6x^2 - 11x - 7 = 0$

Factors of $-42$   Sums
$-14, 3$ and $14, -3$   $-11, 11$

$6x^2 - 14x + 3x - 7 = 0$
$2x(3x-7) + 1(3x-7) = 0$
$(3x-7)(2x+1) = 0$

$3x - 7 = 0$
$3x = 7$
$x = \dfrac{7}{3}$

$2x + 1 = 0$
$2x = -1$
$x = -\dfrac{1}{2}$

13. $10x + 8 = 3x^2$
$3x^2 - 10x - 8 = 0$
$(3x+2)(x-4) = 0$

$x = 4$
$3x + 2 = 0$
$3x = -2$
$x = -\dfrac{2}{3}$

15. $x^2 + 5x = 3 - x^2$
$2x^2 + 5x - 3 = 0$
$(2x-1)(x+3) = 0$

$2x - 1 = 0$
$2x = 1$
$x = \dfrac{1}{2}$
$x = -3$

17. $x(x-3) = 70$
$x^2 - 3x - 70 = 0$
$(x-10)(x+7) = 0$

$x = 10$
$x - 3 = 7$

The dimensions are 7 x 10.

19. $x(2x-3) = 77$
$2x^2 - 3x - 77 = 0$
$(2x+11)(x-7) = 0$

$x = 7$
$2x - 3 = 2(7) - 3 = 11$

The dimensions are 7 x 11.

21. $\frac{1}{2}x(x+5) = 42$
$x(x+5) = 84$
$x^2 + 5x = 84$
$x^2 + 5x - 84 = 0$
$(x+12)(x-7) = 0$

$x = 7$
$x + 5 = 12$

The dimensions are 7 x 12.

23. $\frac{1}{2}x(2x+1+3x+2) = 27$
$\frac{1}{2}x(5x+3) = 27$
$x(5x+3) = 54$
$5x^2 + 3x = 54$
$5x^2 + 3x - 54 = 0$
$5x^2 + 18x - 15x - 54 = 0$
$x(5x+18) - 3(5x+18) = 0$
$(x-3)(5x+18) = 0$

$x = 3$
$2x + 1 = 2(3) + 1 = 7$
$3x + 2 = 3(3) + 2 = 11$

The dimensions are 3, 7, and 11.

25. $2(x+1)(3x+2) = 48$
$(x+1)(3x+2) = 24$
$3x^2 + 2x + 3x + 2 = 24$
$3x^2 + 5x - 22 = 0$
$3x^2 + 11x - 6x - 22 = 0$
$x(3x+11) - 2(3x+11) = 0$
$(x-2)(3x+11) = 0$

$x = 2$
$x + 1 = 3$
$3x + 2 = 3(2) + 2 = 8$

The dimensions are 2 x 3 x 8.

Factoring

27. $x(x+3) = 40$
$x^2 + 3x = 40$
$x^2 + 3x - 40 = 0$
$(x+8)(x-5) = 0$

$x = 5$
$x + 3 = 8$

The width is 5 feet and the length is 8 feet.

29. $\frac{1}{2}x(x-1) = 21$
$x(x-1) = 42$
$x^2 - x - 42 = 0$
$(x-7)(x+6) = 0$

$x = 7$
$x - 1 = 6$

The base is 6 meters and the height is 7 meters.

31. $x^2 = 2(4x)$
$x^2 = 8x$
$x^2 - 8x = 0$
$x(x-8) = 0$
$x = 8$

The sides are 8 units longs.

33. $2x(x-4) = x^2 + 20$
$2x^2 - 8x = x^2 + 20$
$x^2 - 8x - 20 = 0$
$(x-10)(x+2) = 0$

$x = 10$

The length of a side is 10 feet.

35. $x^2 + (2x-2)^2 = 5^2$
$x^2 + 4x^2 - 8x + 4 = 25$
$5x^2 - 8x + 4 = 25$
$5x^2 - 8x - 21 = 0$
$(5x+7)(x-3) = 0$

$x = 3$
$2x - 2 = 2(3) - 2 = 4$

The dimensions are 3, 4, and 5.

37. $x^2 + (x+3)^2 = (2x-3)^2$
$x^2 + x^2 + 6x + 9 = 4x^2 - 12x + 9$
$2x^2 + 6x + 9 = 4x^2 - 12x + 9$
$2x^2 - 18x = 0$
$2x(x-9) = 0$

$x = 9$
$x + 3 = 12$
$2x - 3 = 2(9) - 3 = 15$

The dimensions are 9, 12, and 15.

39. $x^2 + (x+5)^2 = 25^2$

41. Note: 36 feet = 12 yards
$x^2 + 12^2 = (3x-2)^2$
$x^2 + 144 = 9x^2 - 12x + 4$
$8x^2 - 12x - 140 = 0$
$2x^2 - 3x - 35 = 0$
$(2x+7)(x-5) = 0$
$x = 5$
$3x - 2 = 3(5) - 2 = 13$

The dimensions are 5 yards, 12 yards and 13 yards.

43. $x^2 + (x+7)^2 = (x+8)^2$

$x^2 + x^2 + 14x + 49 = x^2 + 16x + 64$

$2x^2 + 14x + 49 = x^2 + 16x + 64$

$x^2 - 2x - 15 = 0$

$(x-5)(x+3) = 0$

$x = 5$

$x + 7 = 12$

$x + 8 = 13$

The dimensions are 5 feet, 12 feet, and 13 feet.

45. $x(x-3) = 180$

$x^2 - 3x - 180 = 0$

$(x-15)(x+12) = 0$

$x = 15$

$x - 3 = 12$

The dimensions are 12 feet by 15 feet.

47. The area of the overhang $= 4 \cdot 5 - 12$
$= 20 - 12$
$= 8$

Let $x$ = length of overhang.

$5x + 5x + (4-2x)x + (4-2x)x = 8$

$10x + 2x(4-2x) = 8$

$10x + 8x - 4x^2 = 8$

$-4^2 + 18x - 8 = 0$

$2x^2 - 9x + 4 = 0$

$(2x-1)(x-4) = 0$

$x = \dfrac{1}{2}$

The cloth overhang by $\dfrac{1}{2}$ foot.

49.

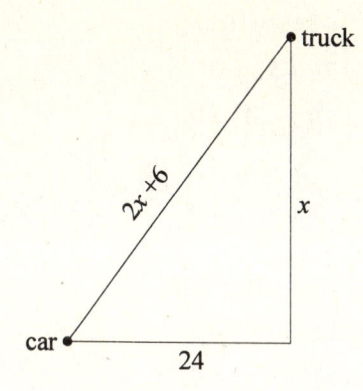

[Figure: right triangle with legs 24 (horizontal, from car) and x (vertical, to truck), hypotenuse 2x+6]

$24^2 + x^2 = (2x+6)^2$
$x^2 + 576 = 4x^2 + 24x + 36$
$3x^2 + 24x - 540 = 0$
$x^2 + 8x - 180 = 0$
$(x-10)(x+18) = 0$

$x = 10$ miles

$\dfrac{10 \text{ miles}}{\frac{1}{2} \text{ hour}} = 20 \text{ mph}$

51.

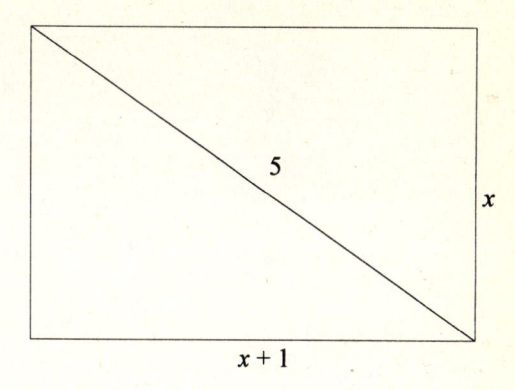

[Figure: rectangle with diagonal 5, width x+1, height x]

$(x+1)^2 + x^2 = 5^2$
$x^2 + 2x + 1 + x^2 = 25$
$2x^2 + 2x - 24 = 0$
$x^2 + x - 12 = 0$
$(x+4)(x-3) = 0$

$x = 3$
$x + 1 = 4$

The dimensions are 3 yards, 4 yards, and 5 yards.

53. $\frac{1}{2}n(n-1) = 300$
$n(n-1) = 600$
$n^2 - n - 600 = 0$
$(n-25)(n+24) = 0$

$n = 25$ students

55. $\frac{n(n+1)}{2} = 55$
$n(n+1) = 110$
$n^2 + n - 110 = 0$
$(n+11)(n-10) = 0$

$n = 10$

57. $n^2 - 20n + 10 = 310$
$n^2 - 20n - 300 = 0$
$(n-30)(n+10) = 0$

$n = 30$ blenders

59. $R = np$
$= (550 - 50p)p$
$= 550p - 50p^2$

$500 = 550p - 50p^2$
$10 = 11p - p^2$
$p^2 - 11p + 10 = 0$
$(p-1)(p-10) = 0$

$p = 1$ or $p = 10$ but $p > 5$ so $p = 10$

$n = 550 - 50(10)$
$= 50$ tickets sold

61. $-2x^2 + 34x - 20 = 84$
$-2x^2 + 34x - 104 = 0$
$x^2 - 17x + 52 = 0$
$(x-13)(x-4) = 0$

$x = 13$ or $x = 4$ but $x > 5$ so $x = 13$ mg

63. $125 = -x^2 - 10x + 200$
$x^2 + 10x - 75 = 0$
$(x+15)(x-5) = 0$
$x = 5$ dollars

65. (a) $0.3x^2 - 3.6x + 36.2 = 36.2$

(b) $0.3x^2 - 3.6x = 0$
$3x^2 - 36x = 0$
$3x(x-12) = 0$

$x = 12$
In 2002 the attendance is predicted to reach 36,200.

67. $x^2 + 22x + 588 = 1428$

69. $x^2 + 22x + 588 = 1428$
$x^2 + 22x - 840 = 0$
$(x-20)(x+42) = 0$

$x = 20 (2010)$

71. $(x+1)^2 + x^2 = 145$
$x^2 + 2x + 1 + x^2 = 145$
$2x^2 + 2x - 144 = 0$
$x^2 + x - 72 = 0$
$(x+9)(x-8) = 0$

$x = 8$
$x + 1 = 9$

The dimensions are $8 \times 8$ and $9 \times 9$.

73.

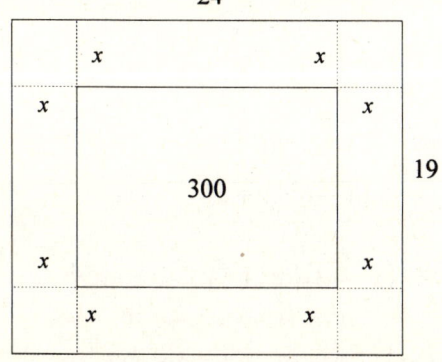

Area of frame and poster $= 19 \cdot 24 = 456$

Area of poster $= 300$

Area of frame $= 456 - 300 = 156$

$2 \cdot 19x + 2(24 - 2x)x = 156$
$38x + (48 - 4x)x = 156$
$38x + 48x - 4x^2 = 156$
$-4x^2 + 86x = 156$
$4x^2 - 86x + 156 = 0$
$2x^2 - 43x + 78 = 0$
$(2x - 39)(x - 2) = 0$

$x = 2$ inches

The frame is 2 inches wide.

75.

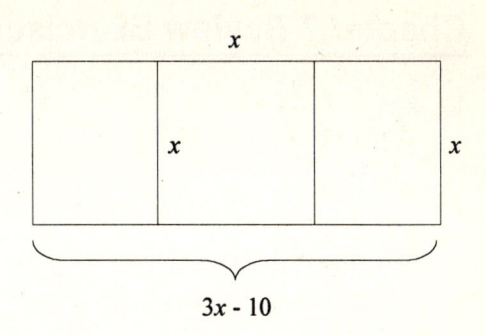

$x(3x - 10) - x^2 = 1000$
$3x^2 - 10x - x^2 = 1000$
$2x^2 - 10x - 1000 = 0$
$x^2 - 5x - 500 = 0$
$(x - 25)(x + 20) = 0$

$x = 25$
$3x - 10 = 3(25) - 10 = 65$

The lobby is 24 feet by 65 feet.

## Chapter 7 Review Exercises

1. $825$
   $= 25 \cdot 33$
   $= 5 \cdot 5 \cdot 3 \cdot 11$

3. (a) $15y^2 = 3 \cdot 5 \cdot y \cdot y$
   $20y^3 = 2 \cdot 2 \cdot 5 \cdot y \cdot y \cdot y$
   $GCF = 5 \cdot y \cdot y = 5y^2$

   (b) $6xz = 2 \cdot 3 \cdot x \cdot z$
   $9x^2z = 3 \cdot 3 \cdot x \cdot x \cdot z$
   $15xz^2 = 3 \cdot 5 \cdot x \cdot z \cdot z$
   $GCF = 3 \cdot x \cdot z = 3xz$

5. (a) $9x^2(3x^2 - 5x - 1)$

   (b) $4x(xy - 1)$

7. $(a^2 + 7)(b - 1)$

9. $x^3(x+3) - 6x(x+3)$
   $= (x^3 - 6x)(x+3)$
   $= x(x^2 - 6)(x+3)$

11. (a) $(4a - 3)(4a + 3)$

    (b) $(pq - 2)(pq + 4)$

13. $(b^2 - 4)(b^2 + 4)$
    $= (b - 2)(b + 2)(b^2 + 4)$

15. (a) $(x - 6)^2$

    (b) $2a(x^2 + 8x + +16)$
    $= 2a(x + 4)^2$

17. (a) $(x + 3)(x^2 - 3x + 9)$

    (b) $3(1 + 8a^3)$
    $= 3(1 + 2a)(1 - 2a + 4a^2)$

19. $(x+1)(x-1)(x-1)^2$
    $= (x+1)(x-1)^3$

21. (a) $(c + 7)(c + 3)$

    (b) $(9 + x)(2 + x)$

23. $y^2 - y - 30$
    $= (y - 6)(y + 5)$

25. $(x^2 - 3)(x^2 + 5)$

27. (a) $3(y^2 - 4yz - 32z^2)$
    $= 3(y - 8z)(y + 4z)$

    (b) Prime

29. $(x-1)(x+1)(x+1)^2(x+2)(x-1)$
    $= (x-1)^2(x+1)^3(x+2)$

31. (a)

    | Factors of 30 | Sums |
    |---|---|
    | 15, 2 | 17 |

    $6x^2 + 15x + 2x + 5$
    $= 3x(2x + 5) + 1(2x + 5)$
    $= (2x + 5)(3x + 1)$

**Factoring**

(b)

| Factors of 24 | Sums |
|---|---|
| $-3, -8$ | $-11$ |

$2x^2 - 3x - 8x + 12$
$= x(2x-3) - 4(2x-3)$
$= (2x-3)(x-4)$

33. (a)

| Factors of 42 | Sums |
|---|---|
| 7, 6 | 13 |

$3a^2 + 7ab + 6ab + 4b^2$
$= a(3a + 7b) + 2b(3a + 7b)$
$= (3a + 7b)(a + 2b)$

(b)

| Factors 18 | Sums |
|---|---|
| $-1, -18$ | $-19$ |

$2x^2 - xy - 18xy + 9y^2$
$= x(2x - y) - 9y(2x - y)$
$= (2x - y)(x - 9y)$

35. $-1(3x^2 - x - 24)$

| Factors of $-72$ | Sums |
|---|---|
| $-9, 8$ | $-1$ |

$-1(3x^2 - 9x + 8x - 24)$
$= -1[3x(x-3) - 8(x-3)]$
$= -1(x-3)(3x-8)$

37. (a)

| Factors of $-3$ | Sums |
|---|---|
| $1, -3$ and $-1, 3$ | $-2, 2$ |

Prime

(b) $2a(24a^2 + 53a + 28)$
$= 2a(24a^2 + 21a + 32a + 28)$
$= 2a[3a(8a+7) + 4(8a+7)]$
$= 2a(8a+7)(3a+4)$

39. $(2x+1)(x-1)(2x-1)^2$

41. $4(x^2 - 4)$
$= 4(x-2)(x+2)$

43. $(t^2 - 4)(t^2 + 4)$
$= (t+2)(t-2)(t^2 + 4)$

45. $a^2(a+1) + 1(a+1)$
$= (a+1)(a^2 + 1)$

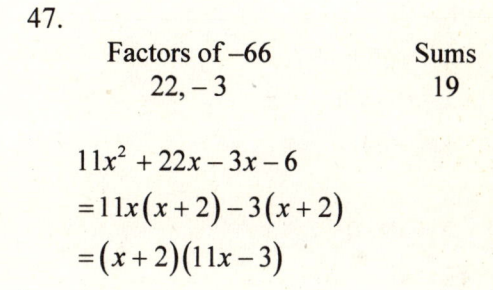

47.

| Factors of $-66$ | Sums |
|---|---|
| $22, -3$ | 19 |

$11x^2 + 22x - 3x - 6$
$= 11x(x+2) - 3(x+2)$
$= (x+2)(11x-3)$

49. $2x(x^2 + 6x + 9)$
$= 2x(x+3)^2$

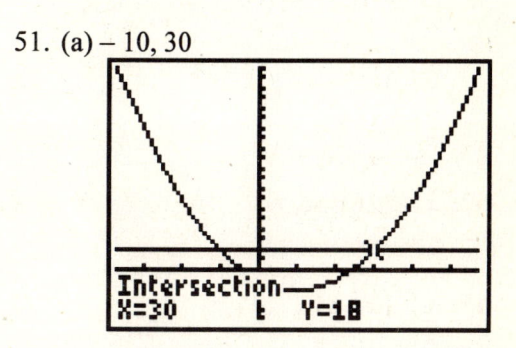

51. (a) $-10, 30$

(b) 10

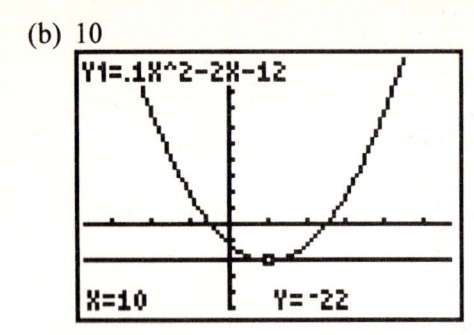

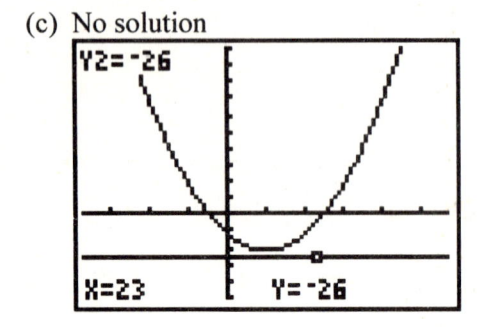

[calculator screen: Y1=.1X^2-2X-12, X=10, Y=-22]

(c) No solution

[calculator screen: Y2=-26, X=23, Y=-26]

53. $3x = 0$
$x = 0$

$x + 0.4 = 0$
$x = -0.4$

$2x - 1 = 0$
$2x = 1$
$x = \dfrac{1}{2}$

55. $(x-7)(x+4) = 0$

$x = 7$
$x = -4$

57. $x^2 - 5x = 36$
$x^2 - 5x - 36 = 0$
$(x-9)(x+4) = 0$

$x = 9$
$x = -4$

59. $x^3 + 3x^2 - 40x = 0$
$x(x^2 + 3x - 40) = 0$
$x(x+8)(x-5) = 0$

$x = 0$
$x = -8$
$x = 5$

61. $x - y = 2$
$x = y + 2$

$xy = 63$
$(x+2)y = 63$
$y^2 + 2y - 63 = 0$
$(y+9)(y-7) = 0$

$x = 7$
$x = 7 + 2 = 9$
The older brother is 9.

63. $x(x+2) = 4(x+1) - 1$
$x^2 + 2x = 4x + 4 - 1$
$x^2 + 2x = 4x + 3$
$x^2 - 2x - 3 = 0$
$(x-3)(x+1) = 0$

$x = 3$
$x + 1 = 4$
$x + 2 = 5$

The largest intger is 5.

**Factoring**

65. $(2x+2)^2 + x^2 = (3x-2)^2$
$4x^2 + 8x + 4 + x^2 = 9x^2 - 12x + 4$
$5x^2 + 8x + 4 = 9x^2 - 12x + 4$
$4x^2 - 20x = 0$
$4x(x-5) = 0$

$x = 0$
$x = 5$
$3x - 2 = 3(5) - 2 = 13$ feet

67. $l = 2w - 1$
$l^2 + w^2 = (2 \cdot 8.5)^2$
$(2w-1)^2 + w^2 = 17^2$
$4w^2 - 4w + 1 + w^2 = 289$
$5w^2 - 4w - 288 = 0$
$5w^2 - 40w + 36w - 288 = 0$
$5w(w-8) + 36(w-8) = 0$
$(5w+36)(w-8) = 0$

$w = 8$
$l = 2(8) - 1 = 15$ inches

69. $l = 7w + 1$
$\dfrac{4}{12} \cdot l \cdot w = 10$
$\dfrac{1}{3}(7w+1)w = 10$
$7w^2 + w = 30$
$7w^2 + w - 30 = 0$

Factors of −210 | Sums
15,− 14 | 1

$7w^2 + 15w - 14w - 30 = 0$
$w(7w+15) - 2(7w+15) = 0$
$(7w+15)(w-2) = 0$

$w = 2$
$l = 7(2) + 1 = 15$ feet

# Chapter 7 Test

1. $2x^4 y^2 (x + 3y - 6xy)$
GCF $= 2x^4 y^2$

3. $x^4(x-1) - 2(x-1)$
$= (x-1)(x^4 - 2)$

5. Sum of two cubes: $y+1(y^2 - y + 1)$

7. Perfect square trinomial: $(2x-5)^2$

9. $(a+2b)(a-5b)$

11. $c(c^2 + 5c - 14)$
$= c(c+7)(c-2)$

13.
Factors of − 180 | Sums
18, − 10 | 8

$12x^2 + 18x - 10x - 15$
$= 6x(2x+3) - 5(2x+3)$
$= (6x-5)(2x+3)$

15.
$$\begin{array}{r}8x-3\phantom{)}\\2x-9{\overline{\smash{)}}16x^2-66x-27}\\\underline{16x^2-72x}\phantom{-27}\\6x-27\\\underline{6x-27}\\0\end{array}$$
Length $= 8x - 3$

17. $x^2(x-1) - 4(x-1)$
$= (x-1)(x^2 - 4)$
$= (x-1)(x-2)(x+2)$

19. $(a^2 - 4)(a^2 + 4)$
$= (a-2)(a+2)(a^2 + 4)$

21. $6x = 0$
$x = 0$

$x - 7 = 0$
$x = 7$

23. $6x^2 - 7x = 5$
$6x^2 - 7x - 5 = 0$
$6x^2 - 10x + 3x - 5 = 0$
$2x(3x - 5) + 1(3x - 5) = 0$
$(3x - 5)(2x + 1) = 0$

$3x - 5 = 0$
$3x = 5$
$x = \dfrac{5}{3}$
$2x + 1 = 0$
$2x = -1$
$x = -\dfrac{1}{2}$

25. The expression $ax^2 + bx + c$ is a perfect square trinomial.

27. $l = w + 3$
$w^2 + l^2 = (w + 6)^2$
$w^2 + (w + 3)^2 = (w + 6)^2$
$w^2 + w^2 + 6w + 9 = w^2 + 12w + 36$
$w^2 + 6w + 9 = 12w + 36$
$w^2 - 6w - 27 = 0$
$(w - 9)(w + 3) = 0$

$w = 9$
$l = w + 3 = 12$ feet

# Chapter 8

# Rational Expressions

## Section 8.1 Introduction to Rational Expressions

1. Because the expression is not defined for −2, the calculator gives an error message.

3. $\dfrac{x}{x+4}$

   $\dfrac{0}{0+4} = \dfrac{0}{4} = 0$

   $\dfrac{4}{4+4} = \dfrac{4}{8} = \dfrac{1}{2}$

   $\dfrac{-4}{-4+4} = \dfrac{-4}{0} =$ undefined

5. $\dfrac{2t-3}{5-t}$

   $\dfrac{2(5)-3}{5-5}$
   $= \dfrac{10-3}{0}$
   $=$ undefined

   $\dfrac{2(-5)-3}{5-(-5)}$
   $= \dfrac{-10-3}{10}$
   $= \dfrac{-13}{10}$

   $\dfrac{2(4)-3}{5-4}$
   $= \dfrac{8-3}{1}$
   $= 5$

7. $r(1) = \dfrac{3(1)}{(1-3(1))^2}$
   $= \dfrac{3}{(1-3)^2}$
   $= \dfrac{3}{(-2)^2}$
   $= \dfrac{3}{4}$

9. $R(-3) = \dfrac{-3+1}{(-3)^2 - 3(-3)}$
   $= \dfrac{-2}{9+9}$
   $= \dfrac{-2}{18}$
   $= \dfrac{-1}{9}$

11. $x + 2 \neq 0$
    $x \neq -2$

13. $4x - x^2 \neq 0$
    $x(4 - x) \neq 0$
    $x \neq 0$ or $4 - x \neq 0$
    $4 \neq x$
    $x \neq 4$

15. $x^2 + 16 \neq 0$
    $x^2 \neq -16$
    Since $x^2 \geq 0$ for all values of $x$, there are no restricted values.

17. $25 - x^2 \neq 0$
    $(5 - x)(5 + x) \neq 0$
    $5 - x \neq 0$ or $5 + x \neq 0$
    $x \neq 5 \qquad x \neq -5$
    So $x \neq \pm 5$.

19. numerator: $4x, -3$
    denominator: $x, -3$

21. numerator: $4, x - 4$
    denominator: $3, x - 4$

23. In the numerator $x$ is a term, not a factor. Only factors can be divided out.

25. $\dfrac{y(y + 4)}{3(y + 4)}$
    $= \dfrac{y}{3}$

27. $\dfrac{5(x + 3)}{x(x - 3)}$

29. $\dfrac{2x^2(2x + 1)}{6x^2}$
    $= \dfrac{(2x + 1)}{3}$

31. $\dfrac{(x + 3)(x + 3)}{(x + 3)(x - 3)}$
    $= \dfrac{x + 3}{x - 3}$

33. $\dfrac{3(x + 2)}{(x - 6)(x + 2)}$
    $= \dfrac{3}{x - 6}$

35. $\dfrac{(w + 4)(w - 2)}{(w - 3)(w - 2)}$
    $= \dfrac{w + 4}{w - 3}$

37. The expressions are opposites because each is $-1$ times the other.

39. $\dfrac{-1(-5 + a)}{a - 5}$
    $= -1 \dfrac{a - 5}{a - 5}$
    $= -1$

41. $\dfrac{3y - 2x}{-1(-2x + 3y)}$
    $= \dfrac{3y - 2x}{-1(3y - 2x)}$
    $= -1$

43. Can't simplify

45. $\dfrac{y - 5}{(5 - y)(2 + 3y)}$
    $= \dfrac{-1(5 - y)}{(5 - y)(2x + 3y)}$
    $= \dfrac{-1}{2x + 3y}$

Rational Expressions

47. $\dfrac{4(3-x)}{x-3}$
$= 4(-1)$
$= -4$

49. $\dfrac{3x(3-x)}{2(x-3)}$
$= \dfrac{3x}{2}(-1)$
$= \dfrac{-3x}{2}$

51. $\dfrac{2z(1-z)}{(z+2)(z-1)}$
$= \dfrac{2z}{z+2}(-1)$
$= -\dfrac{2z}{z+2}$

53. $\dfrac{2(x^2-1)}{3(1-x)}$
$= \dfrac{2(x-1)(x+1)}{3(1-x)}$
$= \dfrac{2(x+1)}{3}(-1)$
$= -\dfrac{2(x+1)}{3}$

55. $\dfrac{10(a-2b)}{15(a-2b)}$
$= \dfrac{10}{15}$
$= \dfrac{2}{3}$

57. $\dfrac{(x-4)(x+2)}{(x+5)(x+2)}$
$= \dfrac{x-4}{x+5}$

59. $\dfrac{(3-x)(3+x)}{x-3}$
$= -1(3+x)$
$= -(x+3)$

61. $\dfrac{(x-4y)(x+4y)}{x+4y}$
$= x-4y$

63. $\dfrac{(y-3)(y+3)}{y-3}$
$= y+3$

65. $\dfrac{5(2-3x)}{3x-2}$
$= 5(-1)$
$= -5$

67. (a) 1

(b) Neither

(c) $-1$

69. No

71. Yes

73. No

75. $\dfrac{6x+2x^2}{2x}$
$= \dfrac{2x(3+x)}{2x}$
$= 3+x$

77. $\dfrac{3x^2+4x}{5x}$
$= \dfrac{x(3x+4)}{5x}$
$= \dfrac{3x+4}{5}$

79. The coefficient of $x$ gives the slope. Because in each case the coefficient of $x$ is positive, the life expectancy is increasing for both men and women.

81. $r(100) = \dfrac{0.33(100) + 48.6}{0.26(100) + 48.6}$

$= \dfrac{33 + 48.6}{26 + 48.6}$

$= \dfrac{81.6}{74.6}$

$\approx 1.09$

The life expectancy of women is about 9% longer than that of men.

83. $\dfrac{(x-1)^5}{[-1(-1+x)]^3}$

$= \dfrac{(x-1)^5}{(-1)^3(x-1)^3}$

$= \dfrac{(x-1)^2(x-1)^3}{-1(x-1)^3}$

$= -(x-1)^2$

85. $\dfrac{x^2(x+3) - 4(x+3)}{(x+2)(x+3)}$

$= \dfrac{(x+3)(x^2-4)}{(x+2)(x+3)}$

$= \dfrac{(x+3)(x-2)(x+2)}{(x+2)(x+2)}$

$= x - 2$

87. $\dfrac{x+y-2}{[(x-2)-y][(x-2)+y]}$

$= \dfrac{x+y-2}{(x-y-2)(x+y-2)}$

$= \dfrac{1}{x-y-2}$

89. $\dfrac{y}{(x-4)(x+1)} = \dfrac{1}{x+1}$

$y = \dfrac{(x-4)(x+1)}{x+1}$

$y = x - 4$

## Section 8.2 Multiplication and Division

1. Multiplying first may create expressions that are difficult to factor. If we divide out the common factors first, then the remaining factors are easier to multiply.

3. $\dfrac{\cancel{8}}{\cancel{8}\cancel{x^2}} \cdot \dfrac{\cancel{8} \cdot 7 \cdot \cancel{x} \cdot x^2}{\cancel{8} \cdot 4}$

$= \dfrac{7x^2}{4}$

5. $\dfrac{2x+1}{\cancel{x-5}} \cdot \dfrac{-3(\cancel{-5+x})}{2x-1}$

$= \dfrac{-3(2x+1)}{2x-1}$

7. $\cancel{y+3} \cdot \dfrac{y-2}{4(\cancel{y+3})}$

$= \dfrac{y-2}{4}$

9. $\dfrac{4(\cancel{x-3})}{3x^2} \cdot \dfrac{6x}{2(\cancel{x-3})}$

$= \dfrac{24x}{6x^2}$

$= \dfrac{4}{x}$

Rational Expressions

11. $\dfrac{(w-4)(w+6)}{9w^5} \cdot \dfrac{3w}{w-4}$

$= \dfrac{3w(w+6)}{9w^5}$

$= \dfrac{w+6}{3w^4}$

13. $\dfrac{6x\cancel{(x+3)}}{\cancel{(x-3)}\cancel{(x+3)}} \cdot \dfrac{\cancel{(x-3)}(x-3)}{3x}$

$= \dfrac{6x(x-3)}{3x}$

$= 2(x-3)$

15. $\dfrac{\cancel{2}(3x-2)}{x+1} \cdot \dfrac{1+x}{\cancel{2}(2-3x)}$

$= \dfrac{(3x-2)}{-1(-2+3x)}$

$= -1$

17. $\dfrac{\cancel{(y+1)}\cancel{(y-1)}}{8\cancel{(y+1)}} \cdot \dfrac{y\cancel{(y-1)}}{\cancel{(y-1)}\cancel{(y-1)}}$

$= \dfrac{y}{8}$

19. $\dfrac{3x\cancel{(x-4)}}{\cancel{(x-4)}\cancel{(x+3)}} \cdot \dfrac{\cancel{x+3}}{15x}$

$= \dfrac{3x}{15x}$

$= \dfrac{1}{5}$

21. Change the operation to multiplication and take the reciprocal of the divisor.

23. $\dfrac{1}{x}$

25. $a+3$

27. $\dfrac{x-1}{x+2}$

29. $\dfrac{30a^4b^3}{b} \cdot \dfrac{1}{12a^2b^3}$

$= \dfrac{30a^4b^3}{12a^2b^4}$

$= \dfrac{5a^2}{2b}$

31. $\dfrac{6\cancel{(2x+1)}}{x^2} \cdot \dfrac{3x}{9\cancel{(2x+1)}}$

$= \dfrac{18x}{9x^2}$

$= \dfrac{2}{x}$

33. $\dfrac{x-12}{-16} \cdot \dfrac{8}{12-x}$

$= \dfrac{8\cancel{(x-12)}}{16[-1\cancel{(-12+x)}]}$

$= \dfrac{8}{16}$

$= \dfrac{1}{2}$

35. $\dfrac{x\cancel{(1-3x)}}{\cancel{(1-3x)}} \cdot \dfrac{1}{-3x^2}$

$= \dfrac{x}{-3x^2}$

$= \dfrac{-1}{3x}$

37. $\dfrac{3(x-y)}{(2x-5y)\cancel{(2x-5y)}} \cdot \dfrac{7\cancel{(2x-5y)}}{12(y-x)}$

$= \dfrac{21\cancel{(x-y)}}{12(2x-5y)[-1\cancel{(-y+x)}]}$

$= \dfrac{-7}{4(2x-5y)}$

39. $\dfrac{4h}{\cancel{h-4}} \cdot \dfrac{\cancel{(h-4)}(h+1)}{24h^2}$

$= \dfrac{4h(h+1)}{24h^2}$

$= \dfrac{h+1}{6h}$

41. $\dfrac{\cancel{(y-4)}\cancel{(y-4)}}{\cancel{(y-4)}\cancel{(y+1)}} \cdot \dfrac{\cancel{(y+1)}\cancel{(y+4)}}{\cancel{(y+4)}\cancel{(y-4)}}$

$= 1$

43. $\dfrac{3x^2 \cancel{(x+2)}}{(2x-3)\cancel{(x+2)}} \cdot \dfrac{1}{3x(2x-5)}$

$= \dfrac{3x^2}{3x(2x-3)(2x-5)}$

$= \dfrac{x}{(2x-3)(2x-5)}$

45. $\dfrac{4d}{8d(d^2-4)} \cdot (d-3)(d+2)$

$= \dfrac{(d-3)\cancel{(d+2)}}{2\cancel{(d+2)}(d-2)}$

$= \dfrac{d-3}{2(d-2)}$

47. $\dfrac{\cancel{(z-5)}\cancel{(z+2)}}{4z\cancel{(z+2)}} \cdot \dfrac{3(5-z)}{\cancel{(z-5)}(z-5)}$

$= \dfrac{-3\cancel{(z-5)}}{4z\cancel{(z-5)}}$

$= -\dfrac{3}{4z}$

49. $\dfrac{b^3\cancel{(a+b)}}{a^2(2a-b)} \cdot \dfrac{a(b-2a)}{b^4\cancel{(b+a)}}$

$= \dfrac{-ab^3\cancel{(2a-b)}}{a^2 b^4 \cancel{(2a-b)}}$

$= -\dfrac{1}{ab}$

51. $\dfrac{x^2 y^4}{(x-3)\cancel{(x+1)}} \cdot \dfrac{\cancel{(x+1)}(x+2)}{xy^5}$

$= \dfrac{x^2 y^4 (x+2)}{xy^5 (x-3)}$

$= \dfrac{x(x+2)}{y(x-3)}$

53. $\dfrac{(a+1)\cancel{(a+4)}}{4a^3} \cdot \dfrac{a\cancel{(a-4)}}{\cancel{(a+4)}\cancel{(a-4)}}$

$= \dfrac{a(a+1)}{4a^3}$

$= \dfrac{a+1}{4a^2}$

55. $\dfrac{3\cancel{(k-2)}}{\cancel{(k+7)}\cancel{(k-2)}} \cdot \dfrac{\cancel{(k+7)}\cancel{(k-6)}}{9\cancel{(k-6)}}$

$= \dfrac{3}{9}$

$= \dfrac{1}{3}$

57. $\dfrac{\cancel{(x+8)}\cancel{(x-3)}}{\cancel{(x+3)}\cancel{(x-3)}} \cdot \dfrac{(x+6)\cancel{(x+3)}}{(x-8)\cancel{(x+8)}}$

$= \dfrac{x+6}{x-8}$

**Rational Expressions**

59. $\dfrac{(x+4)\cancel{(x-2)}}{(x+8)\cancel{(x-3)}} \cdot \dfrac{\cancel{(x+7)}\cancel{(x-3)}}{\cancel{(x+7)}\cancel{(x-2)}}$

    $= \dfrac{x+4}{x+8}$

61. $\dfrac{x^2}{x+2} \cdot \dfrac{4x+8}{x}$

    $= \dfrac{x^2}{\cancel{x+2}} \cdot \dfrac{4\cancel{(x+2)}}{x}$

    $= \dfrac{4x^2}{x}$

    $= 4x$

63. $x \div \dfrac{5x}{2x+1}$

    $= x \cdot \dfrac{2x+1}{5x}$

    $= \dfrac{2x+1}{5}$

65. (a) No, since $(x-3)^2 \geq 0$ for all values of $x$, $(x-3)^2 + 19 > 0$.

    (b) $N = 10 + \dfrac{590(7-3)}{(7-3)^2 + 19}$

    $= 10 + \dfrac{590(4)}{16+19}$

    $= 10 + \dfrac{2360}{35}$

    $= 77.4285$ thousand

    $= 77,429$

67. $\dfrac{3\cancel{(2x+3)}}{x-1} \cdot \dfrac{y}{x\cancel{(2x+3)}} = \dfrac{-3}{x}$

    $\dfrac{3y}{x(x-1)} = \dfrac{-3}{x}$

    $\dfrac{3y}{x-1} = -3$

    $3y = -3(x-1)$

    $y = -1(x-1)$

    $= -x+1$

    $= 1-x$

69. $\dfrac{x(y+3)-2(y+3)}{(x-2)(x-4)} \div \dfrac{(y+2)(y+3)}{y(x-4)+2(x-4)}$

$= \dfrac{(\cancel{x-2})(\cancel{y+3})}{(\cancel{x-2})(\cancel{x-4})} \cdot \dfrac{(\cancel{x-4})(\cancel{y+2})}{(\cancel{y+2})(\cancel{y+3})}$

$= 1$

71. $\dfrac{(\cancel{x-2})(x^2+2x+4)}{(\cancel{x+2})(\cancel{x-2})} \cdot \dfrac{(\cancel{x+4})(\cancel{x+2})}{(\cancel{x+4})(x-3)} \div \dfrac{x^2+2x+4}{x^2(x-3)}$

$= \dfrac{\cancel{x^2+2x+4}}{\cancel{x-3}} \cdot \dfrac{x^2(\cancel{x-3})}{\cancel{x^2+2x+4}}$

$= x^2$

## Section 8.3 Addition and Subtraction (Like Denominators)

1. In order for us to add fractions, the denominators must be the same.
Reducing $\dfrac{3}{9}$ first would result in different denominators.

3. $\dfrac{18}{4x}$

$= \dfrac{9}{2x}$

5. $\dfrac{5x-5}{x}$

7. $\dfrac{4x-4}{1-x}$

$= \dfrac{-4(\cancel{-x+1})}{\cancel{1-x}}$

$= -4$

9. $\dfrac{2x-3}{2x+3}$

11. $\dfrac{y^2+4y}{y+4}$

$= \dfrac{y(\cancel{y+4})}{(\cancel{y+4})}$

$= y$

13. $\dfrac{4y^2+3y-1}{1-4y}$

$= \dfrac{(4y-1)(y+1)}{1-4y}$

$= \dfrac{(\cancel{4y-1})(y+1)}{-1(\cancel{4y-1})}$

$= -1(y+1)$

$= -y-1$

15. $\dfrac{x^2+7x+10}{x^2-4}$

$= \dfrac{(x+2)(x+5)}{(x+2)(x-2)}$

$= \dfrac{x+5}{x-2}$

# Rational Expressions

17. $\dfrac{4x+12}{(x+2)(x+3)}$

    $= \dfrac{4\cancel{(x+3)}}{x+2\cancel{(x+3)}}$

    $= \dfrac{4}{x+2}$

19. $\dfrac{1}{x}$

21. $\dfrac{t^3+8-8-(-5t^3)}{3t}$

    $= \dfrac{6t^3}{3t}$

    $= 2t^2$

23. $\dfrac{5x+2-(5x-2)}{x}$

    $= \dfrac{4}{x}$

25. $\dfrac{3x-4}{4-3x}$

    $= -1$

27. $\dfrac{x-30-(6-19x)}{5x-9}$

    $= \dfrac{20x-36}{5x-9}$

    $= \dfrac{4\cancel{(5x-9)}}{\cancel{(5x-9)}}$

    $= 4$

29. $\dfrac{8-x-(2+x)}{x-3}$

    $= \dfrac{6-2x}{x-3}$

    $= \dfrac{-2\cancel{(-3+x)}}{\cancel{x-3}}$

    $= -2$

31. $\dfrac{2x-1-(2-x)}{x^2-7x+6}$

    $= \dfrac{3x-3}{(x-6)(x-1)}$

    $= \dfrac{3\cancel{(x-1)}}{(x-6)\cancel{(x-1)}}$

    $= \dfrac{3}{x-6}$

33. $\dfrac{2x^2-3x-(2x+7)}{2x-7}$

    $= \dfrac{2x^2-5x-7}{2x-7}$

    $= \dfrac{\cancel{(2x-7)}(x+1)}{\cancel{(2x-7)}}$

    $= x+1$

35. $\dfrac{3x-1-(2x-1)}{2x^2-5x}$

    $= \dfrac{x}{x(2x-5)}$

    $= \dfrac{1}{2x-5}$

37. Insert parentheses to write the numerator as $x-(2a+3)$.

39. $\dfrac{9x-5}{x}$

41. $\dfrac{2x-7}{5}$

43. $\dfrac{2z+2+5z-5}{3z^2}$

    $= \dfrac{7z-3}{3z^2}$

45. $\dfrac{5x-4y-(8x-9y)}{y}$
$=\dfrac{-3x+5y}{y}$

47. $\dfrac{2x-3y+2y-x}{x-y}$
$=\dfrac{x-y}{x-y}$
$=1$

49. $\dfrac{4x+10-(7x+7)}{x-1}$
$=\dfrac{-3x+3}{x-1}$
$=\dfrac{-3(\cancel{x-1})}{(\cancel{x-1})}$
$=-3$

51. $\dfrac{4t+20}{4(t+5)}$
$=\dfrac{4(t+5)}{4(t+5)}$
$=1$

53. $\dfrac{x+15-(1-x)}{x+7}$
$=\dfrac{2x+14}{x+7}$
$=\dfrac{2(\cancel{x+7})}{(\cancel{x+7})}$
$=2$

55. $\dfrac{3x-6-(x+2)}{(x+2)(x-4)}$
$=\dfrac{2x-8}{(x+2)(x-4)}$
$=\dfrac{2(\cancel{x-4})}{(x+2)(\cancel{x-4})}$
$=\dfrac{2}{x+2}$

57. $\dfrac{5x+4-(x-x^2)}{x^2-4}$
$=\dfrac{x^2+4x+4}{x^2-4}$
$=\dfrac{(\cancel{x+2})(x+2)}{(\cancel{x+2})(x-2)}$
$=\dfrac{x+2}{x-2}$

59. $\dfrac{2x^2-5x}{2x^2-x-10}$
$=\dfrac{x(\cancel{2x-5})}{(\cancel{2x-5})(x+2)}$
$=\dfrac{x}{x+2}$

61. $2y-5+y=3x+3$
$y=x+8$

63. $y-(2x+1)=x^2-1$
$y=x^2-1+2x+1$
$y=x^2+2x$

## Rational Expressions

65. $\dfrac{7x+16}{x+3} - \dfrac{x-2}{x+3}$

   $= \dfrac{6x+18}{x+3}$

   $= \dfrac{6\cancel{(x+3)}}{\cancel{x+3}}$

   $= 6$

67. $\dfrac{2z+2-(4z-3)+5z-6}{3z}$

   $= \dfrac{3z-1}{3z}$

69. (a) $C_2$

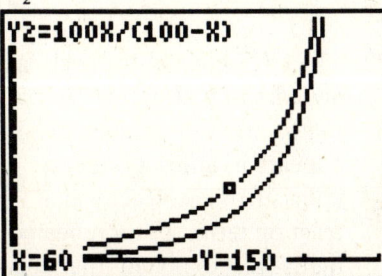

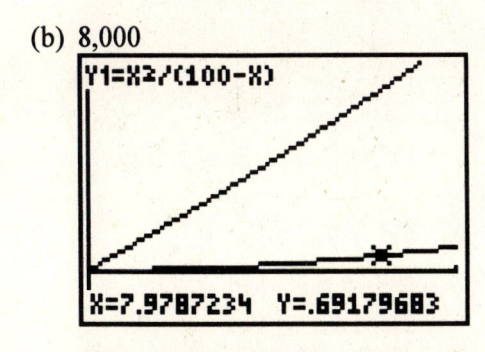

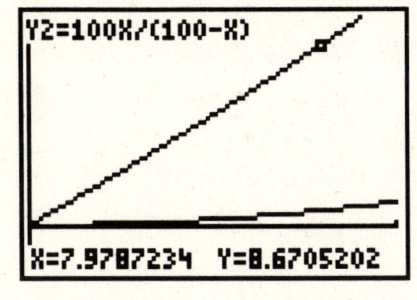

   Y2=100X/(100-X)
   X=60    Y=150

   (b) 8,000

   Y1=X2/(100-X)
   X=7.9787234  Y=.69179683

   Y2=100X/(100-X)
   X=7.9787234  Y=8.6705202

   (c) $\dfrac{100x}{100-x} - \dfrac{x^2}{100-x}$

   $= \dfrac{100x - x^2}{100-x}$

   $= \dfrac{x(100-x)}{100-x}$

   $= x$

   At $x = 8$, the function value is 8. This corresponds with 8000 votes.

71. $\dfrac{t^2+t-12}{t+3} - \dfrac{2t}{t+3}$

   $= \dfrac{t^2-t-12}{t+3}$

   $= \dfrac{(t-4)\cancel{(t+3)}}{\cancel{(t+3)}}$

   $= t - 4$

73. (a) 1 since men and women make up 100 % of the workforce

   (b) $\dfrac{0.77x + 52.42}{1.90x + 85.29} + \dfrac{1.13x + 32.87}{1.90x + 85.29}$

   $= \dfrac{1.90x + 85.29}{1.90x + 85.29}$

   $= 1$

75. $\dfrac{x^2 - 1 - (2x-5)}{x^3 + 8}$

   $= \dfrac{\cancel{x^2 - 2x + 4}}{(x+2)\cancel{(x^2 - 2x + 4)}}$

   $= \dfrac{1}{x+2}$

77. $\dfrac{x-3}{x-5} + \dfrac{x}{x+1} \cdot \dfrac{x+1}{x-5}$

$= \dfrac{x-3}{x-5} + \dfrac{x}{x-5}$

$= \dfrac{2x-3}{x-5}$

79. $\dfrac{3x-1}{x^2-x} \cdot \dfrac{x^2-1}{6x-2}$

$= \dfrac{\cancel{3x-1}}{x\cancel{(x-1)}} \cdot \dfrac{\cancel{(x-1)}(x+1)}{2\cancel{(3x-1)}}$

$= \dfrac{x+1}{2x}$

## Section 8.4 Least Common Denominators

1. When both denominators are in factored form, it is easier to determine the factors that are in the new denominator.

3. $\dfrac{3y}{2x} \cdot \dfrac{3y}{3y} = \dfrac{9y^2}{6xy}$

5. $\dfrac{-5}{12y^2} \cdot \dfrac{5y^3}{5y^3} = \dfrac{-25y^3}{60y^5}$

7. $\dfrac{4}{3x} \cdot \dfrac{3x+2}{(3x+2)} = \dfrac{12x^2+8}{9x^2+6x}$

9. $\dfrac{2x-1}{x+1} \cdot \dfrac{3x}{3x} = \dfrac{6x^2-3x}{3x^2+3x}$

11. $\dfrac{y}{y-3} \cdot \dfrac{-1}{-1} = \dfrac{-y}{-y+3} = \dfrac{-y}{3-y}$

13. $\dfrac{-6}{x-3} \cdot \dfrac{x-3}{x-3} = \dfrac{-6x+18}{(x-3)^2}$

15. $\dfrac{x-5}{4x+12} \cdot \dfrac{2x}{2x} = \dfrac{2x^2-10x}{8x^2+24x}$

17. $\dfrac{-3}{x+5} \cdot \dfrac{x-1}{x-1} = \dfrac{-3x+3^2}{x^2+4x-5}$

19. $\dfrac{y-6}{y-4} \cdot \dfrac{y+6}{y+6} = \dfrac{y^2-36}{y^2+2y-24}$

21. $\dfrac{2x}{6x-15} \cdot \dfrac{(2x+5)}{(2x+5)} = \dfrac{4x^2+10x}{12x^2-75}$

23. The GCF consists of each factor that appears in both terms. For a repeated factor, use the smallest exponent that appears in at least on term. The LCM consists of each factor that appears in at least on term. For a repeated factor, use the largest exponent that appears on the factor.

25. $24x : 2 \cdot 2 \cdot 2 \cdot 3 \cdot x$
$8y : 2 \cdot 2 \cdot 2 \cdot y$
LCM: $2 \cdot 2 \cdot 2 \cdot 3 \cdot x \cdot y = 24xy$

27. $5x^3 = 5 \cdot x \cdot x \cdot x$
$10x^2 = 2 \cdot 5 \cdot x \cdot x$
LCM: $2 \cdot 5 \cdot x \cdot x \cdot x = 10x^3$

29. LCM: $b(b+2)$

31. $8x : 2^3 x$
$4x - 8 : 4(x-8)$
LCM: $8x(x-8)$

33. $2x - 7$ since $-1(2x-7) = 7 - 2x$

35. $x^2 - 49 : (x-7)(x+7)$
LCM: $(x-7)(x+7)$

37. $x^2 - 64 : (x-8)(x+8)$
    $3x - 24 : 3(x-8)$
    LCM $: 3(x-8)(x+8)$

39. $x^2 - 6x + 5 : (x-1)(x-5)$
    $x^2 + 4x - 5 : (x+5)(x-1)$
    LCM $: (x-1)(x+5)(x-5)$

41. $35 : 7 \cdot 5$
    $42y : 7 \cdot 6y$
    LCD $: 7 \cdot 6y \cdot 5 = 210y$

43. $2x^4 y^3$

45. $(x-7)(x-8)$

47. $5x^2 : 5 \cdot x^2$
    $x^2 - 2x : x(x-2)$
    LCD $: 5x^2(x-2)$

49. $4x^2 - 1 : (2x-1)(2x+1)$
    $1 - 2x : -1(2x-1)$
    LCD $: (2x-1)(2x+1)$

51. $x^2 - 5x - 6 : (x-6)(x+1)$
    LCD $: (x-6)(x+1)$

53. $x^2 - 1 : (x-1)(x+1)$
    $x^2 + 2x - 3$
    LCD $: (x-1)(x+1)(x+3)$

55. $x^2 + 6x + 9 : (x+3)^2$
    $x^2 + 2x - 3 : (x+3)(x-1)$
    LCD $: (x+3)^2 (x-1)$

57. LCD $: 3ab^2$
    $\dfrac{a^2}{3ab^2}, \dfrac{3b^3}{3ab^2}$

59. LCD: $x(x+4)$

$$\frac{2(x+4)}{x(x+4)} = \frac{2x+8}{x(x+4)}, \frac{x^2}{x(x+4)}$$

61. LCD: $x+3$

$$\frac{(2x+5)(x+3)}{x+3} = \frac{2x^2+11+15}{x+3}, \frac{x}{x+3}$$

63. LCD: $(x+1)(x-2)$

$$\frac{x(x+1)}{(x-2)(x+1)} = \frac{x^2+x}{(x-2)(x+1)}, \frac{(x+2)(x-2)}{(x+1)(x-2)} = \frac{x^2-4}{(x+1)(x-2)}$$

65. LCD: $2(x+3) \cdot 5$

$$\frac{x}{2(x+3)} \cdot \frac{5}{5} = \frac{5x}{10(x+3)}, \frac{x+1}{5(x+3)} \cdot \frac{2}{2} = \frac{2x+2}{10(x+3)}$$

67. LCD: $x(x+3)^2$

$$\frac{3x}{x(x+3)^2}, \frac{2}{x(x+3)} \cdot \frac{x+3}{x+3} = \frac{2x+6}{x(x+3)^2}$$

69. LCD: $(2x-5)(x-2)$

$$\frac{x+3}{(2x-5)(x-2)}, \frac{3}{x-2} \cdot \frac{2x-5}{2x-5} = \frac{6x-15}{(x-2)(2x-5)}$$

71. LCD: $(2x+1)^2(x+1)$

$$\frac{x}{(2x+1)^2} \cdot \frac{x+1}{x+1} = \frac{x^2+x}{(2x+1)^2(x+1)}, \frac{x+2}{(2x+1)(x+1)} \cdot \frac{(2x+1)}{(2x+1)} = \frac{2x^2+5x+2}{(2x+1)^2(x+1)}$$

73. The LCM of 3 and 4 is 12 so every 12 seconds the handlers will discharge luggage simultaneously. Two minutes is 120 seconds so the handlers will discharge simultaneously 10 times.

75. The LCM is $4 \cdot 7 = 28$. The cars will pass each other simultaneously at 28 minutes.

77. $14 : 2 \cdot 7$
    $33 : 3 \cdot 11$
    $78 : 2 \cdot 3 \cdot 13$
    LCM: $2 \cdot 3 \cdot 7 \cdot 11 \cdot 13 = 6006$

**Rational Expressions**

79. $x^3 + 1 : (x+1)(x^2 - x + 1)$

$x^2 + 2x + 1 : (x+1)^2$

$x^3 - x^2 + x : x(x^2 - x + 1)$

LCD $: x(x+1)^2(x^2 - x + 1)$

81. $a^4 - b^4$
$= (a^2 + b^2)(a^2 - b^2)$
$= (a^2 + b^2)(a+b)(a-b)$

$a^3 b + ab^3 = ab(a^2 + b^2)$

LCD $: ab(a^2 + b^2)(a+b)(a-b) = (a^4 - b^4)ab$

$\dfrac{2}{a^4 - b^4} = \dfrac{2ab}{(a^4 - b^4)ab}, \dfrac{1(a^2 - b^2)}{ab(a^2 + b^2)(a^2 - b^2)}$

## Section 8.5 Addition and Subtraction (Unlike Denominators)

1. The denominators have no common factors.

3. $\dfrac{8x - 6}{8 - 3x} = \dfrac{y}{3x - 8} \cdot \dfrac{(-1)}{(-1)} = \dfrac{-y}{8 - 3x}$

   So $y = -(8x - 6) = -8x + 6$

   $\dfrac{5x - 1}{3x - 8} + \dfrac{-8x + 6}{3x - 8} = \dfrac{-3x + 5}{3x - 8}$

5. $\dfrac{x}{8} - \dfrac{9}{8} = \dfrac{x - 9}{8}$

7. $\dfrac{-9}{x} + \dfrac{5}{x} = \dfrac{-4}{x}$

9. $\dfrac{x^2}{x - 9} + \dfrac{-81}{x - 9}$
   $= \dfrac{x^2 - 81}{x - 9}$
   $= \dfrac{(x+9)(x-9)}{x-9}$
   $= x + 9$

11. $\dfrac{x - 6}{x^2 - 36} - \dfrac{6 - x}{x^2 - 36}$
    $= \dfrac{2x - 12}{x^2 - 36}$
    $= \dfrac{2(x - 6)}{(x+6)(x-6)}$
    $= \dfrac{2}{x + 6}$

13. $\dfrac{1}{x - 1} + \dfrac{1}{x}$
    $\dfrac{x}{x} \cdot \dfrac{1}{x - 1} + \dfrac{1}{x} \cdot \dfrac{x - 1}{x - 1}$
    $= \dfrac{x}{x(x-1)} + \dfrac{(x-1)}{x(x-1)}$
    $= \dfrac{2x - 1}{x(x-1)}$

15. $\dfrac{x}{1} + \dfrac{5}{x}$
    $= \dfrac{x \cdot x + 1 \cdot 5}{1 \cdot x}$
    $= \dfrac{x^2 + 5}{x}$

17. $\dfrac{8}{1} - \dfrac{7}{x}$

$= \dfrac{8 \cdot x - 1 \cdot 7}{1 \cdot x}$

$= \dfrac{8x - 7}{x}$

19. $\dfrac{(2x+3) \cdot 7 - 4(x-5)}{4 \cdot 7}$

$= \dfrac{14x + 21 - 4x + 20}{28}$

$= \dfrac{10x + 41}{28}$

21. $\dfrac{2}{1} + \dfrac{8}{x-5}$

$= \dfrac{2(x-5) + 1 \cdot 8}{1 \cdot (x-5)}$

$= \dfrac{2x - 10 + 8}{x - 5}$

$= \dfrac{2x - 2}{x - 5}$

23. $\dfrac{3(x+7) + 4x}{x(x+7)}$

$= \dfrac{3x + 21 + 4x}{x(x+7)}$

$= \dfrac{7x + 21}{x(x+7)}$

25. $\dfrac{x \cdot x - (x-2)(x-2)}{x(x-2)}$

$= \dfrac{x^2 - (x^2 - 4x + 4)}{x(x-2)}$

$= \dfrac{4x - 4}{x(x-2)}$

27. $\dfrac{8(x-3) + 2(x+5)}{(x+5)(x-3)}$

$= \dfrac{8x - 24 + 2x + 10}{(x+5)(x-3)}$

$= \dfrac{10x - 14}{(x+5)(x-3)}$

29. $\dfrac{x(x+4) - 3(3x-2)}{(3x-2)(x+4)}$

$= \dfrac{x^2 + 4x - 9x + 6}{(3x-2)(x+4)}$

$= \dfrac{x^2 - 5x + 6}{(3x-2)(x+4)}$

31. $\dfrac{(x+1)(x-4) - (x+5)(x-2)}{(x+5)(x-4)}$

$= \dfrac{x^2 - 4x + x - 4 - (x^2 - 2x + 5x - 10)}{(x+5)(x-4)}$

$= \dfrac{x^2 - 3x - 4 - x^2 - 3x + 10}{(x+5)(x-4)}$

$= \dfrac{-6x + 6}{(x+5)(x-4)}$

33. $\dfrac{(y+3)(3+y) + (y-3)(3-y)}{(3-y)(3+y)}$

$= \dfrac{3y + y^2 + 9 + 3y + 3y - y^2 - 9 + 3y}{(3-y)(3+y)}$

$= \dfrac{12y}{(3-y)(3+y)}$

35. The factions have the same denominator. The next step is to add the numerators.

37. $\dfrac{8}{x} \cdot \dfrac{x}{x} + \dfrac{2}{x^2}$

$= \dfrac{8x}{x^2} + \dfrac{2}{x^2}$

$= \dfrac{8x + 2}{x^2}$

Rational Expressions

39. $\dfrac{5}{5} \cdot \dfrac{4}{x+2} - \dfrac{3}{5x+10}$

$= \dfrac{20-3}{5x+10}$

$= \dfrac{17}{5(x+2)}$

41. $\dfrac{x-5}{x-5} \cdot \dfrac{x}{x+5} - \dfrac{50}{(x-5)(x+5)}$

$= \dfrac{x^2 - 5x - 50}{(x-5)(x+5)}$

$= \dfrac{(x-10)\cancel{(x+5)}}{(x-5)\cancel{(x+5)}}$

$= \dfrac{x-10}{x-5}$

43. $\dfrac{x+9}{x+9} \cdot \dfrac{5}{x+9} - \dfrac{x}{(x+9)^2}$

$= \dfrac{5x + 45 - x}{(x+9)^2}$

$= \dfrac{4x + 45}{(x+9)^2}$

45. $\dfrac{3}{x+2} + \dfrac{5}{(x+2)(x+5)}$

$= \dfrac{x+5}{x+5} \cdot \dfrac{3}{x+2} + \dfrac{5}{(x+2)(x+5)}$

$= \dfrac{3x + 15 + 5}{(x+5)(x+2)}$

$= \dfrac{3x + 20}{(x+5)(x+2)}$

47. $\dfrac{4}{x+3} - \dfrac{2}{(x+3)(x+1)}$

$= \dfrac{4(x+1) - 2}{(x+3)(x+1)}$

$= \dfrac{4x + 4 - 2}{(x+3)(x+1)}$

$= \dfrac{4x + 2}{(x+3)(x+1)}$

49. $\dfrac{x-9}{x-9} \cdot \dfrac{2}{x} - \dfrac{3}{x(x-9)}$

$= \dfrac{2x - 18 - 3}{x(x-9)}$

$= \dfrac{2x - 21}{x(x-9)}$

51. $\dfrac{2}{2} \cdot \dfrac{4}{5x} - \dfrac{3}{2x} \cdot \dfrac{5}{5}$

$= \dfrac{8 - 15}{10x}$

$= \dfrac{-7}{10x}$

53. $\dfrac{2}{2} \cdot \dfrac{x+2}{15} + \dfrac{3x-5}{6} \cdot \dfrac{5}{5}$

$= \dfrac{2x + 4 + 15x - 25}{30}$

$= \dfrac{17x - 21}{30}$

55. $\dfrac{x}{x} \cdot \dfrac{x}{2(x+2)} - \dfrac{2}{x(x+2)} \cdot \dfrac{2}{2}$

$= \dfrac{x^2 - 4}{2x(x+2)}$

$= \dfrac{(x-2)\cancel{(x+2)}}{2x\cancel{(x+2)}}$

$= \dfrac{x-2}{2x}$

57. $\dfrac{x}{6(x+3)} + \dfrac{5}{4(x+3)}$

$= \dfrac{2}{2} \cdot \dfrac{x}{6(x+3)} + \dfrac{5}{4(x+3)} \cdot \dfrac{3}{3}$

$= \dfrac{2x+15}{12(x+3)}$

59. $\dfrac{4}{y(2y+3)} - \dfrac{2y-5}{3(2y+3)}$

$= \dfrac{3}{3} \cdot \dfrac{4}{y(2y+3)} - \dfrac{2y-5}{3(2y+3)} \cdot \dfrac{y}{y}$

$= \dfrac{12 - 2y^2 + 5y}{3y(2y+3)}$

$= \dfrac{(4-y)\cancel{(2y+3)}}{3y\cancel{(2y+3)}}$

$= \dfrac{4-y}{3y}$

61. $\dfrac{x}{x(x-5)} - \dfrac{1}{x+7}$

$= \dfrac{(x+7)x - 1 \cdot (x)(x-5)}{x(x-5)(x+7)}$

$= \dfrac{x^2 + 7x - x^2 + 5x}{x(x-5)(x+7)}$

$= \dfrac{12x}{x(x-5)(x+7)}$

$= \dfrac{12}{(x-5)(x+7)}$

63. $\dfrac{3}{(x+6)(x+3)} + \dfrac{1}{(x-6)(x+6)}$

$= \dfrac{3(x-6) + 1 \cdot (x+3)}{(x+6)(x+3)(x-6)}$

$= \dfrac{3x - 18 + x + 3}{(x+6)(x+3)(x-6)}$

$= \dfrac{4x - 15}{(x+6)(x+3)(x-6)}$

65. $\dfrac{x}{(x+5)(x+4)} - \dfrac{15}{(x+8)(x+5)}$

$= \dfrac{x(x+8) - 15(x+4)}{(x+4)(x+5)(x+8)}$

$= \dfrac{x^2 + 8x - 15x - 60}{(x+4)(x+5)(x+8)}$

$= \dfrac{x^2 - 7x - 60}{(x+4)(x+5)(x+8)}$

$= \dfrac{(x-12)\cancel{(x+5)}}{(x+4)\cancel{(x+5)}(x+8)}$

$= \dfrac{x-12}{(x+4)(x+8)}$

67. $\dfrac{x}{(x+3)(x+5)} + \dfrac{15}{(x+5)(x-1)}$

$= \dfrac{x(x-1) + 15(x+3)}{(x+3)(x+5)(x-1)}$

$= \dfrac{x^2 - x + 15x + 45}{(x+3)(x+5)(x-1)}$

$= \dfrac{x^2 + 14x + 45}{(x+3)(x+5)(x-1)}$

$= \dfrac{(x+9)\cancel{(x+5)}}{(x+3)\cancel{(x+5)}(x-1)}$

$= \dfrac{x+9}{(x+3)(x-1)}$

Rational Expressions

69. $\dfrac{x(x+2)}{(x+3)(x-1)} - \dfrac{4x-1}{4(x-1)}$

$= \dfrac{x(x+2) \cdot 4 - (x+3)(4x-1)}{4(x-1)(x+3)}$

$= \dfrac{4x^2 + 8x - (4x^2 + 11x - 3)}{4(x-1)(x+3)}$

$= \dfrac{-3x+3}{4(x-1)(x+3)}$

$= \dfrac{-3\cancel{(x-1)}}{4\cancel{(x-1)}(x+3)}$

$= \dfrac{-3}{4(x+3)}$

71. $\dfrac{2(x+3)}{x(x+2)} - \dfrac{x+5}{(x+2)(x-1)}$

$= \dfrac{2(x+3)(x-1) - (x+5)x}{x(x+2)(x-1)}$

$= \dfrac{2(x^2+2x-3) - x^2 - 5x}{x(x+2)(x-1)}$

$= \dfrac{2x^2 + 4x - 6 - x^2 - 5x}{x(x+2)(x-1)}$

$= \dfrac{x^2 - x - 6}{x(x+2)(x-1)}$

$= \dfrac{(x-3)\cancel{(x+2)}}{x\cancel{(x+2)}(x-1)}$

$= \dfrac{x-3}{x(x-1)}$

73. $\dfrac{2x}{x+1} - \dfrac{x}{x+2}$

$= \dfrac{2x(x+2) - x(x+1)}{(x+1)(x+2)}$

$= \dfrac{2x^2 + 4x - x^2 - x}{(x+1)(x+2)}$

$= \dfrac{x^2 + 3x}{(x+1)(x+2)}$

75. $\dfrac{20x}{x+1} + \dfrac{5x}{x-1}$

$= \dfrac{20x(x-1) + 5x(x+1)}{(x+1)(x-1)}$

$= \dfrac{20x^2 - 20x + 5x^2 + 5x}{(x+1)(x-1)}$

$= \dfrac{25x^2 - 15x}{(x+1)(x-1)}$

77. (a) $g(x) = \dfrac{34(x+3)}{x+3} - \dfrac{179}{x+3}$

$= \dfrac{34x + 102 - 179}{x+3}$

$= \dfrac{34x - 77}{x+3}$

(b) $f(15) = 34 - \dfrac{179}{15+3}$

$= 24.06$

$g(15) = \dfrac{34(15) - 77}{15+3}$

$= 24.06$

In 2001, the SUV market share is predicated to be 24.1%.

79. $\dfrac{3x+6(x+4)+12}{x(x+4)}$

$= \dfrac{3x+6x+24+12}{x(x+4)}$

$= \dfrac{9x+36}{x(x+4)}$

$= \dfrac{9\cancel{(x+4)}}{x\cancel{(x+4)}}$

$= \dfrac{9}{x}$

81. $\dfrac{4}{(x-3)(x+1)} + \dfrac{2}{(x+1)(x+2)} - \dfrac{5}{(x-3)(x+2)}$

$\dfrac{4(x+2)+2(x-3)-5(x+1)}{(x-3)(x+1)(x+2)}$

$= \dfrac{4x+8+2x-6-5x-5}{(x-3)(x+1)(x+2)}$

$= \dfrac{x-3}{(x-3)(x+1)(x+2)}$

$= \dfrac{1}{(x+1)(x+2)}$

83. $\dfrac{a+3}{a-2}\left[\dfrac{6a}{(a-2)(a+2)} + \dfrac{(a-1)(a-2)}{(a-2)(a+2)}\right]$

$= \dfrac{a+3}{a-2} - \left(\dfrac{6a+a^2-3a+2}{(a-2)(a+2)}\right)$

$= \dfrac{a+3}{a-2} - \left(\dfrac{a^2+3a+2}{(a-2)(a+2)}\right)$

$= \dfrac{a+3}{a-2} - \left(\dfrac{\cancel{(a+2)}(a+1)}{\cancel{(a+2)}(a-2)}\right)$

$= \dfrac{a+3}{a-2} - \dfrac{a+1}{a-2}$

$= \dfrac{a+3-(a+1)}{a-2}$

$= \dfrac{2}{a-2}$

## Section 8.6 Complex Fractions

1. Simplifying the complex fraction can be accomplished in just one step by multiplying the numerator and the denominator by the LCD $x$.

3. $\dfrac{\dfrac{3}{a}}{\dfrac{2}{b}}$

$= \dfrac{3}{a} \cdot \dfrac{b}{2}$

$= \dfrac{3b}{2a}$

5. $\dfrac{\dfrac{3}{t^4}}{\dfrac{18}{t}}$

$= \dfrac{3}{t^4} \cdot \dfrac{t}{18}$

$= \dfrac{3t}{18t^4}$

$= \dfrac{1}{6t^3}$

Rational Expressions

7. $\dfrac{a}{a-3} \div \dfrac{a^2}{3}$

$= \dfrac{a}{a-3} \cdot \dfrac{3}{a^2}$

$= \dfrac{3}{a(a-3)}$

9. $\dfrac{y+5}{y-7} \div \dfrac{y-5}{7-y}$

$= \dfrac{y+5}{y-7} \cdot \dfrac{7-y}{y-5}$

$= \dfrac{(y+5)(7-y)}{(y-5)(y-7)}$

$= -\dfrac{y+5}{y-5}$

11. $\dfrac{x^2}{x^2-9} \div \dfrac{2x}{x+3}$

$= \dfrac{x^2}{(x+3)(x-3)} \cdot \dfrac{(x+3)}{2x}$

$= \dfrac{x}{2(x-3)}$

13. $\dfrac{3x+6}{1} \div \dfrac{x+2}{x^2+9}$

$= \dfrac{3(x+2)}{1} \cdot \dfrac{x^2+9}{(x+2)}$

$= 3(x^2+9)$

15. $\dfrac{9x^2}{6(x+1)} \div \dfrac{63x^3}{(x-1)(x+1)}$

$= \dfrac{\cancel{9x^2}}{6\cancel{(x+1)}} \cdot \dfrac{(x-1)\cancel{(x+1)}}{\cancel{9x^2} \cdot 7x}$

$= \dfrac{x-1}{42x}$

17. $\dfrac{4\left(1+\dfrac{1}{2}\right)}{4\left(2-\dfrac{3}{4}\right)}$

$= \dfrac{4+2}{8-3}$

$= \dfrac{6}{5}$

19. $\dfrac{12\left(\dfrac{3}{4}-\dfrac{1}{3}\right)}{12\left(\dfrac{5}{6}+\dfrac{1}{2}\right)}$

$= \dfrac{9-4}{10+6}$

$= \dfrac{5}{16}$

21. $\dfrac{9x\left(\dfrac{3}{x}-\dfrac{5}{9}\right)}{9x\left(\dfrac{4}{3}-\dfrac{4}{x}\right)}$

$= \dfrac{27-5x}{12x-36}$

23. $\dfrac{x^2\left(4-\dfrac{1}{x^2}\right)}{x^2\left(\dfrac{2}{x}-\dfrac{1}{x^2}\right)}$

$= \dfrac{4x^2-1}{2x-1}$

$= \dfrac{\cancel{(2x-1)}(2x+1)}{\cancel{(2x-1)}}$

$= 2x+1$

25.
$$\frac{ab\left(\dfrac{a}{b}\right)}{ab\left(\dfrac{2}{a}+b\right)}$$
$$=\frac{a^2}{2b+ab^2}$$

27.
$$\frac{x\left(\dfrac{1}{x}-4\right)}{x\left(\dfrac{2}{x}+3\right)}$$
$$=\frac{1-4x}{2+3x}$$

29.
$$\frac{x(1)}{x\left(\dfrac{2}{x}+y\right)}$$
$$=\frac{x}{2+xy}$$

31.
$$\frac{xy\left(\dfrac{1}{x}+\dfrac{1}{y}\right)}{xy\left(\dfrac{1}{x}+\dfrac{1}{xy}\right)}$$
$$=\frac{y+x}{y+1}$$

33.
$$\frac{x^2\left(4-\dfrac{5}{x}\right)}{x^2\left(\dfrac{4}{x}-\dfrac{5}{x^2}\right)}$$
$$=\frac{4x^2-5x}{4x-5}$$
$$=\frac{x\cancel{(4x-5)}}{\cancel{(4x-5)}}$$
$$=x$$

35.
$$\frac{xy\left(\dfrac{1}{xy}\right)}{xy\left(\dfrac{1}{x}+\dfrac{1}{y}\right)}$$
$$=\frac{1}{y+x}$$

37.
$$\frac{ab(a+b)}{ab\left(\dfrac{1}{a}+\dfrac{1}{b}\right)}$$
$$=\frac{a^2b+ab^2}{b+a}$$
$$=\frac{ab\cancel{(a+b)}}{\cancel{a+b}}$$
$$=ab$$

39.
$$\frac{6\left(\dfrac{a}{3}+\dfrac{b}{2}\right)}{6\left(\dfrac{a}{3}-\dfrac{b}{2}\right)}$$
$$=\frac{2a+3b}{2a-3b}$$

41.
$$\frac{(x+4)\left(1+\dfrac{2}{x+4}\right)}{(x+4)\left(\dfrac{3}{x+4}\right)}$$
$$=\frac{x+4+2}{3}$$
$$=\frac{x+6}{3}$$

**Rational Expressions**

43.
$$\frac{(a-2)\left(2+\frac{7}{a-2}\right)}{(a-2)\left(2-\frac{1}{a-2}\right)}$$
$$=\frac{2(a-2)+7}{2(a-2)-1}$$
$$=\frac{2a-4+7}{2a-4-1}$$
$$=\frac{2a+3}{2a-5}$$

45.
$$\frac{(y+2)\left(\frac{y^2}{y+2}-y\right)}{(y+2)\left(\frac{y}{y+2}\right)}$$
$$=\frac{y^2-y(y+2)}{y}$$
$$=\frac{y^2-y^2-2y}{y}$$
$$=\frac{-2y}{y}$$
$$=-2$$

47. (a)
$$\frac{t\left(1+\frac{20}{t}\right)}{t\left(\frac{1}{t}+1\right)}$$
$$=\frac{t+20}{1+t}$$

(b)
$$\frac{0+20}{1+0}=20 \text{ owls}$$

It is not valid because the original expression is not defined if $t$ is 0.

49.
$$\frac{\frac{3t}{2}}{\frac{5t}{3}+\frac{5}{2}}$$
$$=\frac{6\left(\frac{3t}{2}\right)}{6\left(\frac{5t}{3}+\frac{5}{2}\right)}$$
$$=\frac{9t}{10t+15}$$

51. $r(x)=\dfrac{1167x-4580}{1.8x+77.2}$

53.

| $x$ | $r(x)$ | actual |
|---|---|---|
| 7 | 40.0 | 41.0 |
| 9 | 63.4 | 60.4 |
| 10 | 74.5 | 76.7 |

The estimate is most accurate for 1997.

55.
$$\frac{\frac{1}{x}+1}{\frac{1}{x^2}-1}$$
$$=\frac{x^2\left(\frac{1}{x}+1\right)}{x^2\left(\frac{1}{x^2}-1\right)}$$
$$=\frac{x+x^2}{1-x^2}$$
$$=\frac{x(1+x)}{(1-x)(1+x)}$$
$$=\frac{x}{1-x}$$

**57.**

$$\frac{x(x+1)\left(\dfrac{2}{x}+\dfrac{3}{x+1}\right)}{x(x+1)\left(\dfrac{1}{x+1}-\dfrac{2}{x}\right)}$$

$$= \frac{2(x+1)+3x}{x-2(x+1)}$$

$$= \frac{2x+2+3x}{x-2x-2}$$

$$= \frac{5x+2}{-x-2}$$

**59.**

$$\frac{16c\left(\dfrac{c}{16}-\dfrac{1}{c}\right)}{16c\left(\dfrac{1}{2}+\dfrac{c+6}{c}\right)}$$

$$= \frac{c^2-16}{8c+16(c+6)}$$

$$= \frac{(c+4)(c-4)}{8[c+2(c+6)]}$$

$$= \frac{(c+4)(c-4)}{8(c+2c+12)}$$

$$= \frac{(c+4)(c-4)}{8(3c+12)}$$

$$= \frac{(c+4)(c-4)}{24(c+4)}$$

$$= \frac{c-4}{24}$$

**61.**

$$f\left(\frac{2}{t}\right) = \frac{\dfrac{2}{t}}{1+\dfrac{2}{t}}$$

$$= \frac{t\left(\dfrac{2}{t}\right)}{t\left(1+\dfrac{2}{t}\right)}$$

$$= \frac{2}{t+2}$$

## Section 8.7 Equations with Rational Expressions

1. One of the apparent solutions may be a restricted value.

   (a) No

   (b) No

   (c) Yes

3. $7 = \dfrac{7}{x}$

   $7x = 7$

   $x = 1$

Rational Expressions

5. $3(2x-5) = 2(x-3)$
$6x - 15 = 2x - 6$
$4x - 9 = 0$
$4x = 9$
$x = \dfrac{9}{4}$

(a) No

(b) Yes

(c) No

7. 8

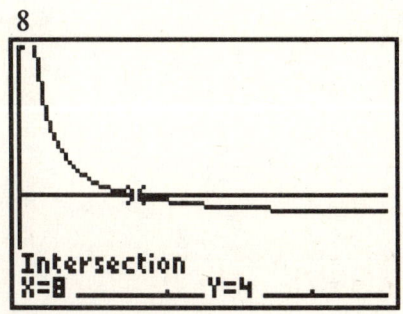

9. $-3$

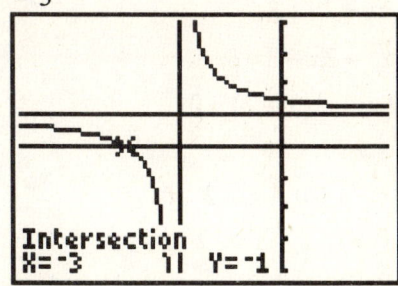

11. 2, 10

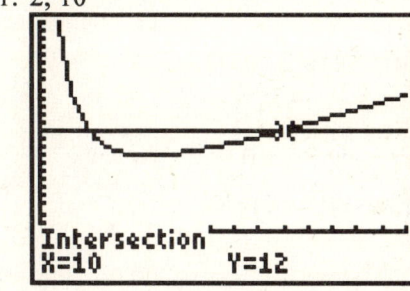

13. $-4, 6$

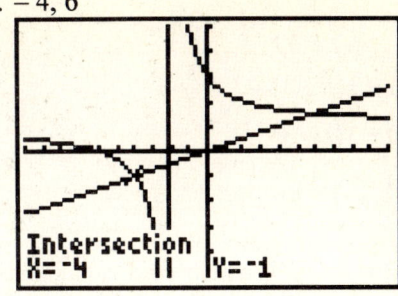

15. $x = 0, x = 2$

17. $x = 0, -1$

19. $\dfrac{2x+1}{(x-6)(x+1)} = 1 + \dfrac{1}{x}$
$x = 6, x = -1, x = 0$

21. $\dfrac{t}{3} = -1$
$t = -3$

23. $8x\left(\dfrac{7}{x}\right) = 8x\left(\dfrac{4}{x} + \dfrac{15}{8}\right)$
$56 = 32 + 15x$
$24 = 15x$
$x = \dfrac{24}{15}$
$x = \dfrac{8}{5}$

25. $4(x-4) = 3x$
$4x - 16 = 3x$
$x - 16 = 0$
$x = 16$

27. $t+3\left(\dfrac{3}{t+3}\right)=(t+3)\left(3+\dfrac{3}{t+3}\right)$

$3=3(t+3)+3$

$0=3(t+3)$

$0=t+3$

$t=-3$

But $t=-3$ is a restricted value, so there is no solution.

29. $-2(1-2x)=3(3x+2)$

$-2+4x=9x+6$

$-2=5x+6$

$-8=5x$

$x=-\dfrac{8}{5}$

31. $\dfrac{x-1+x}{x(x-1)}=\dfrac{2}{x+2}$

$\dfrac{2x-1}{x(x-1)}=\dfrac{2}{x+2}$

$(2x-1)(x+2)=2x(x-1)$

$2x^2+3x-2=2x^2-2x$

$3x-2=-2x$

$-2=-5x$

$x=\dfrac{2}{5}$

33. $\dfrac{4(t-2)-3t}{t(t-2)}=\dfrac{9}{t(2-t)}$

$\dfrac{4t-8-3t}{t(t-2)}=\dfrac{-9}{t(t-2)}$

$t-8=-9$

$t=-1$

35. $x\left(14x+\dfrac{6}{x}\right)=25x$

$14x^2+6=25x$

$14x^2-25x+6=0$

$(2x-3)(7x-2)=0$

$2x-3=0 \quad 7x-2=0$

$2x=3 \quad\quad 7x=2$

$x=\dfrac{3}{2} \quad\quad x=\dfrac{2}{7}$

37. $\dfrac{x}{x+2}=\dfrac{4(x+2)-2x}{x(x+2)}$

$\dfrac{x}{x+2}=\dfrac{4x+8-2x}{x(x+2)}$

$\dfrac{x}{x+2}=\dfrac{2x+8}{x(x+2)}$

$\dfrac{x^2}{x(x+2)}=\dfrac{2x+8}{x(x+2)}$

$x^2=2x+8$

$x^2-2x-8=0$

$(x-4)(x+2)=0$

$x-4=0 \quad x+2=0$

$x=4 \quad\quad x=-2$

But $x=-2$ is a restricted value, so $x=4$.

39. $3t(t-1)=2(t+1)$

$3t^2-3t=2t+2$

$3t^2-5t-2=0$

$(3t+1)(t-2)=0$

$3t+1=0 \quad t-2=0$

$3t=-1 \quad\quad t=2$

$t=-\dfrac{1}{3}$

**Rational Expressions**

41. $\dfrac{8(t-1)+3t}{t(t-1)} = 3$

$8t - 8 + 3t = 3t(t-1)$

$11t - 8 = 3t^2 - 3t$

$0 = 3t^2 - 14t + 8$

$0 = (3t-2)(t-4)$

$3t - 2 = 0 \qquad t - 4 = 0$

$3t = 2 \qquad\quad t = 4$

$t = \dfrac{2}{3}$

43. $\dfrac{x+2}{x-12} - \dfrac{1}{x} = \dfrac{14}{x(x-12)}$

$\dfrac{x(x+2) - (x-12)}{x(x-12)} = \dfrac{14}{x(x-12)}$

$x^2 + 2x - x + 12 = 14$

$x^2 + x - 2 = 0$

$(x+2)(x-1) = 0$

$x = -2, 1$

45. $\dfrac{x}{x} \cdot \dfrac{x+8}{x+2} + \dfrac{12}{x(x+2)} = \dfrac{2}{x} \cdot \dfrac{x+2}{x+2}$

$\dfrac{x(x+8) + 12}{x(x+2)} = \dfrac{2(x+2)}{x(x+2)}$

$x^2 + 8x + 12 = 2x + 4$

$x^2 + 6x + 8 = 0$

$(x+4)(x+2) = 0$

$x = -4, -2$

But $x = -2$ is a restricted value, so $x = -4$.

47. $\dfrac{3}{(x+1)(x+3)} = \dfrac{2}{x+1} - 1 \cdot \dfrac{x+1}{x+1}$

$\dfrac{3}{(x+1)(x+3)} = \dfrac{2-(x+1)}{(x+1)}$

$\dfrac{3}{(x+1)(x+3)} = \dfrac{1-x}{x+1} \cdot \dfrac{x+3}{x+3}$

$3 = (1-x)(x+3)$

$3 = x - x^2 + 3 - 3x$

$3 = -x^2 - 2x + 3$

$x^2 + 2x = 0$

$x(x+2) = 0$

$x = 0, \; x = -2$

49. In (i), we rewrite each fraction with the LCD and then add the numerators. In (ii), we multiply both sides (each term) of the equation by the LCD to clear the fractions in the equation.

51. $3t\left(\dfrac{4}{t} - 1\right) = 3t\left(\dfrac{1}{3}\right)$

$12 - 3t = t$

$12 = 4t$

$t = 3$

53. $6y\left(\dfrac{5}{3} + \dfrac{y}{6}\right) = 6y\left(\dfrac{4}{y}\right)$

$10y + y^2 = 24$

$y^2 + 10y - 24 = 0$

$(y+12)(y-2) = 0$

$y = -12, \; y = 2$

55. $\dfrac{2x+8}{x+4} = 1$

$\dfrac{2(x+4)}{x+4} = 1$

$2 = 1$

No solution

57. $3x^2 = 4 - 11x$
$3x^2 + 11x - 4 = 0$
$(3x - 1)(x + 4) = 0$

$3x - 1 = 0 \quad x + 4 = 0$
$3x = 1 \qquad x = -4$
$x = \dfrac{1}{3}$

59. $\dfrac{x(x-3)}{x-3} - \dfrac{6}{x-3} = \dfrac{2x}{x-3}$
$x^2 - 3x - 6 = 2x$
$x^2 - 5x - 6 = 0$
$(x - 6)(x + 1) = 0$

$x = 6, -1$

61. $\dfrac{7}{2x-5} + \dfrac{2(2x-5)}{(2x-5)} = \dfrac{-3}{2x-5}$
$7 + 4x - 10 = -3$
$4x - 3 = -3$
$4x = 0$
$x = 0$

63. $\dfrac{3(2x-1) - 4(x+1)}{(x+1)(2x-1)} = \dfrac{5}{(2x-1)(x+1)}$
$6x - 3 - 4x - 4 = 5$
$2x - 7 = 5$
$2x = 12$
$x = 6$

65. $\dfrac{(x-2)^2 \cdot 9}{(x-2)^2} = \left(8 + \dfrac{1}{(x-2)}\right)(x-2)^2$
$9 = 8(x-2)^2 + (x-2)$
$9 = 8(x^2 - 4x + 4) + x - 2$
$9 = 8x^2 - 32x + 32 + x - 2$
$0 = 8x^2 - 31x + 21$
$0 = (8x - 7)(x - 3)$

$8x - 7 = 0 \quad x - 3 = 0$
$8x = 7 \qquad x = 3$
$x = \dfrac{7}{8}$

67. $12x\left(\dfrac{5}{4} - \dfrac{4}{3}\right) = 12x\left(\dfrac{7}{x}\right)$
$15x - 16x = 84$
$-x = 84$
$x = -84$

69. $\dfrac{\cancel{(x-8)}\cancel{(x+5)}}{3\cancel{(x+5)}} \cdot \dfrac{9x\cancel{(x+8)}}{\cancel{(x-8)}\cancel{(x+8)}}$
$= 3x$

71. $(x+4)^2 = 25$
$x^2 + 8x + 16 = 25$
$x^2 + 8x - 9 = 0$
$(x+9)(x-1) = 0$
$x = -9, \; x = 1$

73. $\dfrac{x}{x} \cdot \dfrac{8}{x} + \dfrac{2}{x^2}$
$= \dfrac{8x + 2}{x^2}$

75. $\dfrac{-3x - 12}{x + 4}$
$= \dfrac{-3(x+4)}{x+4}$
$= -3$

77. $\dfrac{2(t+5)+3(t-5)}{(t+5)(t-5)} = \dfrac{10}{(t-5)(t+5)}$

$2t+10+3t-15=10$

$5t-5=10$

$5t=15$

$t=3$

79. $\dfrac{x-12}{-16} \cdot \dfrac{8}{12-x}$

$= \dfrac{\cancel{12-x}}{16} \cdot \dfrac{8}{\cancel{12-x}}$

$= \dfrac{8}{16}$

$= \dfrac{1}{2}$

81. $(x^2+x)\left(9+\dfrac{3}{x+1}\right) = \dfrac{-4}{x^2+x} \cdot (x^2+x)$

$9(x^2+x)+3x=-4$

$9x^2+9x+3x=-4$

$9x^2+12x+4=0$

$(3x+2)^2=0$

$3x+2=0$

$3x=-2$

$x=\dfrac{-2}{3}$

83. $d=rt$

$r=\dfrac{d}{t}$

85. $\dfrac{100}{52.9-0.11x} - \dfrac{100}{60.4-0.16x} = 0.21$

87. $\dfrac{3}{2x-1} - \dfrac{5}{(2x-1)(4x+3)} = \dfrac{4}{4x+3}$

$\dfrac{3(4x+3)-5}{(2x-1)(4x+3)} = \dfrac{4(2x-1)}{(4x+3)(2x-1)}$

$12x+9-5=8x-4$

$4x+4=-4$

$4x=-8$

$x=-2$

89. $\dfrac{x}{x-4} - \dfrac{4}{x+2} = \dfrac{24}{(x-4)(x+2)}$

$\dfrac{x(x+2)-4(x-4)}{(x-4)(x+2)} = \dfrac{24}{(x-4)(x+2)}$

$x^2+2x-4x+16=24$

$x^2-2x-8=0$

$(x-4)(x+2)=0$

$x=4,\ x=-2$

But $x=4$ and $x=-2$ are restricted values, so there is no solution.

91. $(x^2-2)(x^2+1) = 2(2x^2-3)$

$x^4-x^2-2=4x^2-6$

$x^4-5x^2+4=0$

$(x^2-4)(x^2-1)=0$

$(x-2)(x+2)(x-1)(x+1)=0$

$x=2, x=-2, x=1, x=-1$

93. (a) Cross-multiplication results in the possible solution 1, which is an extraneous solution. Therefore, the equation has no solution.

(b) Multiplication by the LCD results in the contradiction 1=2. Again, the equation has no solution.

## Section 8.8 Applications

1. $y = 3x$
$\dfrac{1}{x} + \dfrac{1}{y} = \dfrac{7}{6}$
$\dfrac{1}{x} + \dfrac{1}{3x} = \dfrac{7}{6}$
$\dfrac{3}{3x} + \dfrac{1}{3x} = \dfrac{7}{6}$
$\dfrac{4}{3x} = \dfrac{7}{6}$
$24 = 21x$
$\dfrac{24}{21} = x$
$\dfrac{8}{7} = x$

$y = 3\left(\dfrac{8}{7}\right)$
$= \dfrac{24}{7}$

3. $x - 5\left(\dfrac{1}{x}\right) = 4$
$x - \dfrac{5}{x} = 4$
$x^2 - 5 = 4x$
$x^2 - 4x - 5 = 0$
$(x-5)(x+1) = 0$
$x = 5, \ x = -1$

5. $\dfrac{1}{x} - \dfrac{1}{2x} = 5$
$2x\left(\dfrac{1}{x} - \dfrac{1}{2x}\right) = 5 \cdot 2x$
$2 - 1 = 10x$
$1 = 10x$
$x = \dfrac{1}{10}$

7. $\dfrac{1}{x} - \dfrac{1}{x+2} = \dfrac{1}{4}$
$\dfrac{x+2-x}{x(x+2)} = \dfrac{1}{4}$
$\dfrac{2}{x(x+2)} = \dfrac{1}{4}$
$8 = x(x+2)$
$8 = x^2 + 2x$
$x^2 + 2x - 8 = 0$
$(x+4)(x-2) = 0$
$x = -4, \ x = 2$

9. $y = x + 3$
$\dfrac{1}{x} - \dfrac{1}{y} = \dfrac{3}{4}$
$\dfrac{1}{x} - \dfrac{1}{x+3} = \dfrac{3}{4}$
$\dfrac{x+3-x}{x(x+3)} = \dfrac{3}{4}$
$\dfrac{3}{x(x+3)} = \dfrac{3}{4}$
$12 = 3x(x+3)$
$12 = 3x^2 + 9x$
$4 = x^2 + 3x$
$0 = x^2 + 3x - 4$
$0 = (x+4)(x-1)$
$x = -4$
$y = -4 + 3$
$= -1$
or
$x = 1$
$y = 1 + 3$
$= 4$

Rational Expressions

11. $t = \dfrac{d}{r}$ where $r$ is the speed of the boat in still water.

$t = \dfrac{5}{r-1}$

$t = \dfrac{9}{r+1}$

$\dfrac{5}{r-1} = \dfrac{9}{r+1}$

$5(r+1) = 9(r-1)$

$5r + 5 = 9r - 9$

$14 = 4r$

$\dfrac{14}{4} = r$

$r = 3.5$ mph

13. $d = rt$

icy roads: $63 = \dfrac{1}{2}r(t)$

dry roads: $84 = r(5 - t)$

$126 = rt$

$84 = 5r - rt$

$84 = 5r - (126)$

$210 = 5r$

$r = 42$ mph

15. Let $t =$ jogging time
Let $r =$ jogging speed

Jogging: $9 = rt \Rightarrow t = \dfrac{9}{r}$

Cycling: $30 = (r+9)\left(t + \dfrac{1}{2}\right)$

$30 = (r+9)\left(\dfrac{9}{r} + \dfrac{1}{2}\right)$

$30 = 9 + \dfrac{1}{2}r + \dfrac{81}{r} + \dfrac{9}{2}$

$2r(30) = 2r\left(9 + \dfrac{1}{2}r + \dfrac{81}{r} + \dfrac{9}{2}\right)$

$60r = 18r + r^2 + 162 + 9r$

$0 = r^2 + 27r - 60r + 162$

$0 = r^2 - 33r + 162$

$0 = (r - 27)(r - 6)$

$r = 27$ or $r = 6$

It is unreasonable for a person to jog 27 mph, so the correct answer is 6 mph.

17. Let $t$ = amount of time against wind
Let $r$ = plane speed

With wind: $360 = (r+20)(8-t)$

Against wind: $400 = (r-20)t$

$$t = \frac{400}{r-20}$$

$$360 = (r+20)\left(8 - \frac{400}{r-20}\right)$$

$$360 = 8r - \frac{400r}{r-20} + 160 - \frac{8000}{r-20}$$

$$200 = 8r + \frac{-400r - 8000}{r-20}$$

$$200(r-20) = 8r(r-20) - 400r - 8000$$

$$200r - 4000 = 8r^2 - 160r - 400r - 8000$$

$$0 = 8r^2 - 760r - 4000$$

$$0 = r^2 - 95r - 500$$

$$0 = (r-100)(r+5)$$

$$r = 100 \text{ mph}$$

19. Let $r$ = rate on first leg
$t$ = time on first leg

First leg: $25 = rt \Rightarrow \frac{25}{t} = r$

Second leg: $30 = (r+5)\left(t - \frac{1}{2}\right)$

$$30 = \left(\frac{25}{t} + 5\right)\left(t - \frac{1}{2}\right)$$

$$2t(30) = 2t\left(\frac{5}{t} + 5\right)\left(t - \frac{1}{2}\right)$$

$$60t = (50 + 10t)\left(t - \frac{1}{2}\right)$$

$$60t = 50t - 25 + 10t^2 - 5t$$

$$60t = 10t^2 + 45t - 25$$

$$0 = 10t^2 - 15t - 25$$

$$0 = 2t^2 - 3t - 5$$

$$0 = (2t-5)(t+1)$$

$$2t - 5 = 0$$

$$2t = 5$$

$$t = 2.5 \text{ hours}$$

$$r = \frac{25}{t}$$

$$= \frac{25}{2.5}$$

$$= 10 \text{ mph}$$

21. $\frac{1}{10} + \frac{1}{x} = \frac{1}{6}$

$$\frac{x+10}{10x} = \frac{1}{6}$$

$$6(x+10) = 10x$$

$$6x + 60 = 10x$$

$$60 = 4x$$

$$x = 15 \text{ hours}$$

Rational Expressions

23. $\dfrac{1}{x} + \dfrac{1}{3x} = \dfrac{1}{1.5}$

$3x\left(\dfrac{1}{x} + \dfrac{1}{3}x\right) = 3x\left(\dfrac{1}{1.5}\right)$

$3 + 1 = 2x$

$4 = 2x$

$x = 2$ hours

25. $\dfrac{1}{10} + \dfrac{1}{15} = \dfrac{1}{x}$

$\dfrac{15 + 10}{150} = \dfrac{1}{x}$

$\dfrac{25}{150} = \dfrac{1}{x}$

$\dfrac{150}{25} = x$

$x = 6$ hours

27. $\dfrac{1}{t} + \dfrac{1}{t+25} = \dfrac{1}{30}$

$\dfrac{t+25+t}{t(t+25)} = \dfrac{1}{30}$

$30(2t+25) = t(t+25)$

$60t + 750 = t^2 + 25t$

$t^2 + 25t - 60t - 750 = 0$

$t^2 - 35t - 750 = 0$

$(t-50)(t+15) = 0$

Fast copier: $t = 50$ minutes
Slow copier: $t + 25 = 75$ minutes

29. physics rate: $x$ problems per hour
math rate: $x + 5$ math problems per hour

Let $t$ = time to complete 45 problem physics assignment.

physics: $45 = xt$

$t = \dfrac{45}{x}$

math: $40 = (x+5)(t-1)$

$40 = (x+5)\left(\dfrac{45}{x} - 1\right)$

$40x = (x+5)(45-x)$

$40x = 45x - x^2 + 225 - 5x$

$x^2 - 225 = 0$

$(x-15)(x+15) = 0$

$x = 15$ physics problems per hour
$x + 5 = 20$ math problems per hour

31. $\dfrac{1}{5} + \dfrac{1}{x} = \dfrac{1}{2}$

$\dfrac{x+5}{5x} = \dfrac{1}{2}$

$2(x+5) = 5x$

$2x + 10 = 5x$

$10 = 3x$

$x = \dfrac{10}{3}$ hours

33. $m = \dfrac{k}{n}$

$4 = \dfrac{k}{5}$

$k = 20$

35. $m = \dfrac{k}{n}$

$5.1 = \dfrac{k}{-3.2}$

$k = -16.32$

37. $m = \dfrac{k}{n}$

$9 = \dfrac{k}{2}$

$k = 18$

$m = \dfrac{18}{n}$

$m = \dfrac{18}{3}$

$= 6$ hours

39. $c = \dfrac{k}{x}$

$252 = \dfrac{k}{4}$

$k = 1008$

$c = \dfrac{1008}{x}$

$168 = \dfrac{1008}{x}$

$x = \dfrac{1008}{168}$

$x = 6$ people

41. $d = \dfrac{k}{T}$

$2300 = \dfrac{k}{20}$

$46,000 = k$

$d = \dfrac{46,000}{T}$

$d = \dfrac{46,000}{46}$

$= 1000$ tires

43. $C = \dfrac{k}{p}$

$200 = \dfrac{k}{1000}$

$200,000 = k$

$C = \dfrac{200,000}{p}$

$C = \dfrac{200,000}{800}$

$= 250$ people

45. $p =$ price of bird feed
$5p =$ price of salad

$\dfrac{4}{p} =$ amount of bird feed

$\dfrac{2}{5p} =$ amount of salad

$\dfrac{4}{p} + \dfrac{2}{5p} = 22$

$5p\left(\dfrac{4}{p} + \dfrac{2}{5p}\right) = 5p(22)$

$20 + 2 = 110p$

$\dfrac{22}{110} = p$

$p = 0.20$

The unit price is $0.20 per pound.

**Rational Expressions**

47. $x$ = hours worked by first person
$x + 1$ = hours worked by second person

$\dfrac{56}{x}$ = hourly wage of first worker

$\dfrac{54}{x+1}$ = hourly wage of second worker

$\dfrac{56}{x} - \dfrac{54}{x+1} = 1$
$56(x+1) - 54x = x(x+1)$
$56x + 56 - 54x = x^2 + x$
$2x + 56 = x^2 + x$
$x^2 - x - 56 = 0$
$(x-8)(x+7) = 0$

first: $x = 8$ hours
second: $x + 1 = 9$ hours

49. The number of one–parent family groups is increasingly at a faster rate than the number of two–parent families.

51. $\dfrac{0.03x + 25}{0.28x + 6.9} = 2$

53. $r$ = motorcycle speed
$r + 10$ = car speed

$\dfrac{125}{r+10}$ = car travel time

$\dfrac{120}{r}$ = motorcycle travel time

$\dfrac{120}{r} - \dfrac{125}{r+10} = 0.5$
$120(r+10) - 125r = 0.5r(r+10)$
$120r + 1200 - 125r = 0.5r^2 + 5r$
$1200 - 5r = 0.5r^2 + 5r$
$0 = 0.5r^2 + 10r - 1200$
$0 = r^2 + 20r - 2400$
$0 = (r+60)(r-40)$

$r = 40$ mph
car: $r+10 = 50$ mph

55. $t$ = time to make 15 dozen donuts
$2t$ = time to make 24 dozen cookies

$\dfrac{15}{t}$ = number of dozen donuts per hour

$\dfrac{24}{2t}$ = number of dozen cookies per hour

$\dfrac{15}{t} - \dfrac{24}{2t} = 2$

$\dfrac{15}{t} - \dfrac{12}{t} = 2$

$\dfrac{3}{t} = 2$

$3 = 2t$

donuts: $t = 1.5$ hours

cookies: $2t = 2(1.5) = 3$ hours

donuts: $\dfrac{15}{t} = \dfrac{15}{1.5} = 10$ dozen

cookies: $\dfrac{24}{2t} = \dfrac{24}{2(1.5)} = 8$ dozen

57. $x$ = hours required for farmer to bale hay alone

$x + 4$ = hours required for neighbor to bale hay alone

$\dfrac{1}{x}$ = farmer baling rate

$\dfrac{1}{x+4}$ = neighbor baling rate

$3\left(\dfrac{1}{x}\right) + 2\left(\dfrac{1}{x} + \dfrac{1}{x+4}\right) = \dfrac{19}{24}$

$\dfrac{3}{x} + \dfrac{2}{x} + \dfrac{2}{x+4} = \dfrac{19}{24}$

$\dfrac{5}{x} + \dfrac{2}{x+4} = \dfrac{19}{24}$

$5(x+4) + 2x = \dfrac{19}{24}(x(x+4))$

$120(x+4) + 48x = 19(x^2 + 4x)$

$120x + 480 + 48x = 19x^2 + 76x$

$168x + 480 = 19x^2 + 76x$

$0 = 19x^2 - 92x - 480$

$0 = (x-8)(19x+60)$

$x = 8$ hours

## Chapter 8 Review Exercises

1. (a) $\dfrac{3+6}{3^2 - 5(3)}$

$= \dfrac{9}{9-15}$

$= \dfrac{9}{-6}$

$= -1.5$

(b) $\dfrac{5+6}{5^2 - 5(5)}$

$= \dfrac{11}{0}$

$=$ undefined

3. (ii)

5. $\dfrac{(\cancel{x-2})(x+2)}{(\cancel{x-2})(x-2)}$

$= \dfrac{x+2}{x-2}$

7. $\dfrac{a-4b}{(4b-a)(4b+a)}$

$= \dfrac{-1(4b-a)}{(4b-a)(4b+a)}$

$= \dfrac{-1}{4b+a}$

# Rational Expressions

**9.** Cannot simplify

**11.** $\dfrac{x+2}{x-2} = -1$

$x+2 = -1(x-2)$

$x+2 = -x+2$

$x = -x$

$1 = -1$

No

**13.** By dividing out common factors and then multiplying, we ensure that the result is already in simplified form.

**15.** $\dfrac{(x-4)(x+3)}{(x+3)(x+4)} \cdot \dfrac{x+4}{(x-4)(x+4)}$

$= \dfrac{1}{x+4}$

**17.** $\dfrac{3}{1} \div x+2$

$= \dfrac{3}{1} \cdot \dfrac{1}{x+2}$

**19.** $x=2, x=3, x=-5$

**21.** (i) $\dfrac{x-y}{x-y} = 1$ if $x \neq y$

(ii) $\dfrac{x}{x-y} + \dfrac{y}{-(-y+x)}$

$= \dfrac{x-y}{x-y} = 1$ if $x \neq y$

**23.** $\dfrac{2x-5+12-x}{(x+7)(x-3)}$

$= \dfrac{x+7}{(x+7)(x-3)}$

$= \dfrac{1}{x-3}$

**25.** $\dfrac{x^2 - 3x - (x^2 - 4x - 2)}{x^2 + 2x}$

$= \dfrac{x+2}{x(x+2)}$

$= \dfrac{1}{x}$

**27.** $3x^2 - 5x + 1 - y = 2x^2 - 7x + 4$

$x^2 + 2x - 3 = y$

**29.** $\dfrac{2x+1}{x(2x-1)}$

$= \dfrac{7(2x+1)}{7x(2x-1)}$

$= \dfrac{14x+7}{14x^2 - 7x}$

**31.** $(x-3)(x+3)$

$3(x+3)$

LCM: $3(x+3)(x-3)$

**33.** $3x^2(x+3)$

**35.** $6y-15 = 3(2y-5)$

$4y-10 = 2(2y-5)$

LCD: $6(2y-5)$

$\dfrac{2y}{12y-30}, \dfrac{9y}{12y-30}$

**37.** $\dfrac{x}{x+3} + \dfrac{4}{x}$

$= \dfrac{x^2 + 4(x+3)}{x(x+3)}$

$= \dfrac{x^2 + 4x + 12}{x(x+3)}$

39. $\dfrac{x^2}{x-5} - \dfrac{25}{x-5}$

$= \dfrac{x^2 - 25}{x-5}$

$= \dfrac{(x+5)(x-5)}{x-5}$

$= x+5$

41. $\dfrac{2}{x+3} + \dfrac{5}{(x+3)(x-3)}$

$= \dfrac{2(x-3)+5}{(x+3)(x-3)}$

$= \dfrac{2x-6+5}{(x+3)(x-3)}$

$= \dfrac{2x-1}{(x+3)(x-3)}$

43. $\dfrac{5}{(x+6)(x-4)} - \dfrac{1}{(x+6)(x+3)}$

$= \dfrac{5(x+3) - 1(x-4)}{(x+6)(x-4)(x+3)}$

$= \dfrac{5x+15-x+4}{(x+6)(x-4)(x+3)}$

$= \dfrac{4x+19}{(x+6)(x-4)(x+3)}$

45. Both are correct.

47.

$\dfrac{a^2 b^2 \left( \dfrac{3}{a} + \dfrac{5}{b^2} \right)}{a^2 b^2 \left( \dfrac{4}{b} - \dfrac{1}{a^2} \right)}$

$= \dfrac{3b^2 a + 5a^2}{4a^2 b - b^2}$

49. $\dfrac{2(a-b)-b}{a-b} \cdot \dfrac{1}{2a-3b}$

$= \dfrac{2a-2b-b}{a-b} \cdot \dfrac{1}{2a-3b}$

$= \dfrac{2a-3b}{a-b} \cdot \dfrac{1}{2a-3b}$

$= \dfrac{1}{a-b}$

51. The left side of the equation is not defined if $x$ is 3.

53. $\dfrac{5}{6} + \dfrac{2}{2x-1} = \dfrac{5}{3(2x-1)}$

$6(2x-1)\left(\dfrac{5}{6} + \dfrac{2}{2x-1}\right) = 6(2x-1)\left(\dfrac{5}{3(2x-1)}\right)$

$5(2x-1) + 12 = 2(5)$

$10x - 5 + 12 = 10$

$10x + 7 = 10$

$10x = 3$

$x = \dfrac{3}{10}$

55. $\dfrac{2x + 20 - (8 - 2x)}{x+3} = 1$

$\dfrac{4x+12}{x+3} = 1$

$\dfrac{4\cancel{(x+3)}}{\cancel{x+3}} = 1$

$4 = 1$

No solution

Rational Expressions

57. $x - 8\left(\dfrac{1}{x}\right) = 2$

$x - \dfrac{8}{x} = 2$

$x^2 - 8 = 2x$

$x^2 - 2x - 8 = 0$

$(x-4)(x+2) = 0$

$x = 4, -2$

Since $x < 0, x = -2$ is the number.

59. $x$ = number of hours for carpet layer to do job alone.

$x + 2$ = number of hours for nephew to do job alone.

$\dfrac{1}{x} + \dfrac{1}{x+2} = \dfrac{1}{\frac{12}{5}}$

$\dfrac{x+2+x}{x(x+2)} = \dfrac{5}{12}$

$\dfrac{2x+2}{x(x+2)} = \dfrac{5}{12}$

$12(2x+2) = 5x(x+2)$

$24x + 24 = 5x^2 + 10x$

$5x^2 - 14x - 24 = 0$

$(5x+6)(x-4) = 0$

Carpet layer: $x = 4$ hours
Nephew: $x + 2 = 6$ hours

## Chapter 8 Test

1. $\dfrac{x-4}{-(x^2-8x+15)}$

$= \dfrac{x-4}{-(x-3)(x-5)}$

The restricted values are $x = 3, 5$.

3. $\dfrac{(x+y)(x-y)}{(y-x)(y-x)}$

$= \dfrac{(x+y)\cancel{(x-y)}}{-\cancel{(x-y)}(y-x)}$

$= -\dfrac{x+y}{y-x}$

$= \dfrac{x+y}{x-y}$

5. $\dfrac{2\cancel{x}}{\cancel{x+2}} \cdot \dfrac{x\cancel{(x+2)}}{-4\cancel{(-x+2)}}$

$= -\dfrac{x}{4}$

7. $\dfrac{1-3x}{x(x+6)} + \dfrac{4x+5}{x(x+6)}$

$= \dfrac{\cancel{x+6}}{x\cancel{(x+6)}}$

$= \dfrac{1}{x}$

9. $\dfrac{2x-(x+4)}{y}$

$= \dfrac{x-4}{y}$

11. We use the largest exponent to determine the LCM, whereas, the smallest exponent is used to determine the GCF.

13. $x^2 - 4 = (x-2)(x+2)$
    $x^2 + 6x + 8 = (x+2)(x+4)$
    LCM $= (x-2)(x+2)(x+4)$

15. $\dfrac{x(x+5)+3(x-8)}{(x-8)(x+5)}$
    $= \dfrac{x^2+5x+3x-24}{(x-8)(x+5)}$
    $= \dfrac{x^2+8x-24}{(x-8)(x+5)}$

17. $\dfrac{2}{(x+1)^2} - \dfrac{x}{(x+1)(x+2)}$
    $= \dfrac{2(x+2)-x(x+1)}{(x+1)^2(x+2)}$
    $= \dfrac{2x+4-x^2-x}{(x+1)^2(x+2)}$
    $= \dfrac{-x^2+x+4}{(x+1)^2(x+2)}$

19. $\dfrac{\dfrac{xy+1}{y}}{\dfrac{1-xy}{x}}$

    $= \dfrac{xy+1}{y} \cdot \dfrac{x}{1-xy}$
    $= \dfrac{x(xy+1)}{y(1-xy)}$
    $= \dfrac{x^2y+x}{y-xy^2}$

21. $\dfrac{x+7}{x+2} + \dfrac{10}{x(x+2)} = \dfrac{2}{x}$
    $\dfrac{x(x+7)+10}{x(x+2)} = \dfrac{2(x+2)}{x(x+2)}$
    $x^2+7x+10 = 2x+4$
    $x^2+5x+6 = 0$
    $(x+2)(x+3) = 0$
    $x = -2, -3$

    But $x = -2$ is a restricted value so $x = -3$.

23. $x =$ time for owner
    $x+3 =$ time for employee

    $\dfrac{1}{x} =$ owner rate

    $\dfrac{1}{x+3} =$ employee rate

    $2\left(\dfrac{1}{x}\right) + 4\left(\dfrac{1}{x+3}\right) = \dfrac{90}{100}$
    $\dfrac{2}{x} + \dfrac{4}{x+3} = \dfrac{9}{10}$
    $\dfrac{2(x+3)+4x}{x(x+3)} = \dfrac{9}{10}$
    $\dfrac{2x+6+4x}{x(x+3)} = \dfrac{9}{10}$
    $\dfrac{6x+6}{x(x+3)} = \dfrac{9}{10}$
    $10(6x+6) = 9x(x+3)$
    $60x+60 = 9x^2+27x$
    $9x^2+27x-60x-60 = 0$
    $9x^2-33x-60 = 0$
    $3x^2-11x-20 = 0$
    $(x-5)(3x+4) = 0$
    $x = 5$ hours

## Cumulative Test, Chapters 6 – 8

1. (a) $-x^6 y^3$

   (b) $\dfrac{3 \cdot 4 \cdot a^3 \cdot a^2}{3 \cdot 3 \cdot a^2}$
   
   $= \dfrac{4a^3}{3}$

   (c) $\dfrac{8a^3}{27b^6}$

3. $x^2 + 3 - 2x + 8 + x^2 + 7x - 4$
   $= 2x^2 + 5x + 7$

5. (a) $(2x+1)(2x+1)$
   $= 4x^2 + 4x + 1$

   (b) $4x^2 - 1$

7. In (i), the base is 2, and the expression is evaluated as $-(2^{-2})$. In (ii), the base is $-2$, so the expression is evaluated as $(-2)^{-2}$.

9. $6 \cdot 10^9 \cdot 50$
   $= 300 \cdot 10^9$
   $= 3 \cdot 10^{11}$

11. (a) $3x^2 y(2xy - 3)$

    (b) $a(2x+y) - 2b(2x+y)$
    $= (a - 2b)(2x + y)$

13. (a) $2(x^2 - 11x + 18)$
    $= 2(x-9)(x-2)$

    (b) $(x^2 + 5)(x^2 - 3)$

15. $x(x^4 - 1)$
    $= x(x^2 - 1)(x^2 + 1)$
    $= x(x-1)(x+1)(x^2 + 1)$

17. $x =$ short leg
    $2x + 4 =$ long leg
    $3x - 4 =$ hypotenuse

    $x^2 + (2x + 4)^2 = (3x - 4)^2$
    $x^2 + 4x^2 + 16x + 16 = 9x^2 - 24x + 16$
    $5x^2 + 16x + 16 = 9x^2 - 24x + 16$
    $4x^2 - 40x = 0$
    $4x(x - 10) = 0$

    short: $x = 10$ inches
    long: $2(10) + 4 = 24$ inches

19. $\dfrac{(x+5)(x-5)}{x(x+5)} \cdot \dfrac{2x}{(x-5)(x+3)}$
    $= \dfrac{2}{x+3}$

21. $2x^2 = 2 \cdot x \cdot x$
    $4x^2 - 6x = 2x(2x - 3)$
    LCM: $2x^2(2x - 3)$

23.
$$\frac{\frac{1}{x+1}}{\frac{1+x}{x}}$$
$$=\frac{1}{x+1} \cdot \frac{x}{1+x}$$
$$=\frac{x}{(x+1)^2}$$

25. $t$ = time in car
$t + \frac{36}{60}$ = time in train
$\frac{20}{t}$ = car speed
$\frac{30}{t+\frac{36}{60}}$ = train speed
$\frac{20}{t} - \frac{30}{t+\frac{36}{60}} = 20$
$20\left(t+\frac{36}{60}\right) - 30t = 20t\left(t+\frac{36}{60}\right)$
$20t + 12 - 30t = 20t^2 + 12t$
$12 - 10t = 20t^2 + 12t$
$0 = 20t^2 + 22t - 12$
$0 = 10t^2 + 11t - 6$
$0 = (2t+3)(5t-2)$

$5t - 2 = 0$
$5t = 2$
$t = \frac{2}{5}$
train speed = $\frac{30}{\frac{2}{5}+\frac{36}{60}}$
$= \frac{30}{\frac{2}{5}+\frac{3}{5}}$
$= 30$ mph

# Chapter 9

# Radical Expressions

## Section 9.1 Radicals

1.  (a) The numbers are the squares of the natural numbers. The next three numbers are 25, 36, and 49.

    (b) The numbers are the cubes of the natural numbers. The next three numbers are 125, 216, and 343.

3.  $\sqrt{25} = 5, -5$

5.  $\sqrt{\dfrac{9}{49}}$
    $= \dfrac{\sqrt{9}}{\sqrt{49}}$
    $= \dfrac{3}{7}, -\dfrac{3}{7}$

7.  3

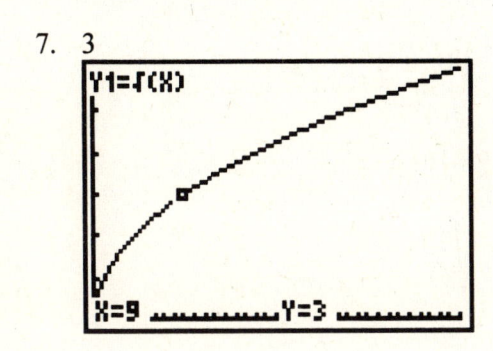

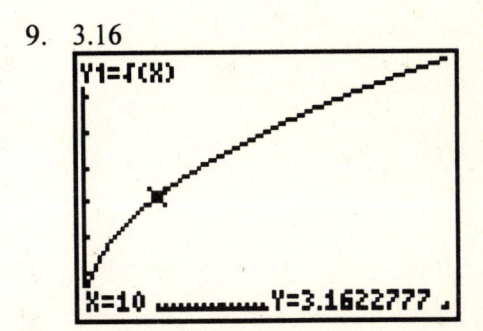

9.  3.16

11. 5

13. 3

15. –9

17. $5(6)$
    $= 30$

19. 100, Rational

21. Not real since there is a negative beneath the radical.

23. Irrational

25. 15.81

27. 2.04

29. 7.94

31. Because $\sqrt{x^2}$ represents a nonnegative number, $y_1 \geq 0$. In $y_2 = x$, $y_2$ represents any real number.

Copyright © Houghton Mifflin Company. All rights reserved.

33. 30

35. $\left(-\sqrt{8}\right)\left(-\sqrt{8}\right)$
$= (-1)^2 \left(\sqrt{8}\right)^2$
$= 1 \cdot 8$
$= 8$

37. $5^2 \left(\sqrt{3}\right)^2$
$= 25 \cdot 3$
$= 75$

39. $2x$

41. $\sqrt{49}$
$= 7$

43. $\sqrt{144}$
$= 12$

45. $-\sqrt{25}$
$= -5$

47. $5t$

49. $5y$

51. $x + 5$

53. $\sqrt{\left(t^3\right)^2}$
$= t^3$

55. $\sqrt{\left(x^6\right)^2}$
$= x^6$

57. $\sqrt{2^6 y^8}$
$= \sqrt{\left(2^3 y^4\right)^2}$
$= 8y^4$

59. $\sqrt{\left[(8ab)^5\right]^2}$
$= (8ab)^5$

61. If $n$ is odd, the $\sqrt[n]{k}$ is defined for all values of $k$. If $n$ is even, then $\sqrt[n]{k}$ is a real number only if $k \geq 0$.

63. $\sqrt[3]{-27}$
$= \sqrt[3]{(-3)^3}$
$= -3$

65. $\sqrt[4]{16}$
$= \sqrt[4]{2^4}$
$= 2, -2$

67. $\sqrt[5]{-32}$
$= \sqrt[5]{(-2)^5}$
$= -2$

69. $\sqrt[6]{-1}$
Not a real number

71. $\sqrt[3]{27}$
$= \sqrt[3]{3^3}$

73. $\sqrt[4]{81}$
$= \sqrt[4]{3^4}$
$= 3$

75. $\sqrt[4]{10,000}$
$\sqrt[4]{10^4}$
$= 10$

77. $\sqrt[4]{\dfrac{2^4}{3^4}}$
$= \dfrac{2}{3}$

79. $\sqrt[3]{\left(t^5\right)^3}$
$= t^5$

**Radical Expressions**

81. $\sqrt[6]{(y^2)^6}$
    $= y^2$

83. $\sqrt[7]{(y^2)^7}$
    $= y^2$

85. $\sqrt[3]{8} = 2$
    $\sqrt[3]{-8} = -2$
    $-\sqrt[3]{8} = -2$
    $-\sqrt[3]{-8} = -(-2) = 2$
    $\sqrt[3]{8} = -\sqrt[3]{-8}$ and $\sqrt[3]{-8} = -\sqrt[3]{8}$

87. 86.46%

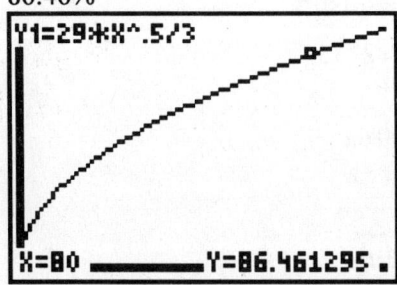

89. $r = \sqrt{\dfrac{6100}{\pi(3.1)}}$
    $= \sqrt{626.35}$
    $= 25.03$

    $d = 2r$
    $= 2(25.03)$
    $= 50.06$
    $\approx 50$ feet

91. (a) The number of miles (in millions) traveled daily in 2002.

    (b) $62.5\sqrt[5]{15} + 5$
    $\quad -62.5\sqrt[5]{20}$
    $= 113.79$
    $\approx 114$ million miles

93. $\sqrt{t^8}$
    $= \sqrt{(t^4)^2}$
    $= t^4$

95. $\sqrt{\sqrt{16}}$
    $= \sqrt{4}$
    $= 2$

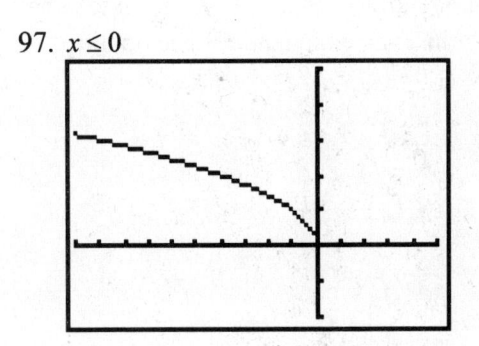

97. $x \leq 0$

## Section 9.2 Product Rule for Radicals

1. (a) The radical sign was omitted in the result: $\sqrt{2}\sqrt{3} = \sqrt{6}$.

   (b) A radical sign is missing on the 4:
   $\sqrt{20} = \sqrt{4}\sqrt{5} = 2\sqrt{5}$.

3. $\sqrt{5 \cdot 7} = \sqrt{35}$

5. $\sqrt{2a \cdot 5b} = \sqrt{10ab}$

7. $3 \cdot 2\sqrt{7 \cdot 5} = 6\sqrt{35}$

9. $\sqrt{100}\sqrt{x^8}$
   $= 10x^4$

11. $\sqrt{x^6} \cdot \sqrt{y^2}$
    $= x^3 y$

13. $\sqrt{25}\sqrt{2}$
    $= 5\sqrt{2}$

15. $\sqrt{4}\sqrt{5}$
    $= 2\sqrt{5}$

17. $\sqrt{16}\sqrt{3}$
    $= 4\sqrt{3}$

19. Write $\sqrt{b^n}$ as $\sqrt{b^{n-1}}\sqrt{b}$. Then take half of the even exponent $n-1$ to obtain $b^{(n-1)/2}\sqrt{b}$.

21. $\sqrt{x^2} \cdot \sqrt{x}$
    $= x\sqrt{x}$

23. $\sqrt{x^{10}}\sqrt{x}$
    $= x^5\sqrt{x}$

25. $\sqrt{4x^6}\sqrt{6}$
    $= 2x^3\sqrt{6}$

27. $\sqrt{49y^{10}}\sqrt{2}$
    $= 7y^5\sqrt{2}$

29. $\sqrt{25x^6}\sqrt{x}$
    $= 5x^3\sqrt{x}$

31. $\sqrt{16y^{14}}\sqrt{y}$
    $= 4y^7\sqrt{y}$

33. $\sqrt{4x^4}\sqrt{7x}$
    $= 2x^2\sqrt{7x}$

35. $\sqrt{25y^{14}}\sqrt{5y}$
    $= 5y^7\sqrt{5y}$

37. $\sqrt{x^8 y^2}\sqrt{y}$
    $= x^4 y\sqrt{y}$

39. $\sqrt{4x^2 y^6}\sqrt{3y}$
    $= 2xy^3\sqrt{3y}$

41. $\sqrt{64}$
    $= 8$

43. $\sqrt{60}$
    $= \sqrt{4}\sqrt{15}$
    $= 2\sqrt{15}$

45. $\sqrt{16x^2}$
    $= 4x$

47. $\sqrt{a^5}$
    $= \sqrt{a^4}\sqrt{a}$
    $= a^2\sqrt{a}$

49. $\sqrt{12t^{10}}$
    $= \sqrt{4t^{10}}\sqrt{3}$
    $= 2t^5\sqrt{3}$

51. $\sqrt[3]{8}\sqrt[3]{5}$
    $= 2\sqrt[3]{5}$

53. $\sqrt[5]{32}\sqrt[5]{2}$
    $= 2\sqrt[5]{2}$

55. $\sqrt[3]{x^6}\sqrt[3]{x}$
    $= x^2\sqrt[3]{x}$

57. $\sqrt[4]{x^4} \cdot \sqrt[4]{x^3}$
    $= x\sqrt[4]{x^3}$

59. $\sqrt[3]{8a^3}\sqrt[3]{3a^2}$
   $= 2a\sqrt[3]{3a^2}$

61. $\sqrt{49 \cdot 10^6}$
   $= 7 \cdot 10^3$

63. $\sqrt{81 \cdot 10^{-4}}$
   $= 9 \cdot 10^{-2}$

65. $\sqrt{7} = \sqrt{10} - \sqrt{3}$
   $2.64 = 3.16 - 1.73$
   $2.64 = 1.43$
   False

67. $5\sqrt{2} = \sqrt{50}$
   $= \sqrt{25}\sqrt{2}$
   $= 5\sqrt{2}$
   True

69. To simplify the radical, take the square root of the coefficient and take half of the exponent: $4x^8$.

71. $\sqrt{2t} + 15\sqrt{5t}$
   $= \sqrt{10t^2 + 75t}$

73. (a) 1.58

   (b) $T = \sqrt{4d^2}\sqrt{5d}$
   $= 2d\sqrt{5d}$
   $T = 2(0.5)\sqrt{5(0.5)}$
   $= \sqrt{2.5}$
   $= 1.58$

75. The graph suggests a steady increase in the amount of chicken that will be eaten.

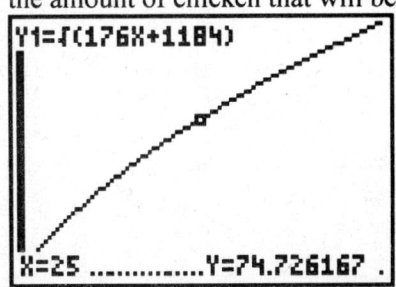

77. 1980–2000

79. $\sqrt{2(x^2 + 8x + 16)}$
   $= \sqrt{2(x+4)^2}$
   $= (x+4)\sqrt{2}$

81. $\sqrt{(2x)^2 \, 3x}$
   $= \sqrt{4x^2 \cdot 3x}$
   $= \sqrt{12x^3}$

83. $\sqrt[2]{(3t^n)^2} = 3t^n$

## Section 9.3 Quotient Rule for Radicals

1. All three approaches are correct, but (ii) will require the fewest steps.

3. $\sqrt{\dfrac{125}{5}}$
   $= \sqrt{25}$
   $= 5$

5. $\sqrt{\dfrac{20x^5}{5x}}$
   $= \sqrt{4x^4}$
   $= 2x^2$

7. $\sqrt{\dfrac{48x}{3x^3}}$
   $= \sqrt{16x^{-2}}$
   $= 4x^{-1}$
   $= \dfrac{4}{x}$

9. $\sqrt{\dfrac{90y^7}{2y}}$
   $= \sqrt{45y^6}$
   $= \sqrt{9y^6}\sqrt{5}$
   $= 3y^3\sqrt{5}$

11. $\dfrac{9}{x^5}$

13. $\dfrac{y^6}{x^5}$

15. $\sqrt{\dfrac{x^6}{25}} \cdot \sqrt{x}$
    $= \dfrac{x^3}{5}\sqrt{x}$

17. $\dfrac{\sqrt{2}}{x^2}$

19. $\dfrac{\sqrt{11y}}{x^3}$

21. (a) The radicand is a fraction. This violates Condition 2 for a simplified square root radical.

    (b) The expression is still not simplified, because the denominator contains a radical. This violates Condition 3 for a simplified square root radical.

23. $\sqrt{\dfrac{75a^9}{3a}}$
    $= \sqrt{25a^8}$
    $= 5a^4$

25. $\sqrt{\dfrac{y^6}{49}}$
    $= y^3 \dfrac{}{7}$

27. $\sqrt{x^4 y^{10}}$
    $= x^2 y^5$

29. $\sqrt{\dfrac{12x^5}{25y^8}}$
    $= \sqrt{\dfrac{4x^4}{25y^8}}\sqrt{3x}$
    $= \dfrac{2x^2}{5y^4}\sqrt{3x}$
    $= \dfrac{2x^2\sqrt{3x}}{5y^4}$

31. $\sqrt{\dfrac{3y^3}{25}}$

$= \sqrt{\dfrac{y^2}{25}}\sqrt{3y}$

$= \dfrac{y}{5}\sqrt{3y}$

$= \dfrac{y\sqrt{3y}}{5}$

33. $\dfrac{4}{\sqrt{5}} \cdot \dfrac{\sqrt{5}}{\sqrt{5}}$

$= \dfrac{4\sqrt{5}}{5}$

35. $\dfrac{\sqrt{5}}{\sqrt{6}} \cdot \dfrac{\sqrt{6}}{\sqrt{6}}$

$= \dfrac{\sqrt{30}}{6}$

37. $\dfrac{3}{\sqrt{15}} \cdot \dfrac{\sqrt{15}}{\sqrt{15}}$

$= \dfrac{3\sqrt{15}}{15}$

$= \dfrac{\sqrt{15}}{5}$

39. $\dfrac{3}{\sqrt{x}} \cdot \dfrac{\sqrt{x}}{\sqrt{x}}$

$= \dfrac{3\sqrt{x}}{x}$

41. $\dfrac{9}{\sqrt{3y}} \cdot \dfrac{\sqrt{3y}}{\sqrt{3y}}$

$= \dfrac{9\sqrt{3y}}{3y}$

$= \dfrac{3\sqrt{3y}}{y}$

43. $\dfrac{\sqrt{5}}{\sqrt{24}} \cdot \dfrac{\sqrt{24}}{\sqrt{24}}$

$= \dfrac{\sqrt{120}}{24}$

$= \dfrac{2\sqrt{30}}{24}$

$= \dfrac{\sqrt{30}}{12}$

45. $\dfrac{y}{\sqrt{5}} \cdot \dfrac{\sqrt{5}}{\sqrt{5}}$

$= \dfrac{\sqrt{5}y}{5}$

47. $\dfrac{10t}{5\sqrt{3}} \cdot \dfrac{\sqrt{3}}{\sqrt{3}}$

$= \dfrac{10t\sqrt{3}}{5 \cdot 3}$

$= \dfrac{2t\sqrt{3}}{3}$

49. $\dfrac{x\sqrt{x}}{\sqrt{5}} \cdot \dfrac{\sqrt{5}}{\sqrt{5}}$

$= \dfrac{x\sqrt{5x}}{5}$

51. $\dfrac{\sqrt{5}}{\sqrt{4x}} \cdot \dfrac{\sqrt{4x}}{\sqrt{4x}}$

$= \dfrac{\sqrt{20x}}{4x}$

$= \dfrac{2\sqrt{5x}}{4x}$

$= \dfrac{\sqrt{5}}{2x}$

53. $\sqrt{\dfrac{3}{5x^2}}$

$= \dfrac{\sqrt{3}}{x\sqrt{5}} \cdot \dfrac{\sqrt{5}}{\sqrt{5}}$

$= \dfrac{\sqrt{15}}{5x}$

55. $\sqrt{\dfrac{x^3}{12}}$

$= \dfrac{x\sqrt{x}}{2\sqrt{3}} \cdot \dfrac{\sqrt{3}}{\sqrt{3}}$

$= \dfrac{x\sqrt{3x}}{6}$

57. $\sqrt{\dfrac{48x^3y}{2x^4y^7}}$

$= \sqrt{24x^{-1}y^{-6}}$

$= \sqrt{4y^{-6}}\sqrt{6x^{-1}}$

$= 2y^{-3}\sqrt{\dfrac{6}{x}}$

$= \dfrac{2\sqrt{6}}{y^3\sqrt{x}} \cdot \dfrac{\sqrt{x}}{x}$

$= \dfrac{2\sqrt{6x}}{xy^3}$

59. $\dfrac{\sqrt[3]{w^2}}{\sqrt[3]{27}}$

$= \dfrac{\sqrt[3]{w^2}}{3}$

61. $\dfrac{\sqrt[4]{x^3}}{\sqrt[4]{16}}$

$= \dfrac{\sqrt[4]{x^3}}{2}$

63. $\dfrac{\sqrt[3]{54}}{\sqrt[3]{125}}$

$= \dfrac{\sqrt[3]{27}\sqrt[3]{2}}{5}$

$= \dfrac{3\sqrt[3]{2}}{5}$

65. $\dfrac{\sqrt[3]{x^5}}{\sqrt[3]{64}}$

$= \dfrac{x\sqrt[3]{x^2}}{4}$

67. $\dfrac{\sqrt[4]{16x^8}\,\sqrt[4]{3x^3}}{\sqrt[4]{3x^3}}$

$= 2x^2$

69. $I = \dfrac{300}{\sqrt{5t}} \cdot \dfrac{\sqrt{5t}}{\sqrt{5t}}$

$= \dfrac{300\sqrt{5t}}{5t}$

$= \dfrac{60\sqrt{5t}}{t}$

71. $\dfrac{\sqrt{75y^5}}{\sqrt{3y^2}}$

$= \sqrt{25y^3}$

$= 5y\sqrt{y}$

73. (a) $\dfrac{5.2}{\sqrt{0.5(60)+10}} + 1.73$

$= \dfrac{5.2}{\sqrt{40}} + 1.73$

$= 2.55$

(b) $\dfrac{5.2}{\sqrt{.05x+10}} \cdot \dfrac{\sqrt{0.5x+10}}{0.5x+10} + 1.73$

$= \dfrac{5.2\sqrt{0.5x+10}}{0.5x+10} + 1.73$

# Radical Expressions

(c) $\dfrac{5.2\sqrt{0.5(60)+10}}{0.5(60)+10}+1.73$

$=\dfrac{5.2\sqrt{40}}{40}+1.73$

$=2.55$

$=2.56$

75. $\dfrac{46x}{\sqrt{x}}\cdot\dfrac{\sqrt{x}}{\sqrt{x}}$

$=\dfrac{46x\sqrt{x}}{x}$

$=46\sqrt{x}$

77. $\dfrac{39.4(x+1)}{\sqrt{x+1}}\cdot\dfrac{\sqrt{x+1}}{\sqrt{x+1}}+19.5$

$=\dfrac{39.4\cancel{(x+1)}(\sqrt{x+1})}{\cancel{(x+1)}}+19.5$

$=39.4\sqrt{x+1}+19.5$

79. $\dfrac{2}{\sqrt[3]{9}}\cdot\dfrac{\sqrt[3]{3}}{\sqrt[3]{3}}$

$=\dfrac{2\sqrt[3]{3}}{\sqrt[3]{27}}$

$=\dfrac{2\sqrt[3]{3}}{3}$

81. $\dfrac{1}{\sqrt[3]{t^2}}\cdot\dfrac{\sqrt[3]{t}}{\sqrt[3]{t}}$

$=\dfrac{\sqrt[3]{t}}{\sqrt[3]{t^3}}$

$=\dfrac{\sqrt[3]{t}}{t}$

83. $\dfrac{\sqrt[4]{2}}{\sqrt[4]{27}}\cdot\dfrac{\sqrt[4]{3}}{\sqrt[4]{3}}$

$=\dfrac{\sqrt[4]{6}}{\sqrt[4]{81}}$

$=\dfrac{\sqrt[4]{6}}{3}$

85. $\sqrt{x^{2n}}$

$=x^n$

## Section 9.4 Operations with Radicals

1. Because the terms do not have a common factor. The Distributive Property does not apply, and the terms cannot be combined.

3. $\sqrt{7}(4+3)$

$=7\sqrt{7}$

5. $\sqrt{15}(1-3)$

$=-2\sqrt{15}$

7. $3\sqrt{2}-2\sqrt{2}-4\sqrt{3}+\sqrt{3}$

$=\sqrt{2}-3\sqrt{3}$

9. $\sqrt{x}(5-3)$

$=2\sqrt{x}$

11. $\sqrt{7t}(t+5t)$

$=6t\sqrt{7t}$

13. $\sqrt{4}\sqrt{3} - \sqrt{3}$
    $= 2\sqrt{3} - \sqrt{3}$
    $= \sqrt{3}$

15. $\sqrt{9}\sqrt{3} + \sqrt{25}\sqrt{3}$
    $= 3\sqrt{3} + 5\sqrt{3}$
    $= 8\sqrt{3}$

17. $3\sqrt{16}\sqrt{3} + 2\sqrt{3}\sqrt{25}$
    $= 3 \cdot 4\sqrt{3} + 2 \cdot 5\sqrt{3}$
    $= 12\sqrt{3} + 10\sqrt{3}$
    $= 22\sqrt{3}$

19. $3\sqrt{2t} - \sqrt{49}\sqrt{2t}$
    $= 3\sqrt{2t} - 7\sqrt{2t}$
    $= -4\sqrt{2t}$

21. $b\sqrt{9}\sqrt{10b} + 8\sqrt{b^2}\sqrt{10b}$
    $= 3b\sqrt{10b} + 8b\sqrt{10b}$
    $= 11b\sqrt{10b}$

23. $2\sqrt{36}\sqrt{2} - \sqrt{100}\sqrt{3} + \sqrt{100}\sqrt{2}$
    $= 2 \cdot 6\sqrt{2} - 10\sqrt{3} + 10\sqrt{2}$
    $= 12\sqrt{2} + 10\sqrt{2} - 10\sqrt{3}$
    $= 22\sqrt{2} - 10\sqrt{3}$

25. (a) The Power of a Product Rule states that each factor must be squared. Thus $\left(3\sqrt{x}\right)^2 = 3^2 \left(\sqrt{x}\right)^2 = 9x.$

    (b) The square of a binomial is a special product whose pattern is $(A+B)^2 = A^2 + 2AB + B^2$. Thus,
    $\left(\sqrt{x} + \sqrt{5}\right)^2$
    $= \left(\sqrt{x}\right)^2 + 2 \cdot \sqrt{x} \cdot \sqrt{5} + \left(\sqrt{5}\right)^2$
    $= x + 2\sqrt{5x} + 5.$

27. $2\sqrt{5} - \sqrt{5}\sqrt{3}$
    $= 2\sqrt{5} - \sqrt{15}$

29. $\sqrt{16} + \sqrt{14}$
    $= 4 + \sqrt{14}$

31. $6 - 4\sqrt{6} - 3\sqrt{6} + 2\left(\sqrt{6}\right)^2$
    $= 6 - 7\sqrt{6} + 2(6)$
    $= 6 - 7\sqrt{6} + 12$
    $= 18 - 7\sqrt{6}$

33. $3 \cdot 5 + \sqrt{10} + 6\sqrt{10} + 2(2)$
    $= 15 + 7\sqrt{10} + 4$
    $= 19 + 7\sqrt{10}$

35. $x - 2\sqrt{x} + 7\sqrt{x} - 14$
    $= x + 5\sqrt{x} - 14$

37. $\left(\sqrt{5} + 2\right)\left(\sqrt{5} + 2\right)$
    $= 5 + 2\sqrt{5} + 2\sqrt{5} + 4$
    $= 9 + 4\sqrt{5}$

39. $\left(3 + \sqrt{x}\right)^2$
    $= 9 + 2 \cdot 3\sqrt{x} + x$
    $= 9 + 6\sqrt{x} + x$

41. $2\sqrt{3} - 3$;
    $\left(2\sqrt{3} + 3\right)\left(2\sqrt{3} - 3\right)$
    $= 4 \cdot 3 - 3 \cdot 2\sqrt{3} + 3 \cdot 2\sqrt{3} - 9$
    $= 12 - 9$
    $= 3$

43. $\sqrt{10} + \sqrt{5}$;
    $\left(\sqrt{10} - \sqrt{5}\right)\left(\sqrt{10} + \sqrt{5}\right)$
    $= 10 - 5$
    $= 5$

**Radical Expressions**

45. $\sqrt{x} + 4$;
$(\sqrt{x} - 4)(\sqrt{x} + 4)$
$= x - 16$

47. $\dfrac{12 - \sqrt{81}\sqrt{2}}{3}$
$= \dfrac{12 - 9\sqrt{2}}{3}$
$= 4 - 3\sqrt{2}$

49. $\dfrac{10 + \sqrt{25}\sqrt{2}}{15}$
$= \dfrac{10 + 5\sqrt{2}}{15}$
$= \dfrac{5(2 + \sqrt{2})}{5(3)}$
$= \dfrac{2 + \sqrt{2}}{3}$

51. $\dfrac{1}{\sqrt{3} + 5} \cdot \dfrac{\sqrt{3} - 5}{\sqrt{3} - 5}$
$= \dfrac{\sqrt{3} - 5}{3 - 25}$
$= \dfrac{\sqrt{3} - 5}{-22}$
$= \dfrac{5 - \sqrt{3}}{22}$

53. $\dfrac{14}{\sqrt{3} - \sqrt{2}} \cdot \dfrac{3 + \sqrt{2}}{3 + \sqrt{2}}$
$= \dfrac{14(3 + \sqrt{2})}{9 - 2}$
$= \dfrac{14(3 + \sqrt{2})}{7}$
$= 2(3 + \sqrt{2})$

55. $\dfrac{3}{2\sqrt{7} - \sqrt{5}} \cdot \dfrac{2\sqrt{7} + \sqrt{5}}{2\sqrt{7} + \sqrt{5}}$
$= \dfrac{3(2\sqrt{7} + \sqrt{5})}{2^2 \cdot 7 - 5}$
$= \dfrac{6\sqrt{7} + 3\sqrt{5}}{4 \cdot 7 - 5}$
$= \dfrac{6\sqrt{7} + 3\sqrt{5}}{28 - 5}$
$= \dfrac{6\sqrt{7} + 3\sqrt{5}}{23}$

57. $\dfrac{\sqrt{7}}{2\sqrt{5} - \sqrt{7}} \cdot \dfrac{2\sqrt{5} + \sqrt{7}}{2\sqrt{5} + \sqrt{7}}$
$= \dfrac{\sqrt{7}(2\sqrt{5} + \sqrt{7})}{2^2(5) - 7}$
$= \dfrac{2\sqrt{35} + 7}{20 - 7}$
$= \dfrac{2\sqrt{35} - 7}{13}$

59. $\dfrac{\sqrt{2}}{\sqrt{6} - \sqrt{2}} \cdot \dfrac{\sqrt{6} + \sqrt{2}}{\sqrt{6} + \sqrt{2}}$
$= \dfrac{\sqrt{2}(\sqrt{6} + \sqrt{2})}{6 - 2}$
$= \dfrac{\sqrt{12} + 2}{4}$
$= \dfrac{\sqrt{4}\sqrt{3} + 2}{4}$
$= \dfrac{2\sqrt{3} + 2}{4}$
$= \dfrac{2(\sqrt{3} + 1)}{2(2)}$
$= \dfrac{\sqrt{3} + 1}{2}$

61. $\dfrac{y}{1-\sqrt{y}} \cdot \dfrac{1+\sqrt{y}}{1+\sqrt{y}}$

$= \dfrac{y(1+\sqrt{y})}{1-\sqrt{y}}$

63. $\dfrac{\sqrt{11}+1}{\sqrt{11}-1} \cdot \dfrac{\sqrt{11}+1}{\sqrt{11}+1}$

$= \dfrac{11+2\sqrt{11}+1}{11-1}$

$= \dfrac{12+2\sqrt{11}}{10}$

$= \dfrac{2(6+\sqrt{11})}{2(5)}$

$= \dfrac{6+\sqrt{11}}{5}$

65. $\sqrt{3}(5+\sqrt{2}) = 5\sqrt{3}+\sqrt{6}$

67. $\sqrt{x}(\sqrt{2x}-1) = x\sqrt{2}-\sqrt{x}$

69. $(2\sqrt{3})^2 + 4^2 = (2\sqrt{7})^2$

$4(3)+16 = 4(7)$

$12+16 = 28$

$28 = 28$

Yes

71. $(\sqrt{7})^2 + (\sqrt{7})^2 = (7\sqrt{2})^2$

$7+7 = 7(4)$

$14 = 28$

No

73. $(\sqrt{6})^2 + (2\sqrt{3})^2 = 4^2$

$6+4(3) = 16$

$6+12 = 16$

$18 = 16$

No

75. $(\sqrt{2}+\sqrt{15})(\sqrt{3})$

$= \sqrt{6}+\sqrt{45}$

$= \sqrt{6}+\sqrt{9}\sqrt{5}$

$= \sqrt{6}+3\sqrt{5}$

77. Square: $(\sqrt{3}+1)(\sqrt{3}+1)$

$= 3+2\sqrt{3}+1$

$= 4+2\sqrt{3}$

Rectangle: $(\sqrt{3}+1)(\sqrt{3}+1-3)$

$= (\sqrt{3}+1)(\sqrt{3}-2)$

$= 3-2\sqrt{3}+\sqrt{3}-2$

$= 1-\sqrt{3}$

Total: $(4+2\sqrt{3})+(1-\sqrt{3})$

$= 5+\sqrt{3}$

Actually, this situation can't exist since a rectangle cannot have a negative area.

79. $F(55) = \dfrac{113}{\sqrt{55+15}} + 1$

$= \dfrac{113}{\sqrt{70}} + 1$

$= 14.5\%$

81. $F(t) = \dfrac{113}{\sqrt{t+15}} \cdot \dfrac{\sqrt{t+15}}{\sqrt{t+15}} + 1$

$= \dfrac{113\sqrt{t+15}}{t+15} + 1$

$C(t) = \dfrac{44}{\sqrt{t+15}} \cdot \dfrac{\sqrt{t+15}}{\sqrt{t+15}} + 0.9$

$= \dfrac{44\sqrt{t+15}}{t+15} + 0.9$

83. $\dfrac{\sqrt{1}}{\sqrt{5}} + \sqrt{4}\sqrt{5}$

$= \dfrac{1}{\sqrt{5}} + 2\sqrt{5}$

$= \dfrac{1 \cdot \sqrt{5}}{\sqrt{5} \cdot \sqrt{5}} + 2\sqrt{5}$

$= \dfrac{\sqrt{5}}{5} + 2\sqrt{5}$

$= \sqrt{5}\left(\dfrac{1}{5} + 2\right)$

$= \sqrt{5}\left(\dfrac{11}{5}\right)$

$= \dfrac{11\sqrt{5}}{5}$

85. $\dfrac{\sqrt{2}}{\sqrt{3}} \cdot \dfrac{\sqrt{3}}{\sqrt{3}} + \dfrac{\sqrt{3}}{\sqrt{2}} \cdot \dfrac{\sqrt{2}}{\sqrt{2}}$

$= \dfrac{\sqrt{6}}{3} + \dfrac{\sqrt{6}}{2}$

$= \dfrac{\sqrt{6}}{3} \cdot \dfrac{2}{2} + \dfrac{\sqrt{6}}{2} \cdot \dfrac{3}{3}$

$= \dfrac{2\sqrt{6}}{6} + \dfrac{3\sqrt{6}}{6}$

$= \dfrac{5\sqrt{6}}{6}$

87. $\left(\sqrt[3]{2}+1\right)\left(\sqrt[3]{4} - \sqrt[3]{2} + 1\right)$

$= \sqrt[3]{8} - \sqrt[3]{4} + \sqrt[3]{2} + \sqrt[3]{4} - \sqrt[3]{2} + 1$

$= 2 + 1$

$= 3$

89. $\sqrt{12x^2 + 12x + 3} + \sqrt{3x^2 + 18x + 27}$

$= \sqrt{3}\sqrt{4x^2 + 4x + 1} + \sqrt{3}\sqrt{x^2 + 6x + 9}$

$= \sqrt{3}\sqrt{(2x+1)^2} + \sqrt{3}\sqrt{(x+3)^2}$

$= \sqrt{3}(2x+1) + \sqrt{3}(x+3)$

$= \sqrt{3}(2x + 1 + x + 3)$

$= \sqrt{3}(3x + 4)$

## Section 9.5 Equations with Radicals

1. (a) True. The Multiplication Property of Equality allows us to multiply both sides by the same nonzero number. Because $a = b$, we multiply the left side by $a$ and the right side by $b$ to obtain $a^2 = b^2$. If $a$ and $b$ are 0, then clearly $a^2 = b^2$.

   (b) False. For example, $(3)^2 = (-3)^2$, but $3 \neq -3$.

3. 6

5. 3

7. 3

9. −1, 3

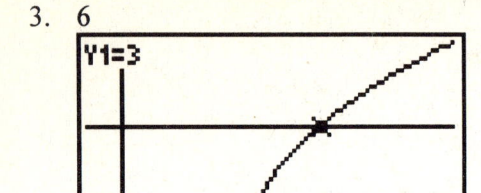

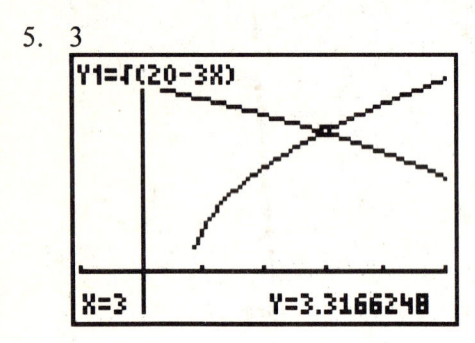

11. 2

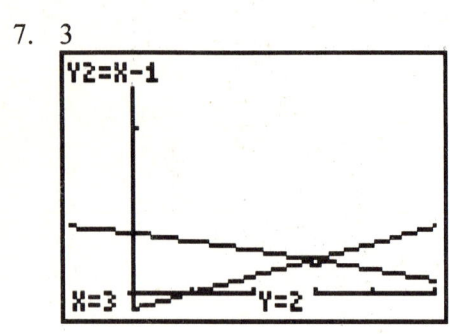

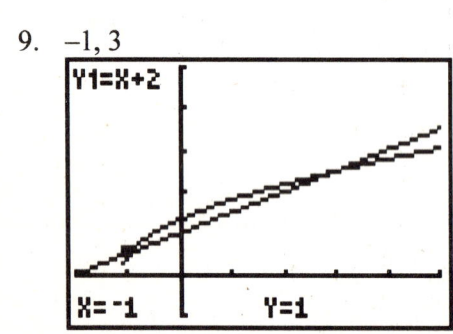

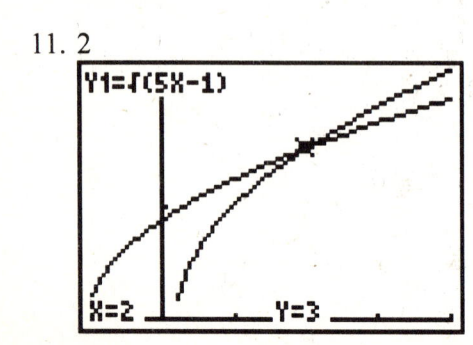

13. $\left(\sqrt{x-5}\right)^2 = 2^2$
    $x - 5 = 4$
    $x = 9$

15. $\left(2\sqrt{x}\right)^2 = \left(\sqrt{x+15}\right)^2$
    $4x = x + 15$
    $3x = 15$
    $x = 5$

17. $\left(\sqrt{x+3}\right)^2 = 7^2$
    $x + 3 = 49$
    $x = 46$

19. $\sqrt{3x+1} = -3 + 7$
    $\left(\sqrt{3x+1}\right)^2 = 4^2$
    $3x + 1 = 16$
    $3x = 15$
    $x = 5$

21. $\sqrt{x+1} = 6 - 8$
    $\sqrt{x+1} = -2$
    No solution since $\sqrt{x+1} \geq 0$.

23. $\left(\sqrt{2x+1}\right)^2 = \left(\sqrt{10-x}\right)^2$
    $2x + 1 = 10 - x$
    $3x + 1 = 10$
    $3x = 9$
    $x = 3$

25. $\left(\sqrt{x-7}\right)^2 = \left(\sqrt{2-2x}\right)^2$
    $x - 7 = 2 - 2x$
    $3x - 7 = 2$
    $3x = 9$
    $x = 3$
    But when $x = 3$, $\sqrt{3-7} = \sqrt{-4}$.
    $\sqrt{-4}$ is undefined.
    There is no solution.

Radical Expressions

27. $\left(\sqrt{2x-24}\right)^2 = \left(\sqrt{x+8}\right)^2$
$2x - 24 = x + 8$
$x - 24 = 8$
$x = 32$

29. $x^2 = \left(\sqrt{x^2 - 2x + 12}\right)^2$
$x^2 = x^2 - 2x + 12$
$0 = -2x + 12$
$2x = 12$
$x = 6$

31. $\left(\sqrt{x^2 + 3x + 3}\right)^2 = x^2$
$x^2 + 3x + 3 = x^2$
$3x + 3 = 0$
$3x = -3$
$x = -1$
However, $\sqrt{(-1)^2 + 3(-1) + 3} = -1$
$1 = -1$
This is a contradiction.
There is no solution.

33. $\left(\sqrt{x^2 + 7}\right)^2 = (x+1)^2$
$x^2 + 7 = x^2 + 2x + 1$
$7 = 2x + 1$
$6 = 2x$
$x = 3$

35. $\left(\sqrt{x^2 + 3x}\right)^2 = (x+1)^2$
$x^2 + 3x = x^2 + 2x + 1$
$3x = 2x + 1$
$x = 1$

37. The solution of $\sqrt{x^2 + 3x} = 2$ are −4 and 1. The first equation is not satisfied by $x = -4$ because $\sqrt{-4}$ is not a real number.

39. $\left(\sqrt{x+3}\right)^2 = (x+3)^2$
$x + 3 = x^2 + 6x + 9$
$0 = x^2 + 5x + 6$
$0 = (x+2)(x+3)$
$x = -2, -3$

41. $x + 2 = 3\sqrt{x}$
$(x+2)^2 = \left(3\sqrt{x}\right)^2$
$x^2 + 4x + 4 = 9x$
$x^2 - 5x + 4 = 0$
$(x-4)(x-1) = 0$
$x = 1, 4$

43. $t^2 = \left(\sqrt{4t-3}\right)^2$
$t^2 = 4t - 3$
$t^2 - 4t + 3 = 0$
$(t-1)(t-3) = 0$
$t = 1, 3$

45. $x + 1 = \sqrt{3x+1}$
$(x+1)^2 = \left(\sqrt{3x+1}\right)^2$
$x^2 + 2x + 1 = 3x + 1$
$x^2 - x = 0$
$x(x-1) = 0$
$x = 0, 1$

47. $x + 1 = 3\sqrt{x-1}$
$(x+1)^2 = \left(3\sqrt{x-1}\right)^2$
$x^2 + 2x + 1 = 9(x-1)$
$x^2 + 2x + 1 = 9x - 9$
$x^2 - 7x + 10 = 0$
$(x-2)(x-5) = 0$
$x = 2, 5$

49. $(x-3)^2 = (\sqrt{x-1})^2$

$x^2 - 6x + 9 = x - 1$

$x^2 - 7x + 10 = 0$

$(x-2)(x-5) = 0$

$x = 2, 5$

But $2 - 3 = \sqrt{2-1}$ leads to a contradiction, $-1 = 1$. So $x = 5$ is the solution.

51. $2\sqrt{x} = 8 - x$

$(2\sqrt{x})^2 = (8-x)^2$

$4x = 64 - 16x + x^2$

$x^2 - 20x + 64 = 0$

$(x-16)(x-4) = 0$

$x = 16, 4$

But $2\sqrt{16} = 8 - 16$

$2(4) = -8$

$8 = -8$

So $x = 4$ is the only solution.

53. $7\sqrt{x} = -x - 10$

$(7\sqrt{x})^2 = (-x-10)^2$

$49x = x^2 + 20x + 100$

$0 = x^2 - 29x + 100$

$0 = (x-25)(x-4)$

$x = 4, 25$

But $7\sqrt{4} = -4 - 10$

$7(2) = -14$

$14 = -14$

and

$7\sqrt{25} = -25 - 10$

$7(5) = -35$

$35 = -35$

So there is no solution.

55. $x^2 = (\sqrt{10-3x})^2$

$x^2 = 10 - 3x$

$x^2 + 3x - 10 = 0$

$(x+5)(x-2) = 0$

$x = -5, 2$

But $-5 = \sqrt{10-3(-5)}$

$-5 = \sqrt{10+15}$

$-5 = \sqrt{25}$

$-5 = 5$

So $x = 2$ is the only solution.

57. $\sqrt{x-2} = x - 4$

$(\sqrt{x-2})^2 = (x-4)^2$

$x - 2 = x^2 - 8x + 16$

$x^2 - 9x + 18 = 0$

$(x-3)(x-6) = 0$

$x = 3, 6$

But $\sqrt{3-2} = 3 - 4$

$\sqrt{1} = -1$

$1 = -1$

So $x = 6$ is the only solution.

**Radical Expressions**

59. $\sqrt{1-3x} = x-1$
$\left(\sqrt{1-3x}\right)^2 = (x-1)^2$
$1-3x = x^2 - 2x + 1$
$0 = x^2 + x$
$0 = x(x+1)$
$x = 0, x = -1$

$\sqrt{1-3(0)} = 0 - 1$
$\sqrt{1} = -1$
$1 = -1$
and $\sqrt{1-3(-1)} = -1-1$
$\sqrt{1+3} = -2$
$\sqrt{4} = -2$
$2 = -2$
There is no solution.

61. $\left(\sqrt{x+2}\right)^2 = \sqrt{x+8}$
$x + 4\sqrt{x} + 4 = x + 8$
$4\sqrt{x} = 4$
$\left(4\sqrt{x}\right)^2 = 4^2$
$16x = 16$
$x = 1$

63. $\left(\sqrt{2t+5}\right)^2 = \left(\sqrt{2t}+1\right)^2$
$2t + 5 = 2t + 2\sqrt{2t} + 1$
$4 = 2\sqrt{2t}$
$2 = \sqrt{2t}$
$(2)^2 = \left(\sqrt{2t}\right)^2$
$4 = 2t$
$t = 2$

65. $\left(\sqrt{3t}+1\right)^2 = \left(\sqrt{1+5t}\right)^2$
$3t + 2\sqrt{3t} + 1 = 1 + 5t$
$2\sqrt{3t} = 2t$
$\sqrt{3t} = t$
$\left(\sqrt{3t}\right)^2 = t^2$
$3t = t^2$
$0 = t^2 - 3t$
$0 = t(t-3)$
$t = 0, 3$

67. $\left(\sqrt{2x}+1\right)^2 = \left(\sqrt{x+7}\right)^2$
$2x + 2\sqrt{2x} + 1 = x + 7$
$x + 2\sqrt{2x} - 6 = 0$
$x - 6 = -2\sqrt{2x}$
$(x-6)^2 = \left(-2\sqrt{2x}\right)^2$
$x^2 - 12x + 36 = 4(2x)$
$x^2 - 12x + 36 = 8x$
$x^2 - 20x + 36 = 0$
$(x-18)(x-2) = 0$
$x = 2, 18$

But $\sqrt{2(18)} + 1 = \sqrt{18+7}$
$\sqrt{36} + 1 = \sqrt{25}$
$6 + 1 = 5$
$7 = 5$
So $x = 2$ is the only solution.

69. $3\sqrt{x} = \sqrt{x+16}$
$\left(3\sqrt{x}\right)^2 = \left(\sqrt{x+16}\right)^2$
$9x = x + 16$
$8x = 16$
$x = 2$

71. $\sqrt{2x+1} = x-7$
$\left(\sqrt{2x+1}\right)^2 = (x-7)^2$
$2x+1 = x^2 - 14x + 49$
$0 = x^2 - 16x + 48$
$0 = (x-12)(x-4)$
$x = 12, 4$

$\sqrt{2(4)+1} = 4-7$
$\sqrt{8+1} = -3$
$\sqrt{9} = -3$
$3 = -3$
So $x = 12$ is the only solution.

73. $d = 1.5\sqrt{h}$
$100 = 1.5\sqrt{h}$
$\dfrac{100}{1.5} = \sqrt{h}$
$\left(\dfrac{100}{1.5}\right)^2 = h$
$h = 4444$ feet

75. $340 = 20\sqrt{2n+7}$
$17 = \sqrt{2n+7}$
$(17)^2 = 2n+7$
$289 = 2n+7$
$282 = 2n$
$n = 141$

77. $\sqrt{x-2} - x + 12 = 8$
$\sqrt{x-2} = x-4$
$\left(\sqrt{x-2}\right)^2 = (x-4)^2$
$x-2 = x^2 - 8x + 16$
$0 = x^2 - 9x + 18$
$0 = (x-3)(x-6)$
$x = 3, 6$
But $\sqrt{3-2} - 3 + 12 = 8$
$\sqrt{1} - 3 + 12 = 8$
$1 - 3 + 12 = 8$
$10 = 8$
So $x = 6$ hours is the only solution.

79. $\sqrt{2n+9} = \sqrt{n+15}$
$\left(\sqrt{2n+9}\right)^2 = \left(\sqrt{n+15}\right)^2$
$2n+9 = n+15$
$n = 6$ years since 1995. So in 2001 the enrollment of the two groups will be the same.

81. (a) $2.3\sqrt{x+1} + 43.1 = 56.9$

(b) $2.3\sqrt{x+1} = 13.8$
$\sqrt{x+1} = 6$
$\left(\sqrt{x+1}\right)^2 = 6^2$
$x+1 = 36$
$x = 35$ years since 1970 (2005).

83. $\sqrt[3]{x} + 5 = 3$
$\sqrt[3]{x} = -2$
$\left(\sqrt[3]{x}\right)^3 = (-2)^3$
$x = -8$

85. $\left(\sqrt[4]{1-x}\right)^4 = \left(\sqrt[4]{3x-7}\right)^4$
    $1 - x = 3x - 7$
    $0 = 4x - 8$
    $8 = 4x$
    $x = 2$

    But $\sqrt[4]{1-2} = \sqrt[4]{-1}$ which is undefined.
    There is no solution.

87. $r^2 = \left(\sqrt{\dfrac{3v}{\pi h}}\right)^2$
    $r^2 = \dfrac{3v}{\pi h}$
    $\pi h r^2 = 3v$
    $h = \dfrac{3v}{\pi r^2}$

## Section 9.6 Applications with Right Triangles

1. The hypotenuse is the longest side of the triangle and is opposite the right angle.

3. $5^2 + 12^2 = x^2$
   $25 + 144 = x^2$
   $x^2 = 169$
   $x = 13$

5. $x^2 + 6^2 = 10^2$
   $x^2 + 36 = 100$
   $x^2 = 64$
   $x = 8$

7. $6^2 + 9^2 = x^2$
   $36 + 81 = x^2$
   $x^2 = 117$
   $x = \sqrt{117}$
   $x \approx 10.82$

9. $x^2 + 2^2 = \left(\sqrt{15}\right)^2$
   $x^2 + 4 = 15$
   $x^2 = 11$
   $x = \sqrt{11}$
   $x \approx 3.32$

11. $\left(2\sqrt{5}\right)^2 + x^2 = 6^2$
    $4(5) + x^2 = 36$
    $20 + x^2 = 36$
    $x^2 = 16$
    $x = 4$

13. $8^2 + 15^2 = c^2$
    $64 + 225 = c^2$
    $c^2 = 289$
    $c = 17$

15. $a^2 + 12^2 = 13^2$
    $a^2 + 144 = 169$
    $a^2 = 25$
    $a = 5$

17. $a^2 + 5^2 = 11^2$
    $a^2 + 25 = 121$
    $a^2 = 96$
    $a = \sqrt{96}$
    $a \approx 9.80$

19. $\left(6\sqrt{2}\right)^2 + b^2 = 10^2$
    $36(2) + b^2 = 100$
    $72 + b^2 = 100$
    $b^2 = 28$
    $b = \sqrt{28}$
    $b \approx 5.29$

21. $\left(\sqrt{7}\right)^2 + \left(\sqrt{7}\right)^2 = c^2$
    $7 + 7 = c^2$
    $c^2 = 14$
    $c = \sqrt{14}$
    $c \approx 3.74$

23. $5^2 + 12^2 = x^2$
    $25 + 144 = x^2$
    $x^2 = 169$
    $x = 13.00$ feet

25. $(120)^2 + x^2 = 200^2$
    $14400 + x^2 = 40000$
    $x^2 = 25600$
    $x = 160.00$ feet

27. $70^2 + x^2 = 200^2$
    $4900 + x^2 = 40000$
    $x^2 = 35100$
    $x = \sqrt{35100}$
    $x \approx 187.35$ feet

29. $40^2 + 60^2 = x^2$
    $1600 + 3600 = x^2$
    $x^2 = 5200$
    $x = \sqrt{5200}$
    $x \approx 72.11$ feet

31. $\left(2(8)\right)^2 + \left(2(6)\right)^2 = x^2$
    $16^2 + 12^2 = x^2$
    $256 + 144 = x^2$
    $x^2 = 400$
    $x = 20.00$ miles

33. $65^2 + \left(60(2)\right)^2 = x^2$
    $4225 + (120)^2 = x^2$
    $4225 + 14400 = x^2$
    $x^2 = 18625$
    $x = \sqrt{18625}$
    $x \approx 136.47$ miles

35. $\left(1.5(120)\right)^2 + 110^2 = x^2$
    $180^2 + 110^2 = x^2$
    $32400 + 12100 = x^2$
    $x^2 = 44500$
    $x = \sqrt{44500}$
    $x \approx 210.95$ miles

37. $5^2 + 12^2 = x^2$
    $25 + 144 = x^2$
    $x^2 = 169$
    $x = \sqrt{169}$
    $x = 13$

39. $4^2 + h^2 = 8^2$
    $16 + h^2 = 64$
    $h^2 = 48$
    $h = \sqrt{48}$
    $h = 6.93$ feet

**Radical Expressions**

41.

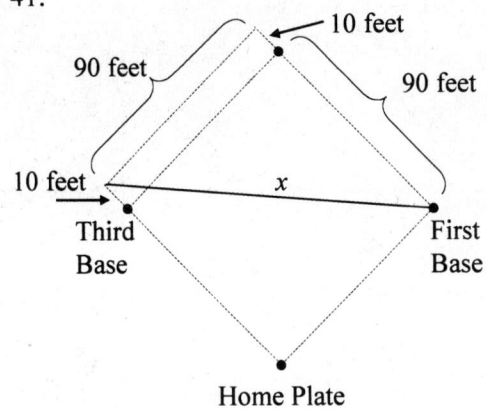

$(90+10)^2 + 90^2 = x^2$
$100^2 + 90^2 = x^2$
$10000 + 8100 = x^2$
$x^2 = 18100$
$x = \sqrt{18100}$
$x \approx 134.54$ feet

43. $x^2 + x^2 = 30^2$
$2x^2 = 900$
$x^2 = 450$
$x = \sqrt{450}$
$x = 21.21$ inches

45. One leg is horizontal and the other is vertical.

47.

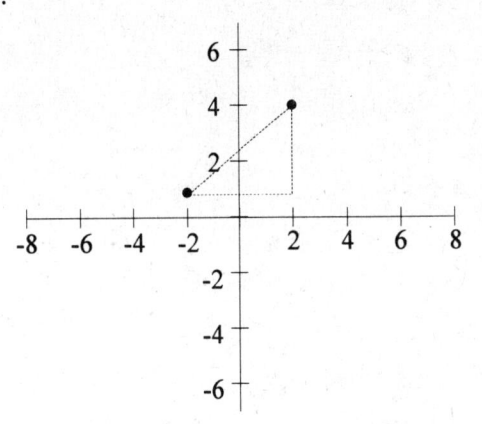

$x^2 = 3^2 + 4^2$
$x^2 = 9 + 16$
$x^2 = 25$
$x = 5$

49. $10 - 2 = 8$
$10 - 4 = 6$

$8^2 + 6^2 = x^2$
$64 + 36 = x^2$
$x^2 = 100$
$x = 10$

51. $7 - 3 = 4$
$-2 - (-8) = -2 + 8 = 6$

$4^2 + 6^2 = x^2$
$16 + 36 = x^2$
$52 = x^2$
$x = \sqrt{52}$
$x \approx 7.21$

53. $y^2 + 24^2 = 25^2$
$y^2 + 576 = 625$
$y^2 = 49$
$y = 7$

$z^2 + 24^2 = 26^2$
$z^2 + 576 = 676$
$z^2 = 100$
$z = 10$

$x = z - y$
$= 10 - 7$
$= 3$ feet

55. $10^2 + 10^2 = x^2$
$100 + 100 = x^2$
$x^2 = 200$
$x = \sqrt{200}$
$x = 14.14$ feet.

This is the length of the diagonal of the square. This is also the diameter of the circle. So the radius of the circle is 7.07 feet.

The area of the circle is $\pi(7.07)^2 = 50\pi$. The area of the square is $10 \cdot 10 = 100$. The area outside the square but inside the circle is $50\pi - 100 = 57.08$ square feet.

57. (a) $x^2 + x^2 = 12^2$
$2x^2 = 144$
$x^2 = 72$
$x = \sqrt{72}$
$x = 8.49$ feet

(b) $x^2 + x^2 = c^2$
$2x^2 = c^2$
$x^2 = \dfrac{c^2}{2}$
$x = \sqrt{\dfrac{c^2}{2}}$
$= \dfrac{c}{\sqrt{2}} \cdot \dfrac{\sqrt{2}}{\sqrt{2}}$
$= \dfrac{c\sqrt{2}}{2}$

## Section 9.7 Rational Exponents

1. In the definition of $b^{\frac{1}{n}}$, the expression $\sqrt[n]{b}$ must be defined. In this case, $\sqrt{-25}$ is not a real number.

3. $100^{\frac{1}{2}}$
$= \sqrt{100}$
$= 10$

# Radical Expressions

5. $\sqrt[3]{8}$
$= \sqrt[3]{2^3}$
$= 2$

7. $(-1)^{\frac{1}{6}}$
Not a real number since $-1 < 0$ and 6 is even.

9. $\left((-2)^5\right)^{\frac{1}{5}}$
$= -2$

11. $\left(2^4\right)^{\frac{1}{4}}$
$= 2$

13. 3.13

15. $-3.11$

17. 3.04

19. The expression in (iv) is not a real number because the base is negative and $n$ is even.

(i) $-\left(2^2\right)^{\frac{3}{2}}$
$= -2^3$
$= -8$

(ii) $\left(2^2\right)^{\frac{-3}{2}}$
$= 2^{-3}$
$= 2^{\frac{1}{3}}$
$= \frac{1}{8}$

(iii) $-\left(2^2\right)^{\frac{-3}{2}}$
$= -2^{-3}$
$= -\frac{1}{2^3}$
$= -\frac{1}{8}$

21. $\left(3^2\right)^{\frac{3}{2}}$
$= 3^3$
$= 27$

23. $\left(3^3\right)^{\frac{2}{3}}$
$= 3^2$
$= 9$

25. $\left((-2)^5\right)^{\frac{2}{5}}$
$= (-2)^2$
$= 4$

27. $\left((-4)^3\right)^{\frac{2}{3}}$
$= (-4)^2$
$= 16$

29. $\left(3^4\right)^{\frac{3}{4}}$
$= 3^3$
$= 27$

31. $\left(\left(\frac{4}{5}\right)^2\right)^{\frac{3}{2}}$
$= \left(\frac{4}{5}\right)^3$
$= \frac{64}{125}$

33. $\left(4^3\right)^{\frac{4}{3}}$
$= 4^4$
$= 256$

35. $\left(4^2\right)^{-\frac{1}{2}}$
$= 4^{-1}$
$= \dfrac{1}{4}$

37. $\left(2^3\right)^{-\frac{2}{3}}$
$= 2^{-2}$
$= \dfrac{1}{2^2}$
$= \dfrac{1}{4}$

39. $\left((-2)^3\right)^{-\frac{5}{3}}$
$(-2)^{-5}$
$= \dfrac{1}{(-2)^5}$
$= -\dfrac{1}{32}$

41. $\left(\left(\dfrac{3}{2}\right)^3\right)^{-\frac{4}{3}}$
$= \left(\dfrac{3}{2}\right)^{-4}$
$= \left(\dfrac{2}{3}\right)^4$
$= \dfrac{16}{81}$

43. $\left(2^2\right)^{-\frac{5}{2}}$
$= 2^{-5}$
$= \dfrac{1}{2^5}$
$= \dfrac{1}{32}$

45. 13.57

47. 0.18

49. 24.68

51. $7^{\frac{3}{5}+\frac{2}{5}}$
$= 7^{\frac{5}{5}}$
$= 7$

53. $a^{-\frac{1}{2}+\frac{3}{2}}$
$= a^{\frac{2}{2}}$
$= a$

55. $y^{\frac{1}{2}+\frac{1}{3}}$
$= y^{\frac{3}{6}+\frac{2}{6}}$
$= y^{\frac{5}{6}}$

57. $5^2$
$= 25$

59. $y^{\frac{6}{3}}$
$= y^2$

61. $a^{\frac{12}{15}}$
$= a^{\frac{4}{5}}$

63. $10^{\frac{4}{3}-\frac{1}{3}}$
$= 10^{\frac{3}{3}}$
$= 10$

65. $x^{1-\frac{1}{2}}$
$= x^{\frac{1}{2}}$

67. $z^{\frac{3}{2}\left(-\frac{1}{2}\right)}$
$= z^{\frac{4}{2}}$
$= z^2$

69. $36^{\frac{1}{2}} x^{\frac{10}{2}}$

$= \left(6^2\right)^{\frac{1}{2}} x^5$

$= 6x^5$

71. $\left(3^2 x^{-6}\right)^{-\frac{1}{2}}$

$= 3^{-\frac{2}{2}} x^{\frac{6}{2}}$

$= 3^{-1} x^3$

$= \frac{1}{3} x^3$

73. $y^{\frac{24}{4}} z^{-\frac{8}{2}}$

$= y^6 z^{-4}$

$= \frac{y^6}{z^4}$

75. $\frac{y^{\frac{3}{3}}}{8^{\frac{1}{3}}}$

$= \frac{y}{2}$

77. $\frac{a^{-2}}{b^{-\frac{18}{3}}}$

$= \frac{a^{-2}}{b^{-6}}$

$= \frac{b^6}{a^2}$

79. $\frac{y^{\frac{12}{2}}}{x^{-\frac{6}{3}}}$

$= \frac{y^4}{x^{-2}}$

$= x^2 y^4$

81. $v = 3.5d^{\frac{1}{2}}$

$= 3.5(230)^{\frac{1}{2}}$

$= 3.5(15 \cdot 17)$

$= 53.08$

$\approx 53$ mph

83. 165 feet

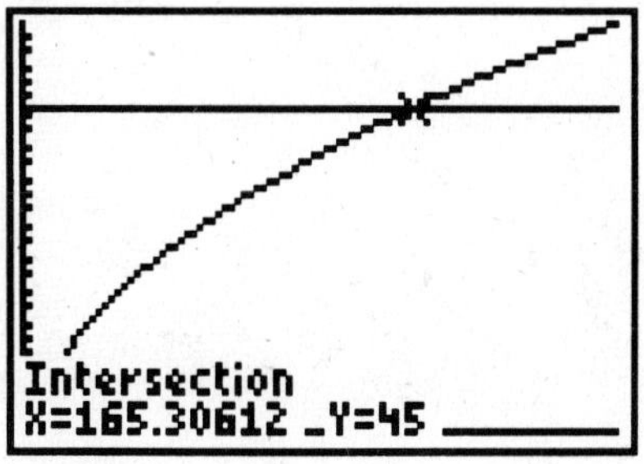
Intersection
X=165.30612 Y=45

85. $t = \frac{d^{\frac{1}{2}}}{4}$

$t = \frac{(450)^{\frac{1}{2}}}{4}$

$= \frac{21.21}{4}$

$= 5.30$

$\approx 5.3$ seconds

87. $2 = \frac{d^{\frac{1}{2}}}{4}$

$8 = d^{\frac{1}{2}}$

$8^2 = d$

$d = 64$ feet

89. $v = 284x^{\frac{1}{5}}$

$= 284\sqrt[5]{x}$

$v = 284(15)^{\frac{1}{5}}$

$= 284(1.72)$

$= 488$ million visitors

91. $R = \dfrac{5\left(284x^{\frac{1}{5}}\right)}{2.8}$

93. $\sqrt[4]{x^2}$
$= x^{\frac{2}{4}}$
$= x^{\frac{1}{2}}$
$= \sqrt{x}$

95. $\left(9^3 x^6\right)^{\frac{1}{12}}$
$= 9^{\frac{3}{12}} x^{\frac{6}{12}}$
$= 9^{\frac{1}{4}} x^{\frac{1}{2}}$
$= \left(3^2\right)^{\frac{1}{4}} x^{\frac{1}{2}}$
$= 3^{\frac{2}{4}} x^{\frac{1}{2}}$
$= 3^{\frac{1}{2}} x^{\frac{1}{2}}$
$= (3x)^{\frac{1}{2}}$
$= \sqrt{3x}$

97. $y^{-\frac{1}{2}} = 7$
$\dfrac{1}{y^{\frac{1}{2}}} = 7$
$1 = 7y^{\frac{1}{2}}$
$\dfrac{1}{7} = y^{\frac{1}{2}}$
$\left(\dfrac{1}{7}\right)^2 = \left(y^{\frac{1}{2}}\right)^2$
$\dfrac{1}{49} = y$

## Chapter 9 Review Exercises

1. (a) $4 \cdot 4 = 16$
   $(-4)(-4) = 16$
   $4, -4$ are square roots of 16

   (b) $\sqrt{16}$
   $= 4$

3. (a) $4.12 - 2.23$
   $= 1.89$

   (b) $\sqrt{17 - 5}$
   $= \sqrt{12}$
   $= 3.46$

5. (a) $\sqrt{(-2)^2}$
   $= \sqrt{4}$
   $= 2$

   (b) $\sqrt{(x+3)^2}$
   $= |x + 3|$

7. (a) $-\left(2^5\right)^{\frac{1}{5}}$
   $= -2$

# Radical Expressions

(b) $\left((-2)^5\right)^{\frac{1}{5}}$
$= -2$

9. $\sqrt{x^2}$
$= |x|$

10. The index and exponent $n$ must be an odd number. If $n$ were an even number, then $\sqrt[7]{a^n} = |a|$.

11. $\sqrt{2x \cdot 3y}$
$= \sqrt{6xy}$

13. (a) $\sqrt{9}\sqrt{2}$
$= 3\sqrt{2}$

(b) $3\sqrt{28}$
$= 3\sqrt{4}\sqrt{7}$
$= 3 \cdot 2\sqrt{7}$
$= 6\sqrt{7}$

15. $\sqrt{25 \cdot a^6 b^{10}}\sqrt{2 \cdot ab}$
$= 5a^3 b^5 \sqrt{2ab}$

17. (a) $\sqrt[3]{x^8}$
$= \sqrt[3]{x^6}\sqrt[3]{x^2}$
$= x^2 \sqrt[3]{x^2}$

(b) $\sqrt[4]{2^4 x^4} \cdot \sqrt[4]{x}$
$= 2x\sqrt[4]{x}$

19. $\sqrt{\dfrac{2x^5}{32x}}$
$= \sqrt{\dfrac{1 \cdot x^4}{16}}$
$= \left(\dfrac{x^4}{16}\right)^{\frac{1}{2}}$
$= \dfrac{x^2}{14}$

21. $\dfrac{\sqrt{3y}}{\sqrt{49}}$
$= \dfrac{\sqrt{3y}}{7}$

23. $\sqrt{\dfrac{48}{49a^{10}}}$
$= \sqrt{\dfrac{16}{49a^{10}}}\sqrt{3}$
$= \dfrac{4}{7a^5}\sqrt{3}$
$= \dfrac{4\sqrt{3}}{7a^5}$

25. $\sqrt{\dfrac{24}{5x^3}}$
$= \sqrt{\dfrac{4}{x^2}}\sqrt{\dfrac{6}{5x}}$
$= \dfrac{2}{x} \cdot \dfrac{\sqrt{6}}{\sqrt{5x}} \cdot \dfrac{\sqrt{5x}}{\sqrt{5x}}$
$= \dfrac{2\sqrt{30x}}{x \cdot 5x}$
$= \dfrac{2\sqrt{30x}}{5x^2}$

27. $\sqrt[3]{\dfrac{8x^3}{3^3}}\sqrt[3]{\dfrac{2x}{1}}$
$= \dfrac{2x\sqrt[3]{2x}}{3}$

29. $\sqrt{7}(1-4)$
    $= -3\sqrt{7}$

31. $\sqrt{36}\sqrt{2} + 3\sqrt{16}\sqrt{2} - \sqrt{16}\sqrt{3}$
    $= 6\sqrt{2} + 3(4)\sqrt{2} - 4\sqrt{3}$
    $= 6\sqrt{2} + 12\sqrt{2} - 4\sqrt{3}$
    $= 18\sqrt{2} - 4\sqrt{3}$

33. $\sqrt{15} + \sqrt{6}$

35. $3\sqrt{14} + 7 - 6\sqrt{2} - 2\sqrt{7}$

37. $(4 - \sqrt{2})(4 + \sqrt{2})$
    $= 16 - 2$
    $= 14$

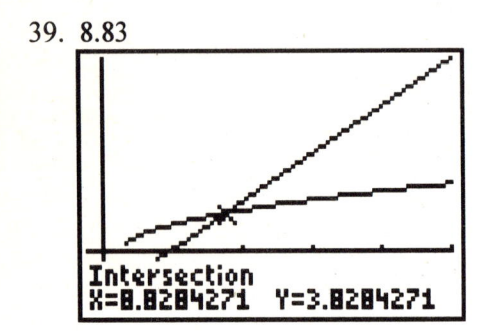

39. 8.83

    Intersection
    X=8.8284271  Y=3.8284271

41. $(\sqrt{3x-4})^2 = (\sqrt{x+6})^2$
    $3x - 4 = x + 6$
    $2x - 10 = 0$
    $2x = 10$
    $x = 5$

43. $\sqrt{6x+19} = x + 2$
    $(\sqrt{6x+19})^2 = (x+2)^2$
    $6x + 19 = x^2 + 4x + 4$
    $0 = x^2 - 2x - 15$
    $0 = (x-5)(x+3)$

    $x = 5, -3$
    But $\sqrt{6(-3)+19} = -3 + 2$
    $\sqrt{1} = -1$
    $1 = -1$
    So $x = 5$ is the only solution.

45. $(\sqrt{x+5})^2 = (5 - \sqrt{x})^2$
    $x + 5 = 25 - 10\sqrt{x} + x$
    $-20 = -10\sqrt{x}$
    $2 = \sqrt{x}$
    $x = 4$

47. $\sqrt{x+1} = x - 1$
    $(\sqrt{x+1})^2 = (x-1)^2$
    $x + 1 = x^2 - 2x + 1$
    $0 = x^2 - 3x$
    $0 = x(x-3)$
    $x = 0, 3$

    But $\sqrt{0+1} = 0 - 1$
    $\sqrt{1} = -1$
    $1 = -1$
    So the only solution is $x = 3$.

# Radical Expressions

49. $4^2 + \left(5\sqrt{2}\right)^2 = x^2$
$16 + 25(2) = x^2$
$16 + 50 = x^2$
$x^2 = 66$
$x = \sqrt{66}$
$x = 8.12$ inches.

51. $30^2 + x^2 = 31^2$
$900 + x^2 = 961$
$x^2 = 61$
$x = \sqrt{61}$
$x = 7.81$ feet

53. (a) $\left(\left(\frac{3}{4}\right)^2\right)^{\frac{1}{2}}$

    (b) $-\left((3)^4\right)^{\frac{1}{4}}$
    $= -3$

55. (a) $\left(4^3\right)^{\frac{5}{3}}$
$= 4^5$
$= 1024$

    (b) $(-16)^{\frac{3}{4}}$
    = Not a real number, since $-16 < 0$ and 4 is even.

57. $25^{\frac{3}{4} - \frac{1}{4}}$
$= 25^{\frac{2}{4}}$
$= 25^{\frac{1}{2}}$
$= \left(5^2\right)^{\frac{1}{2}}$
$= 5$

59. $x^{\frac{1}{3} - \frac{1}{4}}$
$= x^{\frac{4}{12} - \frac{3}{12}}$
$= x^{\frac{1}{12}}$

61. Write the equation as $2\sqrt{x} = 15$ and solve the radical equation.

## Chapter 9 Test

1. (a) $\sqrt{9}$
   $= 3$

   (b) Not a real number, since $\sqrt{-3}$ is not real.

3. Because the radicand is negative, the expression is a real number only if the index is an odd integer.

5. $\sqrt{12a^8}$
   $= \sqrt{4a^8}\sqrt{3}$
   $= 2a^4\sqrt{3}$

7. (a) $\sqrt{4x^3}$
   $= \sqrt{4x^2}\sqrt{x}$
   $= 2x\sqrt{x}$

   (b) $\frac{\sqrt{2}}{\sqrt{7}} \cdot \frac{\sqrt{7}}{\sqrt{7}}$
   $= \frac{\sqrt{14}}{7}$

9. $\sqrt[3]{\dfrac{16x}{2x^4}}$

$= \sqrt[3]{\dfrac{8}{x^3}}$

$= \dfrac{2}{x}$

11. $3 + 2\sqrt{15} - \sqrt{15} - 2\sqrt{5}\sqrt{5}$

$= 3 + \sqrt{15} - 2(5)$

$= 3 + \sqrt{15} - 10$

$= -7 + \sqrt{15}$

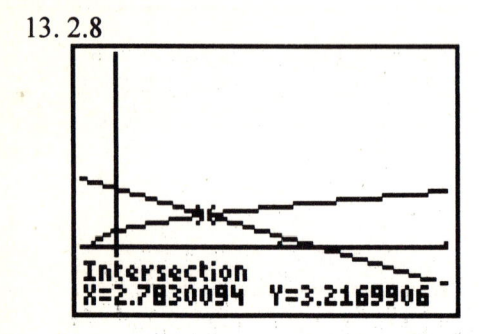

13. 2.8

15. $\left(\sqrt{x^2+1}\right)^2 = (x-1)^2$

$x^2 + 1 = x^2 - 2x + 1$

$1 = -2x + 1$

$0 = -2x$

$x = 0$

But, $\sqrt{0^2 + 1} = 0 - 1$

$\sqrt{1} = -1$

$1 = -1$

So, there is no solution.

17. $5^2 + 11^2 = x^2$

$25 + 121 = x^2$

$146 = x^2$

$x = \sqrt{146}$

$x = 12.1$ inches

19. The base $x$ is the radicand, the denominator of the exponent is the index, and the numerator is the power:

$x^{\frac{2}{3}} = \left(\sqrt[3]{x}\right)^2$.

21. (a) 1.9

(b) 0.4

# Chapter 10

# Quadratic Equations and Functions

## Section 10.1

1. Because the value of $y$ must be 0, we trace to the $x$-intercepts.

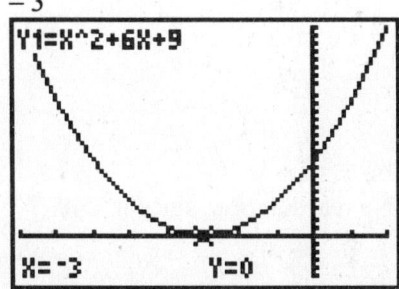

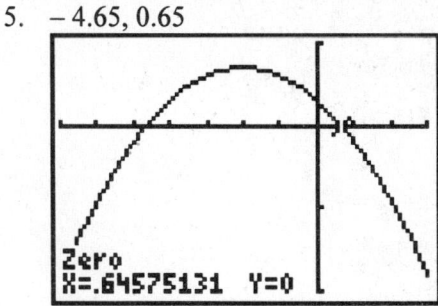

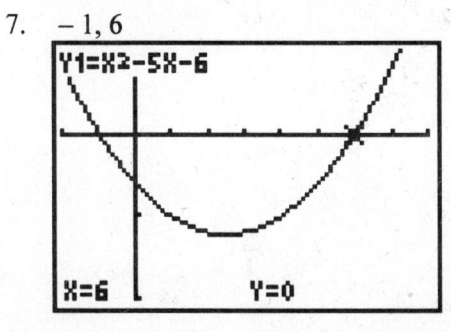

3. $-3$

5. $-4.65, 0.65$

7. $-1, 6$

9. No real number solution.

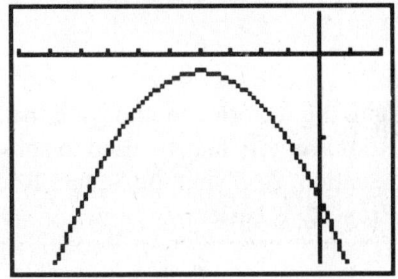

11. $(x+5)(x+1) = 0$
    $x = -5, x = -1$

13. $x^2 - 14x + 49 = 0$
    $(x-7)^2 = 0$
    $x = 7$

15. $3x(2x-1) = 0$
    $3x = 0$
    $x = 0$

    $2x - 1 = 0$
    $2x = 1$
    $x = \dfrac{1}{2}$

17. $x^2 - x - 12 = 0$
    $(x-4)(x+3) = 0$
    $x = 4, x = -3$

325

19. $y^2 - 10y + 25 = 0$
    $(x-5)^2 = 0$
    $x - 5 = 0$
    $x = 5$

21. $t^2 + 4t - 60 = 0$
    $(t+10)(t-6) = 0$
    $t = -10, t = 6$

23. $12x^2 - 28x + 15 = 0$
    $(2x-3)(6x-5) = 0$
    $2x - 3 = 0 \quad 6x - 5 = 0$
    $2x = 3 \quad\quad 6x = 5$
    $x = \dfrac{3}{2} \quad\quad x = \dfrac{5}{6}$

25. Both the factoring method and the Square Root Property may be used to solve the equation. Applying the Square Root Property is easier.

27. $\sqrt{x^2} = \sqrt{49}$
    $|x| = 7$
    $x = \pm 7$

29. $4t^2 - 25 = 0$
    $4t^2 = 25$
    $t^2 = \dfrac{25}{4}$
    $\sqrt{t^2} = \sqrt{\dfrac{25}{4}}$
    $|t| = \dfrac{5}{2}$
    $t = \pm \dfrac{5}{2}$

31. $x^2 = -25$
    No real solution, since $x^2 \geq 0$.

33. $14 = 4z^2$
    $\dfrac{14}{4} = z^2$
    $z^2 = \dfrac{7}{2}$
    $\sqrt{z^2} = \sqrt{\dfrac{7}{2}}$
    $|z| = \dfrac{\sqrt{7}}{\sqrt{2}} \cdot \dfrac{\sqrt{2}}{\sqrt{2}}$
    $|z| = \dfrac{\sqrt{14}}{2}$
    $z = \pm \dfrac{\sqrt{14}}{2}$

35. $x^2 = 125$
    $\sqrt{x^2} = \sqrt{125}$
    $|x| = \sqrt{25}\sqrt{5}$
    $x = \pm 5\sqrt{5}$

37. $w^2 = -7$
    No real solution, since $w^2 > 0$.

39. $\sqrt{(x+5)^2} = \sqrt{9}$
    $|x + 5| = 3$
    $x + 5 = 3$
    $x = -2$

    $x + 5 = -3$
    $x = -8$

## Quadratic Equations and Functions

41. $\sqrt{(w-9)^2} = \sqrt{12}$
$|w-9| = \sqrt{4}\sqrt{3}$
$|w-9| = 2\sqrt{3}$

$w - 9 = 2\sqrt{3}$
$w = 9 + 2\sqrt{3}$

$w - 9 = -2\sqrt{3}$
$w = 9 - 2\sqrt{3}$

43. $(x-6)^2 = -25$

No real solution, since $(x-6)^2 \geq 0$.

45. $\sqrt{(3-x)^2} = \sqrt{\dfrac{49}{16}}$

$|3-x| = \dfrac{7}{4}$

$3 - x = \dfrac{7}{4}$

$-x = \dfrac{7}{4} - 3$

$x = -\dfrac{7}{4} + 3$

$x = -\dfrac{7}{4} + \dfrac{12}{4}$

$x = \dfrac{5}{4}$

$3 - x = -\dfrac{7}{4}$

$-x = -\dfrac{7}{4} - 3$

$-x = -\dfrac{7}{4} - \dfrac{12}{4}$

$-x = -\dfrac{19}{4}$

$x = \dfrac{19}{4}$

47. $(3x-4)^2 = 36$

$\sqrt{(3x-4)^2} = \sqrt{36}$

$|3x-4| = 6$

$3x - 4 = 6$
$3x = 10$
$x = \dfrac{10}{3}$

$3x - 4 = -6$
$3x = -2$
$x = -\dfrac{2}{3}$

49. $9(w+3)^2 - 4 = 0$

$9(w+3)^2 = 4$

$(w+3)^2 = \dfrac{4}{9}$

$\sqrt{(w+3)^2} = \sqrt{\dfrac{4}{9}}$

$|w+3| = \dfrac{2}{3}$

$w + 3 = \dfrac{2}{3}$

$w = \dfrac{2}{3} - 3$

$w = \dfrac{2}{3} - \dfrac{9}{3}$

$w = -\dfrac{7}{3}$

$w + 3 = -\dfrac{2}{3}$

$w = -3 - \dfrac{2}{3}$

$w = -\dfrac{9}{3} - \dfrac{2}{3}$

$w = -\dfrac{11}{3}$

51. Because the left side of (i) is a perfect square trinomial, it is easier to solve with the Square Root Property.

53. $(x-9)^2 = 25$
$\sqrt{(x-9)^2} = \sqrt{25}$
$|x-9| = 5$
$x - 9 = \pm 5$
$x = 9 \pm 5$
$x = 14, 4$

55. $(2x+5)^2 = 9^2$
$\sqrt{(2x+5)^2} = \sqrt{9^2}$
$|2x+5| = 9$
$2x + 5 = \pm 9$
$2x = -5 \pm 9$
$x = \dfrac{-5 \pm 9}{2}$

$x = \dfrac{4}{2} = 2$

$x = -\dfrac{14}{2} = -7$

57. $(x+3)^2 = 10$
$\sqrt{(x+3)^2} = \sqrt{10}$
$|x+3| = \sqrt{10}$
$x + 3 = \pm\sqrt{10}$
$x = -3 \pm \sqrt{10}$

59. $\dfrac{3}{x^2} = 9$
$\dfrac{x^2}{3} = \dfrac{1}{9}$
$x^2 = \dfrac{3}{9}$
$x^2 = \dfrac{1}{3}$
$\sqrt{x^2} = \sqrt{\dfrac{1}{3}}$
$|x| = \dfrac{1}{\sqrt{3}} \cdot \dfrac{\sqrt{3}}{\sqrt{3}}$
$x = \pm\dfrac{\sqrt{3}}{3}$

61. $\dfrac{3}{y^2} = \dfrac{1}{3}$
$\dfrac{1}{y^2} = \dfrac{1}{9}$
$y^2 = 9$
$\sqrt{y^2} = \sqrt{9}$
$|y| = 3$
$y = \pm 3$

63. $x^2 + 12x = 36 + 12x$
$x^2 = 36$
$\sqrt{x^2} = \sqrt{36}$
$|x| = 6$
$x = \pm 6$

65. $4x^2 + 6 = 3x^2 + 15$
$x^2 = 9$
$\sqrt{x^2} = \sqrt{9}$
$|x| = 3$
$x = \pm 3$

**67.** $3(x^2 - 1) = 2x^2 + 5$
$3x^2 - 3 = 2x^2 + 5$
$x^2 = 8$
$\sqrt{x^2} = \sqrt{8}$
$|x| = \sqrt{4}\sqrt{2}$
$x = \pm 2\sqrt{2}$

**69.** $3\left(\dfrac{1}{x}\right) = \dfrac{x-4}{4}$
$\dfrac{3}{x} = \dfrac{x-4}{4}$
$12 = x(x-4)$
$12 = x^2 - 4x$
$0 = x^2 - 4x - 12$
$0 = (x-6)(x+2)$
$x = 6, \; x = -2$

But $x > 0$ so $x = 6$.

**71.** $w =$ width
$3w =$ length
$3w \cdot w =$ area

$3w^2 + 3 = 2w(3w - 4)$
$3w^2 + 3 = 6w^2 - 8w$
$0 = 3w^2 - 8w - 3$
$0 = (3w + 1)(w - 3)$

width: $w = 3$
length: $3w = 3(3) = 9$

**73.** $l =$ length
$l - 2 =$ width
$l(l - 2) =$ area

$(l + 4)(l - 2 + 4) = 2l(l - 2) + 8$
$(l + 4)(l + 2) = 2l^2 - 4l + 8$
$l^2 + 6l + 8 = 2l^2 - 4l + 8$
$0 = l^2 - 10l$
$0 = l(l - 10)$

length: $l = 10$ inches
width: $l - 2 = 10 - 2 = 8$ inches

**75.** $h =$ height
$2h =$ base

$\dfrac{1}{2}h(2h) = 25$
$h^2 = 25$
$\sqrt{h^2} = \sqrt{25}$

height: $h = 5$
base: $2h = 2(5) = 10$ feet

77. $b$ = base
$6b$ = height
$\frac{1}{2}6b(b)$ = area

$3b^2 = 60$
$b^2 = 20$
$\sqrt{b^2} = \sqrt{20}$
$|b| = \sqrt{4}\sqrt{5}$

base: $b = 4.47$
height: $6b = 6(4.47) = 26.83$ inches

If the base of the triangular pennant is touching the ceiling, the pennant will reach 26.83 inches from the ceiling.

79. (a) 16, 78

(b) $x^2 - 94x + 1248 = 0$
$(x - 78)(x - 16) = 0$
$x = 78, x = 16$

People aged 16 or 78 received no assistance.

81. $0.18t^2 + 0.1 = 26$

83. The solution $t = -12$ corresponds to 1978, but the question is about the year after 1990, when the cost will be $26 billion.

85. $3V = \pi r^2 h$
$\dfrac{3V}{\pi h} = r^2$
$r = \sqrt{\dfrac{3V}{\pi h}}$

87. $y^2 = 49 - x^2$
$|y| = \sqrt{49 - x^2}$
$y = \pm\sqrt{49 - x^2}$

89. $(x+3)(x-4) = 0$
$x^2 - 4x + 3x - 12 = 0$
$x^2 - x - 12 = 0$

## Section 10.2 Completing the Square

1. Take half of $-9$, the coefficient of $x$, and square the result.

3. $\left(\dfrac{16}{2}\right)^2$
$= 8^2$
$= 64$

$x^2 + 16x + 64$

5. $\left(\dfrac{-24}{2}\right)^2$
$= (-12)^2$
$= 144$

$x^2 - 24x + 144$

# Quadratic Equations and Functions

7. $\left(-\dfrac{5}{2}\right)^2$

   $=\dfrac{25}{4}$

   $x^2 - 5x + \dfrac{25}{4}$

9. $\left(\dfrac{1}{2} \cdot \dfrac{6}{5}\right)^2$

   $=\left(\dfrac{3}{5}\right)^2$

   $=\dfrac{9}{25}$

   $x^2 + \dfrac{6}{5}x + \dfrac{9}{25}$

11. $x^2 - 4x = 21$

    $x^2 - 4x + \left(-\dfrac{4}{2}\right)^2 = 21 + \left(-\dfrac{4}{2}\right)^2$

    $x^2 - 4x + 4 = 21 + 4$

    $(x-2)^2 = 25$

    $\sqrt{(x-2)^2} = \sqrt{25}$

    $|x-2| = 5$

    $x - 2 = \pm 5$

    $x = 2 \pm 5$

    $x = -3,\ x = 7$

13. $x^2 + 10x = -16$

    $x^2 + 10x + 25 = -16 + 25$

    $(x+5)^2 = 9$

    $\sqrt{(x+5)^2} = \sqrt{9}$

    $|x+5| = 3$

    $x + 5 = \pm 3$

    $x = -5 \pm 3$

    $x = -8,\ x = -2$

15. $x^2 + 5x + \left(\dfrac{5}{2}\right)^2 = 6 + \left(\dfrac{5}{2}\right)^2$

    $x^2 + 5x + \dfrac{25}{4} = 6 + \dfrac{25}{4}$

    $\left(x + \dfrac{5}{2}\right)^2 = \dfrac{24}{4} + \dfrac{25}{4}$

    $\left(x + \dfrac{5}{2}\right)^2 = \dfrac{49}{4}$

    $\sqrt{\left(x + \dfrac{5}{2}\right)^2} = \sqrt{\dfrac{49}{4}}$

    $\left|x + \dfrac{5}{2}\right| = \dfrac{7}{2}$

    $x + \dfrac{5}{2} = \pm \dfrac{7}{2}$

    $x = -\dfrac{5}{2} \pm \dfrac{7}{2}$

    $x = \dfrac{-5 \pm 7}{2}$

    $x = -6,\ x = 1$

17. $x^2 - 9x = -8$

    $x^2 - 9x + \left(\dfrac{9}{2}\right)^2 = -8 + \left(\dfrac{9}{2}\right)^2$

    $\left(x - \dfrac{9}{2}\right)^2 = -\dfrac{32}{4} + \dfrac{81}{4}$

    $\left(x - \dfrac{9}{2}\right)^2 = \dfrac{49}{4}$

    $\sqrt{\left(x - \dfrac{9}{2}\right)^2} = \sqrt{\dfrac{49}{4}}$

    $\left|x - \dfrac{9}{2}\right| = \dfrac{7}{2}$

    $x - \dfrac{9}{2} = \pm \dfrac{7}{2}$

    $x = \dfrac{9}{2} \pm \dfrac{7}{2}$

    $x = \dfrac{9 \pm 7}{2}$

    $x = 8,\ 1$

19. $x^2 - 6x + 9 = 16 + 9$
$(x-3)^2 = 25$
$\sqrt{(x-3)^2} = \sqrt{25}$
$|x-3| = 5$
$x - 3 = \pm 5$
$x = 3 \pm 5$
$x = -2, x = 8$

21. $x^2 + 6x + 9 = 2 + 9$
$(x+3)^2 = 11$
$\sqrt{(x+3)^2} = \sqrt{11}$
$|x+3| = \sqrt{11}$
$x + 3 = \pm\sqrt{11}$
$x = -3 \pm \sqrt{11}$

23. $x^2 - 4x = 2$
$x^2 - 4x + 4 = 2 + 4$
$(x-2)^2 = 6$
$\sqrt{(x-2)^2} = \sqrt{6}$
$|x-2| = \sqrt{6}$
$x - 2 = \pm\sqrt{6}$
$x = 2 \pm \sqrt{6}$

25. $x^2 + 4x = -7$
$x^2 + 4x + 4 = -7 + 4$
$(x+2)^2 = -3$

No real number solution, since $(x+2)^2 \geq 0$.

27. $x^2 - 5x = 5$
$x^2 - 5x + \left(-\dfrac{5}{2}\right)^2 = 5 + \left(-\dfrac{5}{2}\right)^2$
$\left(x - \dfrac{5}{2}\right)^2 = \dfrac{20}{4} + \dfrac{25}{4}$
$\sqrt{\left(x - \dfrac{5}{2}\right)^2} = \sqrt{\dfrac{45}{4}}$
$\left|x - \dfrac{5}{2}\right| = \sqrt{\dfrac{45}{2}}$
$x - \dfrac{5}{2} = \pm \dfrac{3\sqrt{5}}{2}$
$x = \dfrac{5}{2} \pm \dfrac{3\sqrt{5}}{2}$
$x = \dfrac{5 \pm 3\sqrt{5}}{2}$

29. $x^2 + 5x = -2$
$x^2 + 5x + \left(\dfrac{5}{2}\right)^2 = -2 + \left(\dfrac{5}{2}\right)^2$
$\left(x + \dfrac{5}{2}\right)^2 = -\dfrac{8}{4} + \dfrac{25}{4}$
$\sqrt{\left(x + \dfrac{5}{2}\right)^2} = \sqrt{\dfrac{17}{4}}$
$\left|x + \dfrac{5}{2}\right| = \dfrac{\sqrt{17}}{2}$
$x + \dfrac{5}{2} = \pm \dfrac{\sqrt{17}}{2}$
$x = -\dfrac{5}{2} \pm \dfrac{\sqrt{17}}{2}$
$x = \dfrac{-5 \pm \sqrt{17}}{2}$

## Quadratic Equations and Functions

31. $x^2 + 7x = -3$

$x^2 + 7x + \left(\dfrac{7}{2}\right)^2 = -3 + \left(\dfrac{7}{2}\right)^2$

$\left(x + \dfrac{7}{2}\right)^2 = -\dfrac{12}{4} + \dfrac{49}{4}$

$\sqrt{\left(x + \dfrac{7}{2}\right)^2} = \sqrt{\dfrac{37}{4}}$

$\left|x + \dfrac{7}{2}\right| = \dfrac{\sqrt{37}}{2}$

$x + \dfrac{7}{2} = \dfrac{\pm\sqrt{37}}{2}$

$x = -\dfrac{7}{2} \pm \dfrac{\sqrt{37}}{2}$

$x = \dfrac{-7 \pm \sqrt{37}}{2}$

32. $x^2 + 8x = -15$

$x^2 + 8x + 16 = -15 + 16$

$(x+4)^2 = 1$

$\sqrt{(x+4)^2} = \sqrt{1}$

$|x+4| = 1$

$x + 4 = \pm 1$

$x = -4 \pm 1$

$x = -5, -3$

33. Because the equation is in standard form and the left side can be factored, the factoring method is easier.

35. $6x^2 + x = 2$

$x^2 + \dfrac{1}{6}x = \dfrac{2}{6}$

$x^2 + \dfrac{1}{6}x + \left(\dfrac{1}{2} \cdot \dfrac{1}{6}\right)^2 = \dfrac{1}{3} + \left(\dfrac{1}{2} \cdot \dfrac{1}{6}\right)^2$

$\left(x + \dfrac{1}{12}\right)^2 = \dfrac{1}{3} + \dfrac{1}{144}$

$\sqrt{\left(x + \dfrac{1}{12}\right)^2} = \sqrt{\dfrac{49}{144}}$

$\left|x + \dfrac{1}{12}\right| = \dfrac{7}{12}$

$x + \dfrac{1}{12} = \pm\dfrac{7}{12}$

$x = -\dfrac{1}{12} \pm \dfrac{7}{12}$

$x = \dfrac{-1 \pm 7}{12}$

$x = -\dfrac{8}{12} = -\dfrac{2}{3}, \; x = \dfrac{6}{12} = \dfrac{1}{2}$

37. $x^2 + \dfrac{1}{2}x = 3$

$x^2 + \dfrac{1}{2}x + \left(\dfrac{1}{2} \cdot \dfrac{1}{2}\right)^2 = 3 + \left(\dfrac{1}{2} \cdot \dfrac{1}{2}\right)^2$

$\left(x + \dfrac{1}{4}\right)^2 = \dfrac{48}{16} + \dfrac{1}{16}$

$\sqrt{\left(x + \dfrac{1}{4}\right)^2} = \sqrt{\dfrac{49}{16}}$

$\left|x + \dfrac{1}{4}\right| = \dfrac{7}{4}$

$x + \dfrac{1}{4} = \pm\dfrac{7}{4}$

$x = -\dfrac{1}{4} \pm \dfrac{7}{4}$

$x = \dfrac{-1 \pm 7}{4}$

$x = -2, \dfrac{3}{2}$

39. $4x^2 + 7x = -3$

$x^2 + \dfrac{7}{4}x = -\dfrac{3}{4}$

$x^2 + \dfrac{7}{4}x + \left(\dfrac{1}{2} \cdot \dfrac{7}{4}\right)^2 = -\dfrac{3}{4} + \left(\dfrac{1}{2} \cdot \dfrac{7}{4}\right)^2$

$\left(x + \dfrac{7}{8}\right)^2 = -\dfrac{48}{64} + \dfrac{49}{64}$

$\sqrt{\left(x + \dfrac{7}{8}\right)^2} = \sqrt{\dfrac{1}{64}}$

$\left|x + \dfrac{7}{8}\right| = \dfrac{1}{8}$

$x + \dfrac{7}{8} = \pm\dfrac{1}{8}$

$x = \dfrac{-7 \pm 1}{8}$

$x = -1, -\dfrac{3}{4}$

41. $x^2 - x - 6 = 0$

$x^2 - x = 6$

$x^2 - x + \left(\dfrac{1}{2}(-1)\right)^2 = 6 + \left(\dfrac{1}{2}(-1)\right)^2$

$\left(x - \dfrac{1}{2}\right)^2 = \dfrac{24}{4} + \dfrac{1}{4}$

$\sqrt{\left(x - \dfrac{1}{2}\right)^2} = \sqrt{\dfrac{25}{4}}$

$\left|x - \dfrac{1}{2}\right| = \dfrac{5}{2}$

$x - \dfrac{1}{2} = \pm\dfrac{5}{2}$

$x = \dfrac{1 \pm 5}{2}$

$x = 3, -2$

43. $x^2 - \dfrac{1}{2}x = \dfrac{1}{3}$

$x^2 - \dfrac{1}{2}x + \left(\dfrac{1}{2}\left(-\dfrac{1}{2}\right)\right)^2 = \dfrac{1}{3} + \left(\dfrac{1}{2}\left(-\dfrac{1}{2}\right)\right)^2$

$\left(x - \dfrac{1}{4}\right)^2 = \dfrac{1}{3} + \dfrac{1}{16}$

$\sqrt{\left(x - \dfrac{1}{4}\right)^2} = \sqrt{\dfrac{19}{48}}$

$\left|x - \dfrac{1}{4}\right| = \dfrac{\sqrt{19}}{4\sqrt{3}}$

$x - \dfrac{1}{4} = \pm\dfrac{\sqrt{19}}{4\sqrt{3}} \cdot \dfrac{\sqrt{3}}{\sqrt{3}}$

$x - \dfrac{1}{4} = \pm\dfrac{\sqrt{57}}{12}$

$x = \dfrac{3}{12} \pm \dfrac{\sqrt{57}}{12}$

$x = \dfrac{3 \pm \sqrt{57}}{12}$

45. $4x^2 + 4x = -5$

$x^2 + x = -\dfrac{5}{4}$

$x^2 + x + \left(\dfrac{1}{2}(1)\right)^2 = -\dfrac{5}{4} + \left(\dfrac{1}{2}(1)\right)^2$

$\left(x + \dfrac{1}{2}\right)^2 = -\dfrac{5}{4} + \dfrac{1}{4}$

$\left(x + \dfrac{1}{2}\right)^2 = -1$

No real number solution, since $\left(x + \dfrac{1}{2}\right)^2 \geq 0$.

**47.** $2x^2 + 3x = 4$

$x^2 + \dfrac{3}{2}x = 2$

$x^2 + \dfrac{3}{2}x + \left(\dfrac{1}{2} \cdot \dfrac{3}{2}\right)^2 = 2 + \left(\dfrac{1}{2} \cdot \dfrac{3}{2}\right)^2$

$\left(x + \dfrac{3}{4}\right)^2 = \dfrac{32}{16} + \dfrac{9}{16}$

$\sqrt{\left(x + \dfrac{3}{4}\right)^2} = \sqrt{\dfrac{41}{16}}$

$\left|x + \dfrac{3}{4}\right| = \dfrac{\sqrt{41}}{4}$

$x + \dfrac{3}{4} = \pm\dfrac{\sqrt{41}}{4}$

$x = \dfrac{-3 \pm \sqrt{41}}{4}$

**49.** $6x^2 + 9x = -8$

$x^2 + \dfrac{3}{2}x = -\dfrac{4}{3}$

$x^2 + \dfrac{3}{2}x + \left(\dfrac{1}{2}\left(\dfrac{3}{2}\right)\right)^2 = -\dfrac{4}{3} + \left(\dfrac{1}{2}\left(\dfrac{3}{2}\right)\right)^2$

$\left(x + \dfrac{3}{4}\right)^2 = -\dfrac{4}{3} + \dfrac{9}{16}$

$\left(x + \dfrac{3}{4}\right)^2 = -\dfrac{37}{48}$

No real number solution, since

$\left(x + \dfrac{3}{4}\right)^2 \geq 0$.

**51.** $3x^2 + x = 1$

$x^2 + \dfrac{1}{3}x = \dfrac{1}{3}$

$x^2 + \dfrac{1}{3}x + \left(\dfrac{1}{2} \cdot \dfrac{1}{3}\right)^2 = \dfrac{1}{3} + \left(\dfrac{1}{2} \cdot \dfrac{1}{3}\right)^2$

$\left(x + \dfrac{1}{6}\right)^2 = \dfrac{12}{36} + \dfrac{1}{36}$

$\sqrt{\left(x + \dfrac{1}{6}\right)^2} = \sqrt{\dfrac{13}{36}}$

$\left|x + \dfrac{1}{6}\right| = \dfrac{\sqrt{13}}{36}$

$x + \dfrac{1}{6} = \pm\dfrac{\sqrt{13}}{6}$

$x = \dfrac{-1 \pm \sqrt{13}}{6}$

**53.** $x\left(\dfrac{1}{2}x - 1\right) = 60$

$\dfrac{1}{2}x^2 - x = 60$

$x^2 - 2x = 120$

$x^2 - 2x + 1 = 120 + 1$

$(x - 1)^2 = 121$

$\sqrt{(x-1)^2} = \sqrt{121}$

$|x - 1| = 11$

$x - 1 = \pm 11$

$x = 1 \pm 11$

$x = -10, \ x = 12$

But, $x < 0$, so $x = -10$.

55. $x(180-x) = 7200$

$180x - x^2 = 7200$

$x^2 - 180x = -7200$

$x^2 - 180x + 8100 = -7200 + 8100$

$(x-90)^2 = 900$

$\sqrt{(x-90)^2} = \sqrt{900}$

$|x-90| = 30$

$x - 90 = \pm 30$

$x = 90 \pm 30$

$x = 120, 60$

The smaller angle is 60°.

57. $x(x+3) = 180$

$x^2 + 3x = 180$

$x^2 + 3x + \left(\frac{1}{2}(3)\right)^2 = 180 + \left(\frac{1}{2}(3)\right)^2$

$\left(x+\frac{3}{2}\right)^2 = 180 + \frac{9}{4}$

$\sqrt{\left(x+\frac{3}{2}\right)^2} = \sqrt{\frac{729}{4}}$

$\left|x+\frac{3}{2}\right| = \frac{27}{2}$

$x + \frac{3}{2} = \pm\frac{27}{2}$

$x = -\frac{3}{2} \pm \frac{27}{2}$

$x = \frac{24}{2}, x = -\frac{30}{2}$

$x = 12, x = -15$

Width : 12 Blocks
Length : 12 + 3 = 15 blocks
Perimeter = 2 (12) + 2 (15) = 54 blocks

54 lights should be installed.

59. Time upstream = $\dfrac{80}{v-8}$

Time downstream = $\dfrac{60}{v+8}$

$\dfrac{80}{v-8} + \dfrac{60}{v+8} = 5$

$80(v+8) + 60(v-8) = 5(v-8)(v+8)$

$80v + 640 + 60v - 480 = 5(v^2 - 64)$

$140v + 160 = 5v^2 - 320$

$28v + 32 = v^2 - 64$

$0 = v^2 - 28v - 32$

$v^2 - 28v = 32$

$v^2 - 28v + (14)^2 = 32 + (14)^2$

$(v-14)^2 = 228$

$\sqrt{(v-14)^2} = \sqrt{228}$

$|v-14| = 15.1$

$v - 14 = 15.1$

$v \approx 29$ mph

61. $x =$ speed of slow car
$x + 10 =$ speed of fast car

$(x \cdot 1)^2 + ((x+10) \cdot 1)^2 = (90)^2$

$x^2 + x^2 + 20x + 100 = 8100$

$2x^2 + 20x = 8000$

$x^2 + 10x = 4000$

$x^2 + 10x + 25 = 4000 + 25$

$(x+5)^2 = 4025$

$\sqrt{(x+5)^2} = \sqrt{4025}$

$|x+5| = 63.4$

$x + 5 = 63.4$

slow : $x \approx 58$ mph
fast : $x + 10 \approx 68$ mph

Quadratic Equations and Functions

63. (a) $-0.012(x^2 - 136x + 3783) = 4.8$

(b) $x^2 - 136x + 3783 = -400$
$x^2 - 136x = -4183$
$x^2 - 136x + (68)^2 = -4183 + (68)^2$
$(x - 68)^2 = 441$
$\sqrt{(x-68)^2} = \sqrt{441}$
$|x - 68| = 21$
$x - 68 = \pm 21$
$x = 68 \pm 21$
$x = 89, 47$

65. $-0.035(x^2 - 88x + 960) = 34$

67. $x^2 - 88x + 960 = -971.4$
$x^2 - 88x = -1931.4$
$x^2 - 88x + (44)^2 = -1931.4 + (44)^2$
$(x - 44)^2 = 4.6$
$\sqrt{(x-44)^2} = \sqrt{4.6}$
$|x - 44| = 2.1$
$x = 44 \pm 2.1$
$x \approx 42, 46$

69. $4\left(x^2 + \dfrac{3}{4}x + \left(\dfrac{1}{2} \cdot \dfrac{3}{4}\right)^2\right)$
$= 4\left(x^2 + \dfrac{3}{4}x + \dfrac{9}{64}\right)$
$= 4x^2 + 3x + \dfrac{9}{16}$

71. $4x^2 - 8x + 1$
since $\left(\dfrac{-8}{2(4)}\right)^2 = 1$

73. $x^2 + 6x = y - 11$
$x^2 + 6x + 9 = y - 11 + 9$
$(x + 3)^2 = y - 2$
$y - 2 = (x + 3)^2$

## Section 10.3 The Quadratic Formula

1. The Quadratic Formula is derived by completing the square.

3. $a = 2$
$b = -3$
$c = -1$

5. $3x^2 + 0x + 2 = 0$
$a = 3$
$b = 0$
$c = 2$

7. $(x-1)(x-5)=0$
$x=1, x=5$

$x = \dfrac{-(-6) \pm \sqrt{(-6)^2 - 4 \cdot 1 \cdot 5}}{2 \cdot 1}$

$= \dfrac{6 \pm \sqrt{36-20}}{2}$

$= \dfrac{6 \pm \sqrt{16}}{2}$

$= \dfrac{6 \pm 4}{2}$

$= 3 \pm 2$

$x = 1, x = 5$

9. $x(x+7)=0$
$x=0, x=-7$

$x = \dfrac{-7 \pm \sqrt{7^2 - 4 \cdot 1 \cdot 0}}{2 \cdot 1}$

$= \dfrac{-7 \pm \sqrt{49}}{2}$

$= \dfrac{-7 \pm 7}{2}$

$x = 0, -7$

11. $3x^2 + 4x - 4 = 0$
$(3x-2)(x+2) = 0$
$x = \dfrac{2}{3}, x = -2$

$x = \dfrac{-4 \pm \sqrt{4^2 - 4 \cdot 3 \cdot (-4)}}{2 \cdot 3}$

$= \dfrac{-4 \pm \sqrt{16+48}}{6}$

$= \dfrac{-4 \pm \sqrt{64}}{6}$

$= \dfrac{-4 \pm 8}{6}$

$= \dfrac{-2 \pm 4}{3}$

$x = -2, x = \dfrac{2}{3}$

13. $x^2 + 7x + 6 = 0$

$x = \dfrac{-7 \pm \sqrt{7^2 - 4 \cdot 1 \cdot 6}}{2 \cdot 1}$

$= \dfrac{-7 \pm \sqrt{49-24}}{2}$

$= \dfrac{-7 \pm \sqrt{25}}{2}$

$= \dfrac{-7 \pm 5}{2}$

$x = -6, -1$

15. $x = \dfrac{-(-6) \pm \sqrt{(-6)^2 - 4 \cdot 1 \cdot 7}}{2 \cdot 1}$

$= \dfrac{6 \pm \sqrt{36-28}}{2 \cdot 1}$

$= \dfrac{6 \pm \sqrt{8}}{2}$

$= \dfrac{6 \pm 2\sqrt{2}}{2}$

$= 3 \pm \sqrt{2}$

# Quadratic Equations and Functions

17. $x = \dfrac{-3 \pm \sqrt{3^2 - 4 \cdot 1 \cdot 6}}{2 \cdot 1}$

    $= \dfrac{3 \pm \sqrt{9 - 24}}{2}$

    $= \dfrac{-3 \pm \sqrt{-15}}{2}$

    No real number solution, since $\sqrt{-15}$ is undefined.

19. $-3x^2 + 2x + 3 = 0$

    $x = \dfrac{2 \pm \sqrt{2^2 - 4(-3)(3)}}{2(-3)}$

    $= \dfrac{-2 \pm \sqrt{4 + 36}}{-6}$

    $= \dfrac{-2 \pm \sqrt{40}}{-6}$

    $= \dfrac{-2 \pm 2\sqrt{10}}{-6}$

    $= \dfrac{-1 \pm \sqrt{10}}{-3}$

    $= \dfrac{1 \pm \sqrt{10}}{3}$

21. $-x^2 + 0x + 15 = 0$

    $x = \dfrac{-0 \pm \sqrt{0^2 - 4(-1)(15)}}{2 \cdot (-1)}$

    $= \dfrac{0 \pm \sqrt{60}}{-2}$

    $= \dfrac{\pm 2\sqrt{15}}{-2}$

    $= \pm\sqrt{15}$

23. $12x^2 + 30x = 0$

    $2x^2 + 5x = 0$

    $x = \dfrac{-5 \pm \sqrt{5^2 - 4(2)(0)}}{2(2)}$

    $= \dfrac{-5 \pm \sqrt{25}}{4}$

    $= \dfrac{-5 \pm 5}{4}$

    $x = 0, \; x = -\dfrac{5}{2}$

25. $3x^2 - 3x - 2 = 0$

    $x = \dfrac{-(-3) \pm \sqrt{(-3)^2 - 4(3)(-2)}}{2(3)}$

    $= \dfrac{3 \pm \sqrt{9 + 24}}{6}$

    $= \dfrac{3 \pm \sqrt{33}}{6}$

27. Because the graph has only one $x$-intercept, the discriminant is 0.

29. $x = \dfrac{-1 \pm \sqrt{1^2 - 4(10)(-3)}}{2(10)}$

    $= \dfrac{-1 \pm \sqrt{121}}{20}$

    $= \dfrac{-1 \pm 11}{20}$

    $x = -\dfrac{3}{5}, \; x = \dfrac{1}{2}$

31. $x = \dfrac{-(-2) \pm \sqrt{(-2)^2 - 4 \cdot 3 \cdot 1}}{2(3)}$

    $= \dfrac{2 \pm \sqrt{-8}}{6}$

    No real number solution, since $\sqrt{-8}$ is undefined.

33. $x^2 + 6x + 7 = 0$

$x = \dfrac{-6 \pm \sqrt{6^2 - 4(1)(7)}}{2(1)}$

$= \dfrac{-6 \pm \sqrt{36 - 28}}{2}$

$= \dfrac{-6 \pm \sqrt{8}}{2}$

$= \dfrac{-6 \pm 2\sqrt{2}}{2}$

$= -3 \pm \sqrt{2}$

35. $x^2 + 12x + 35 = 0$

$(x + 7)(x + 5) = 0$

$x = -7, x = -5$

37. $x^2 + 5x - 14 = 0$

$(x + 7)(x - 2) = 0$

$x = -7, x = 2$

39. $(x + 9)^2 = 0$

$x = -9$

41. $4x^2 = 11x + 3$

$4x^2 - 11x - 3 = 0$

$(4x + 1)(x - 3) = 0$

$x = -\dfrac{1}{4}, x = 3$

43. $x^2 + 7 = 9x$

$x^2 - 9x + 7 = 0$

$x = \dfrac{-(9) \pm \sqrt{(-9)^2 - 4(1)(7)}}{2(1)}$

$= \dfrac{9 \pm \sqrt{81 - 28}}{2}$

$= \dfrac{9 \pm \sqrt{53}}{2}$

45. $3x^2 + 9x = x - 2$

$3x^2 + 8x + 2 = 0$

$x = \dfrac{-8 \pm \sqrt{8^2 - 4(3)(2)}}{2(3)}$

$= \dfrac{-8 \pm \sqrt{64 - 24}}{6}$

$= \dfrac{-8 \pm \sqrt{40}}{6}$

$= \dfrac{-8 \pm 2\sqrt{10}}{6}$

$= \dfrac{-4 \pm \sqrt{10}}{3}$

47. $x^2 - 5x + 4 = -2$

$x^2 - 5x + 6 = 0$

$(x - 2)(x - 3) = 0$

$x = 2, x = 3$

49. The graph has no $x$-intercepts.

51. $x^2 - 17x + 72 = 0$

$b^2 - 4ac = (-17)^2 - 4(1)(72)$

$= 289 - 288$

$= 1$

There are 2 solutions, since $1 > 0$.

53. $b^2 - 4ac = (-20)^2 - 4(1)(100)$

$= 400 - 400$

$= 0$

There is one solution.

55. $x^2 + 10 = 0$

$b^2 - 4ac = 0^2 - 4(1)(10)$

$= -40$

There are no real solutions, since $-40 < 0$.

## Quadratic Equations and Functions

**57.** $x^2 + 4x + 2 = 0$
$b^2 - 4ac = 4^2 - 4(1)(2)$
$= 16 - 8$
$= 8$
There are two solutions, since $8 > 0$.

**59.** $9x^2 - 24x + 16 = 0$
$b^2 - 4ac = (-24)^2 - 4(9)(16)$
$= 576 - 576$
$= 0$
There is one solution.

**61.** $5x^2 + 5x = -2$
$5x^2 + 5x + 2 = 0$
$b^2 - 4ac = 5^2 - 4(5)(2)$
$= 25 - 40$
$= -15$
There are no real solutions, since $-15 < 0$.

**63.** $w =$ width of garden
$w + 10 =$ length of garden

$2w - 5 =$ length of garden and border
$w + 5 =$ width of garden and border

$(2w - 5)(w + 5) - w(w + 10) = 1025$
$2w^2 + 5w - 25 - (w^2 + 10w) = 1025$
$2w^2 + 5w - 25 - w^2 - 10w = 1025$
$w^2 - 5w = 1050$
$w^2 - 5w + \left(-\dfrac{5}{2}\right)^2 = 1050 + \left(-\dfrac{5}{2}\right)^2$
$\left(w - \dfrac{5}{2}\right)^2 = \dfrac{4200}{4} + \dfrac{25}{4}$
$\left(w - \dfrac{5}{2}\right)^2 = \dfrac{4225}{4}$
$\sqrt{\left(w - \dfrac{5}{2}\right)^2} = \sqrt{\dfrac{4225}{4}}$
$w - \dfrac{5}{2} = \dfrac{65}{2}$
$w = \dfrac{70}{2}$

width of garden: $w = 35$ feet
length of garden: $w + 10 = 45$ feet

65. $w=$ width of material
$2w+1=$ length of material

$w(2w+1) = 9$

$2w^2 + w = 9$

$w^2 + \dfrac{1}{2}w = \dfrac{9}{2}$

$w^2 + \dfrac{1}{2}w + \left(\dfrac{1}{2}\cdot\dfrac{1}{2}\right)^2 = \dfrac{9}{2} + \left(\dfrac{1}{2}\cdot\dfrac{1}{2}\right)^2$

$\left(w + \dfrac{1}{4}\right)^2 = \dfrac{72}{16} + \dfrac{1}{16}$

$\sqrt{\left(w+\dfrac{1}{4}\right)^2} = \sqrt{\dfrac{73}{16}}$

$w + \dfrac{1}{4} = \dfrac{\sqrt{73}}{4}$

$w = \dfrac{\sqrt{73}}{4} - \dfrac{1}{4}$

$= 1.89$ feet

We must deduct $\dfrac{6}{12}$ feet from the width to allow for the folding. The width of the bench is $1.89 - 0.5 = 1.39$ feet.

67.

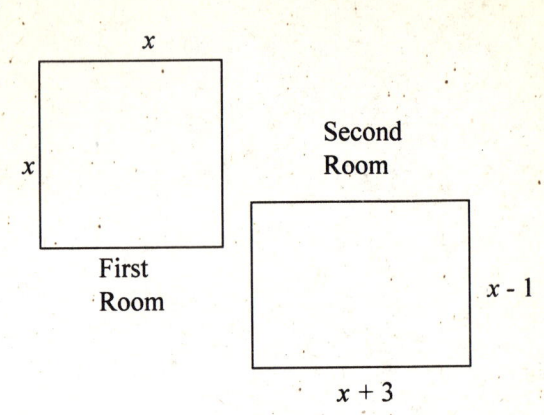

$x^2 + (x+3)(x-1) = 261$

$x^2 + x^2 + 2x - 3 = 261$

$2x^2 + 2x - 264 = 0$

$x^2 + x - 132 = 0$

$(x+12)(x-11) = 0$

$x = 11$ feet

$x + 3 = 11 + 3 = 14$ feet

## Quadratic Equations and Functions

69. $w =$ long leg
    $w + 4 =$ hypotenuse
    $\dfrac{3}{4}w =$ short leg

    $\left(\dfrac{3}{4}w\right)^2 + w^2 = (w+4)^2$

    $\dfrac{9}{16}w^2 + w^2 = w^2 + 8w + 16$

    $\dfrac{9}{16}w^2 = 8w + 16$

    $9w^2 = 128w + 256$

    $w^2 = \dfrac{128}{9}w + \dfrac{256}{9}$

    $w^2 - \dfrac{128}{9}w = \dfrac{256}{9}$

    $w^2 - \dfrac{128}{9}w + \left(\dfrac{128}{18}\right)^2 = \dfrac{256}{9} + \left(\dfrac{128}{18}\right)^2$

    $\left(w - \dfrac{128}{18}\right)^2 = \dfrac{2304}{81} + \dfrac{4096}{81}$

    $\left(w - \dfrac{64}{9}\right)^2 = \dfrac{6400}{81}$

    $\sqrt{\left(w - \dfrac{64}{9}\right)^2} = \sqrt{\dfrac{6400}{81}}$

    $w - \dfrac{64}{9} = \dfrac{80}{9}$

    $w = \dfrac{144}{9}$

    long leg : $w = 16$
    hypotenuse : $w + 4 = 16 + 4 = 20$ feet

71. $x =$ Welltech shares
    $200 - x =$ NGN shares

    $\dfrac{1250}{x} =$ Welltech price

    $\dfrac{4500}{200 - x} =$ NGN price

    $\dfrac{4500}{200 - x} - \dfrac{1250}{x} = 5$

    $4500x - 1250(200 - x) = 5x(200 - x)$

    $4500x - 250{,}000 + 1250x = 1000x - 5x^2$

    $5x^2 + 4750x - 250{,}000 = 0$

    $x^2 + 950x - 50{,}000 = 0$

    $x^2 + 950x = 50{,}000$

    $x^2 + 950x + (475)^2 = 50{,}000 + (475)^2$

    $(x + 475)^2 = 50{,}000 + 225{,}625$

    $(x + 475)^2 = 275{,}625$

    $\sqrt{(x + 475)^2} = \sqrt{275{,}625}$

    $x + 475 = 525$

    $x = 50$ shares

73. $x =$ released
$150 - x =$ admitted

$\dfrac{36{,}000}{x} =$ average released charge

$\dfrac{22{,}500}{150 - x} =$ average admitted charge

$\dfrac{22{,}500}{150 - x} - \dfrac{36{,}000}{x} = 450$

$22{,}500x - 36{,}000(150 - x) = 450x(150 - x)$

$22{,}500x - 5{,}400{,}000 + 36{,}000x = 67{,}500x - 450x^2$

$58{,}500x - 5{,}400{,}000 = -450x^2 + 67{,}500x$

$450x^2 - 9000x - 5{,}400{,}000 = 0$

$x^2 - 20x - 12{,}000 = 0$

$x^2 - 20x = 12{,}000$

$x^2 - 20x + 100 = 12{,}000 + 100$

$(x - 10)^2 = 12{,}100$

$\sqrt{(x-10)^2} = \sqrt{12100}$

$x - 10 = 110$

released : $x = 120$

admitted : $150 - x = 150 - 120 = 30$

75. $x=$ apprentice time to do job alone
$x-1.5=$ electrician time to do job alone

$\dfrac{1}{x} =$ apprentice rate

$\dfrac{1}{x-1.5} =$ electrician rate

$\left(\dfrac{1}{x}+\dfrac{1}{x-1.5}\right)1.8=1$

$\dfrac{1}{x}+\dfrac{1}{x-1.5}=\dfrac{1}{1.8}$

$\dfrac{x-1.5+x}{x(x-1.5)}=\dfrac{1}{1.8}$

$\dfrac{2x-1.5}{x(x-1.5)}=\dfrac{1}{1.8}$

$1.8(2x-1.5)=x(x-1.5)$

$3.6x-2.7=x^2-1.5x$

$x^2-5.1x+2.7=0$

$10x^2-51x+27=0$

$(2x-9)(5x-3)=0$

$2x-9=0 \quad 5x-3=0$
$2x=9 \quad\quad 5x=3$
$x=\dfrac{9}{2} \quad\quad x=\dfrac{3}{5}$

But if $x=\dfrac{3}{5}$, $x-1.5$ is negative.

So $x=\dfrac{9}{2}=4.5$ hours.

77. (a) $-0.029(40)^2+2.02(40)-12.2$
$=-46.4+80.8-12.2$
$=22.2\%$

(b) $22.2=-0.029x^2+2.02x-12.2$

$34.4=-0.029x^2+2.02x$
$0.029x^2-2.02x=-34.4$
$x^2-69.66x=-1186.21$
$x^2-69.66x+1213.13=-1186.21+1213.13$
$(x-34.83)^2=26.92$
$|x-34.83|=5.19$
$x-34.83=\pm 5.19$
$x=34.83\pm 5.19$
$x\approx 40,\ x\approx 30$

At ages 30 and 40, the probability of a person buying life insurance is 22.2%.

79. $405-314=91$ million

81. $1.1t^2-2t=174$
$11t^2-20t=1740$
$t^2-\dfrac{20}{11}t=\dfrac{1740}{11}$
$t^2-\dfrac{20}{11}+\left(\dfrac{20}{22}\right)^2=\dfrac{1740}{11}+\left(\dfrac{20}{22}\right)^2$
$\left(t-\dfrac{10}{11}\right)^2=\dfrac{1740}{11}+\left(\dfrac{10}{11}\right)^2$
$\sqrt{\left(t-\dfrac{10}{11}\right)^2}=\sqrt{\dfrac{19240}{121}}$
$\left|t-\dfrac{10}{11}\right|=12.61$
$t-0.91=12.61$
$t=13.5$ years after 1990 (2003).

The model predicts an increase of 91 million passengers between 2000 and 2003.

83. $x^2 - 4\sqrt{2}x - 4 = 0$
$x^2 - 4\sqrt{2}x = 4$
$x^2 - 4\sqrt{2}x + \left(\dfrac{4\sqrt{2}}{2}\right)^2 = 4 + \left(\dfrac{4\sqrt{2}}{2}\right)^2$
$\left(x - 2\sqrt{2}\right)^2 = 4 + \left(2\sqrt{2}\right)^2$
$\left(x - 2\sqrt{2}\right)^2 = 4 + 8$
$\sqrt{\left(x - 2\sqrt{2}\right)^2} = \sqrt{12}$
$\left|x - 2\sqrt{2}\right| = 2\sqrt{3}$
$x - 2\sqrt{2} = \pm 2\sqrt{3}$
$x = 2\sqrt{2} \pm 2\sqrt{3}$

85. $y^2 + 6y + (5 - x) = 0$
$a = 1$
$b = 6$
$c = 5 - x$

$y = \dfrac{-6 \pm \sqrt{6^2 - 4(1)(5-x)}}{2 \cdot 1}$
$= \dfrac{-6 \pm \sqrt{36 - 4(5-x)}}{2}$
$= \dfrac{-6 \pm \sqrt{36 - 20 + 4x}}{2}$
$= \dfrac{-6 \pm \sqrt{16 + 4x}}{2}$
$= \dfrac{-6 \pm 2\sqrt{4 + x}}{2}$
$= -3 \pm \sqrt{4 + x}$

87. $a = k$
$b = -3$
$c = 3$

$b^2 - 4ac = (-3)^2 - 4 \cdot k \cdot 3$
$= 9 - 12k$

(a) To have two unique solutions, $9 - 12k$ must be positive.
$9 - 12k > 0$
$-12k > -9$
$k < \dfrac{-9}{-12}$
$k < \dfrac{3}{4}$

(b) To have no real number solutions, $9 - 12k$ must be negative.
$9 - 12k < 0$
$9 < 12k$
$\dfrac{9}{12} < k$
$k > \dfrac{3}{4}$

## Section 10.4 Complex Numbers

1. A complete number can be written in the form $a + bi$, where $a$ and $b$ are any real numbers. An imaginary number is a complex number with $b \neq 0$.

3. $\sqrt{-1}\sqrt{15}$
$= i\sqrt{15}$

5. $\sqrt{49}\sqrt{-1}$
$= 7i$

Quadratic Equations and Functions

7. $\sqrt{144}\sqrt{-1}$
$= 12i$

9. $\sqrt{4}\sqrt{5}\sqrt{-1}$
$= 2i\sqrt{5}$

11. $\sqrt{25}\sqrt{-1}\sqrt{3}$
$= 5i\sqrt{3}$

13. (a) 5

(b) 7

15. (a) 0

(b) 10

17. (a) 12

(b) 0

19. (a) $-7$

(b) 15

21. $3 + (2i + 4i)$
$= 3 + 6i$

23. $(3+4)+(-2i+i)$
$= 7 - i$

25. $(7-(-2))+(3i-5i)$
$= 9 - 2i$

27. $(5-3)-(-2i)$
$= 2 + 2i$

29. $(6+6)+(i-i)$
$= 12$

31. $(-3-(-1))+(2i-(-4i))$
$= -2 + 6i$

33. $2i + 3i^2$
$= 2i + 3(-1)$
$= -3 + 2i$

35. $-12i + 2i^2$
$= -12i + 2(-1)$
$= -2 - 12i$

37. $25i^2$
$= 25(-1)$
$= -25$

39. $(5-4i)(5-4i)$
$= 25 - 20i - 20i + 16i^2$
$= 25 - 40i + 16(-1)$
$= 25 - 16 - 40i$
$= 9 - 40i$

41. $6 - 3i - 4i + 2i^2$
$= 6 + 7i + 2(-1)$
$= 6 - 2 - 7i$
$= 4 - 7i$

43. $2 + 4i - i - 2i^2$
$= 2 + 3i - 2(-1)$
$= 2 + 2 + 3i$
$= 4 + 3i$

45. (a) $5 + 3i$

(b) $(5+3i)(5-3i)$

$= 25 - 15i + 15i - 9i^2$
$= 25 - 9(-1)$
$= 25 + 9$
$= 34$

47. (a) $7 - 2i$

(b) $(7 + 2i)(7 - 2i)$
$= 49 - 4i^2$
$= 49 - 4(-1)$
$= 53$

49. (a) $-1 - i$

(b) $(-1 - i)(-1 + i)$
$= (-1)^2 - i^2$
$= 1 - (-1)$
$= 2$

51. In each, we multiply both the numerator and the denominator by the conjugate of the denominator. In (i), multiply by the complex conjugate 1-$i$, and in (ii), multiply by $1 - \sqrt{3}$.

53. $-\dfrac{12}{3} + \dfrac{15i}{3}$
$= -4 + 5i$

55. $\dfrac{3 + 2i}{i} \cdot \dfrac{-i}{-i}$
$= \dfrac{-3i - 2i^2}{-i^2}$
$= \dfrac{-3i - 2(-1)}{-(-1)}$
$= -3i + 2$
$= 2 - 3i$

57. $\dfrac{3i}{1 - 2i} \cdot \dfrac{1 + 2i}{1 + 2i}$
$= \dfrac{3i + 6i^2}{1^2 + 2^2}$
$= \dfrac{3i + 6(-1)}{1 + 4}$
$= \dfrac{3i - 6}{5}$
$= -\dfrac{6}{5} + \dfrac{3}{5}i$

59. $\dfrac{1 + 3i}{1 - 3i} \cdot \dfrac{1 + 3i}{1 + 3i}$
$= \dfrac{1 + 6i + 9i^2}{1^2 + (3)^2}$
$= \dfrac{1 + 6i + 9(-1)}{1 + 9}$
$= \dfrac{-8 + 6i}{10}$
$= -\dfrac{8}{10} + \dfrac{6}{10}i$
$= -\dfrac{4}{5} + \dfrac{3}{5}i$

61. $\dfrac{5}{2 + i} \cdot \dfrac{2 - i}{2 - i}$
$= \dfrac{10 - 5i}{2^2 + 1^2}$
$= \dfrac{10 - 5i}{5}$
$= 2 - i$

63. The imaginary solutions of a quadratic equation with real coefficients are complex conjugates.

65. $y^2 = -4$
$\sqrt{y^2} = \sqrt{-4}$
$|y| = \sqrt{4}\sqrt{-1}$
$y = \pm 2i$

67. $t^2 = -5$
$\sqrt{t^2} = \sqrt{-5}$
$|t| = \sqrt{-1}\sqrt{5}$
$t = \pm i\sqrt{5}$

69. $\sqrt{(x + 5)^2} = \sqrt{-16}$
$|x + 5| = \sqrt{16}\sqrt{-1}$
$x + 5 = \pm 4i$
$x = -5 \pm 4i$

## Quadratic Equations and Functions

71. $\sqrt{(3x-1)^2} = \sqrt{-9}$
$|3x-1| = \sqrt{9}\sqrt{-1}$
$3x - 1 = \pm 3i$
$3x = 1 \pm 3i$
$x = \dfrac{1}{3} \pm \dfrac{3i}{3}$
$x = \dfrac{1}{3} \pm i$

73. $(x+2)^2 = -18$
$\sqrt{(x+2)^2} = \sqrt{-18}$
$|x+2| = \sqrt{9}\sqrt{2}\sqrt{-1}$
$|x+2| = 3i\sqrt{2}$
$x + 2 = \pm 3i\sqrt{2}$
$x = -2 \pm 3i\sqrt{2}$

75. $x = \dfrac{-(-2) \pm \sqrt{(-2)^2 - 4 \cdot 1 \cdot 5}}{2 \cdot 1}$
$x = \dfrac{2 \pm \sqrt{4 - 20}}{2}$
$x = \dfrac{2 \pm \sqrt{-16}}{2}$
$x = \dfrac{2 \pm 4i}{2}$
$x = 1 \pm 2i$

77. $x = \dfrac{-10 \pm \sqrt{10^2 - 4 \cdot 1 \cdot 26}}{2 \cdot 1}$
$x = \dfrac{-10 \pm \sqrt{100 - 104}}{2}$
$x = \dfrac{-10 \pm \sqrt{-4}}{2}$
$x = \dfrac{-10 \pm 2i}{2}$
$x = -5 \pm i$

79. $x^2 + 3x + 4 = 0$
$x = \dfrac{-3 \pm \sqrt{3^2 - 4 \cdot 1 \cdot 4}}{2 \cdot 1}$
$x = \dfrac{-3 \pm \sqrt{9 - 16}}{2}$
$x = \dfrac{-3 \pm i\sqrt{7}}{2}$
$x = -\dfrac{3}{2} \pm i\dfrac{\sqrt{7}}{2}$

81. $x = \dfrac{-(-3) \pm \sqrt{(-3)^2 - 4 \cdot 1 \cdot 9}}{2 \cdot 1}$
$x = \dfrac{3 \pm \sqrt{9 - 36}}{2}$
$x = \dfrac{3 \pm \sqrt{-27}}{2}$
$x = \dfrac{3 \pm 3i\sqrt{3}}{2}$
$x = \dfrac{3}{2} \pm \dfrac{3i\sqrt{3}}{2}$

83. $x^2 + 4x + 7 = 0$
$x = \dfrac{-4 \pm \sqrt{4^2 - 4 \cdot 1 \cdot 7}}{2 \cdot 1}$
$x = \dfrac{-4 \pm \sqrt{16 - 28}}{2}$
$x = \dfrac{-4 \pm \sqrt{-12}}{2}$
$x = \dfrac{-4 \pm 2i\sqrt{3}}{2}$
$x = -2 \pm i\sqrt{3}$

85. $x = \dfrac{-(-4) \pm \sqrt{(-4)^2 - 4 \cdot 4 \cdot 5}}{2 \cdot 4}$

$x = \dfrac{4 \pm \sqrt{16 - 80}}{8}$

$x = \dfrac{4 \pm \sqrt{-64}}{8}$

$x = \dfrac{4 \pm 8i}{8}$

$x = \dfrac{1}{2} \pm i$

87. $x = \dfrac{-12 \pm \sqrt{12^2 - 4 \cdot 9 \cdot 5}}{2 \cdot 9}$

$x = \dfrac{-12 \pm \sqrt{144 - 180}}{18}$

$x = \dfrac{-12 \pm \sqrt{-36}}{18}$

$x = \dfrac{-12 \pm 6i}{18}$

$x = \dfrac{-12}{18} \pm \dfrac{6i}{18}$

$x = -\dfrac{2}{3} \pm \dfrac{1}{3}i$

89. $4x^2 + 2x + 1 = 0$

$x = \dfrac{-2 \pm \sqrt{2^2 - 4 \cdot 4 \cdot 1}}{2 \cdot 4}$

$x = \dfrac{-2 \pm \sqrt{4 - 16}}{8}$

$x = \dfrac{-2 \pm \sqrt{-12}}{8}$

$x = \dfrac{-2 \pm 2i\sqrt{3}}{8}$

$x = -\dfrac{1}{4} \pm \dfrac{i\sqrt{3}}{4}$

91. $x(x + 4) = -8$

$x^2 + 4x + 8 = 0$

$x = \dfrac{-4 \pm \sqrt{4^2 - 4 \cdot 1 \cdot 8}}{2 \cdot 1}$

$x = \dfrac{-4 \pm \sqrt{16 - 32}}{2}$

$x = \dfrac{-4 \pm \sqrt{-16}}{2}$

$x = \dfrac{-4 \pm 4i}{2}$

$x = -2 \pm 2i$

93. $x^2 + 10 = 6x$

$x^2 - 6x + 10 = 0$

$x = \dfrac{-(-6) \pm \sqrt{(-6)^2 - 4 \cdot 1 \cdot 10}}{2 \cdot 1}$

$x = \dfrac{6 \pm \sqrt{36 - 40}}{2}$

$x = \dfrac{6 \pm \sqrt{-4}}{2}$

$x = \dfrac{6 \pm 2i}{2}$

$x = 3 \pm i$

95. $x^2 + 2ix - 5 = 0$

$x = \dfrac{-2i \pm \sqrt{(2i)^2 - 4 \cdot 1 \cdot (-5)}}{2 \cdot 1}$

$x = \dfrac{-2i\sqrt{4i^2 + 20}}{2}$

$x = \dfrac{-2i \pm \sqrt{-4 + 20}}{2}$

$x = \dfrac{-2i \pm \sqrt{16}}{2}$

$x = \dfrac{-2i \pm 4}{2}$

$x = -i \pm 2$

97. $i^3 = i(i^2)$

$= i(-1)$

$= -i$

99. $i^4 = i^2 \cdot i^2$
$= (-1)(-1)$
$= 1$

## Section 10.5 Graphs of Quadratic Equations

1. If $a < 0$, the parabola opens downward and the vertex is the highest point. If $a > 0$, the parabola opens upward and the vertex is the lowest point.

3. Upward, since $a = 1$.

5. Downward, since $a = -1$.

7. Upward, since $a = 2$.

9. Downward, since $a = -1$.

11. The graph is a parabola and always has one $y$-intercept. The graph may have zero, one, or two $x$-intercepts.

13. $y = (x+3)(x-1)$

    $x$-intercepts: $(-3, 0)(1, 0)$
    $y$-intercept: $(0, -3)$
    The value of $c$ is the $y$-coordinate of the $y$-intercept.

15. $y = x^2 - 6x + 9$
    $= (x-3)^2$

    $x$-intercept: $(3, 0)$
    $y$-intercept: $(0, 9)$

17. $y = -2x^2 + 2x$
    $= -2x(x-1)$

    $x$-intercepts: $(0, 0)(1, 0)$
    $y$-intercept: $(0, 0)$

19. $y = x^2 - 3x + 5$

    $x = \dfrac{-(-3) \pm \sqrt{(-3)^2 - 4 \cdot 1 \cdot 5}}{2 \cdot 1}$

    $= \dfrac{3 \pm \sqrt{9 - 20}}{2}$

    $= \dfrac{3 \pm \sqrt{-11}}{2}$

    No $x$-intercepts
    $y$-intercept: $(0, 5)$

21. $y = -4x^2 - 4x - 1$
    $= -(4x^2 + 4x + 1)$
    $= -(2x+1)^2$
    $2x + 1 = 0$
    $2x = -1$
    $x = -\dfrac{1}{2}$

    $x$-intercept: $\left(-\dfrac{1}{2}, 0\right)$
    $y$-intercept: $(0, -1)$

23. $y = -2x^2 + 5x + 12$
    $= -(2x^2 - 5x - 12)$
    $= -(2x+3)(x-4)$
    $0 = 2x + 3$
    $-3 = 2x$
    $-\dfrac{3}{2} = x$

    $x$-intercepts: $\left(-\dfrac{3}{2}, 0\right), (4, 0)$
    $y$-intercept: $(0, 12)$

25. $y = -x^2 + x - 2$

$x = \dfrac{-1 \pm \sqrt{1^2 - 4 \cdot (-1)(-2)}}{2 \cdot (-1)}$

$= \dfrac{-1 \pm \sqrt{1-8}}{-2}$

$= \dfrac{-1 \pm \sqrt{-7}}{-2}$

No $x$-intercepts

$y$-intercept: $(0, -2)$

27. $y = x^2 - 25$
$= (x+5)(x-5)$

$x$-intercepts: $(-5, 0), (5, 0)$

$y$-intercept: $(0, -25)$

29. $x = \dfrac{-6}{2(1)}$

$= -3$

$y = (-3)^2 + 6(-3)$
$= 9 - 18$
$= -9$

vertex: $(-3, -9)$

31. $x = \dfrac{-2}{2(1)}$

$= -1$

$y = (-1)^2 + 2(-1) - 2$
$= 1 - 2 - 2$
$= -3$

vertex: $(-1, -3)$

33. $x = \dfrac{-(-6)}{2(1)}$

$= 3$

$y = 3^2 - 6(3) + 4$
$= 9 - 18 + 4$
$= -5$

vertex: $(3, -5)$

35. $y = -x^2 + 2x$

$x = \dfrac{-2}{2(-1)}$

$= 1$

$y = -(1)^2 + 2(1)$
$= -1 + 2$
$= 1$

vertex: $(1, 1)$

37. Upward, since a>0.

$x = \dfrac{-6}{2(1)}$

$= -3$

$y = (-3)^2 + 6(-3) - 7$
$= 9 - 18 - 7$
$= -16$

vertex: $(-3, -16)$

## Quadratic Equations and Functions

39. Downward, since $a < 0$.

$x = \dfrac{-2}{2(-1)}$
$= 1$

$y = -(1)^2 + 2(1) - 1$
$= -1 + 2 - 1$
$= 0$

vertex: $(1, 0)$

41. Upward, since $a > 0$.

$x = \dfrac{-0}{2(1)}$
$= 0$

$y = 0^2 + 4$
$= 4$

vertex: $(0, 4)$

43. Downward, since $a < 0$.

$x = \dfrac{-(-3)}{2(-1)}$
$= -\dfrac{3}{2}$

$y = -\left(-\dfrac{3}{2}\right)^2 - 3\left(-\dfrac{3}{2}\right) + 2$
$= -\dfrac{9}{4} + \dfrac{9}{2} + 2$
$= -\dfrac{9}{4} + \dfrac{18}{4} + \dfrac{8}{4}$
$= \dfrac{17}{4}$

vertex: $\left(-\dfrac{3}{2}, \dfrac{17}{4}\right)$

45. Upward, since $a > 0$.

$x = \dfrac{-(-12)}{2(2)}$
$= 3$

$y = 2(3)^2 - 12(3)$
$= 2(9) - 36$
$= 18 - 36$
$= -18$

vertex: $(3, -18)$

47. Downward, since $a < 0$.

$x = \dfrac{-6}{2(-3)}$
$= 1$

$y = -3(1)^2 + 6(1) + 1$
$= -3 + 6 + 1$
$= 4$

vertex: $(1, 4)$

49. (a) The parabola opens upward and has no $x$-intercepts.

(b) The parabola opens downward and has two $x$-intercepts.

51. 2, since downward with vertex in Quadrant I.

53. 1, since vertex is on $x$-axis.

55. 0, since vertex in Quadrant II and graph is upward.

57. 0, since downward with vertex below $x$-axis.

59. $y = -2x^2 + 6x$
$b^2 - 4ac = 6^2 - 4 \cdot (-2)(0)$
$= 36$
2 $x$-intercepts

61. $b^2 - 4ac = 10^2 - 4(1)(25)$
$= 100 - 100$
$= 0$
1 $x$-intercept

63. $b^2 - 4ac = 1^2 - 4(2)(-2)$
$= 1 + 16$
$= 17$
2 $x$-intercepts

65. $y = x^2 + x + 3$
$b^2 - 4ac = 1^2 - 4(1)(3)$
$= 1 - 12$
$= -11$
No $x$-intercepts

67. 0, since vertex above $x$-axis.

69. 2, since vertex above $x$-axis.

71. 1, since vertex on $x$-axis.

73. 0, since vertex above $x$-axis and $y$-intercept above vertex.

75. (ii), (iv), (vi), (vii), (viii), (ix)

77. (i), (iii), (vi), (vii), (x)

79. $y = x(x-6)$
$= x^2 - 6x$

(a) $x = \dfrac{-(-6)}{2(1)}$
$= 3$

$y = 3^2 - 6(3)$
$= -9$

vertex: $(3, -9)$

$3(-9) = -27$ is the smallest product.

(b) $7 = x^2 - 6x$
$0 = x^2 - 6x - 7$
$(x-7)(x+1) = 0$
$x = 7, x = -1$

(c) $-10 = x^2 - 6x$
$0 = x^2 - 6x + 10$
$x = \dfrac{-(-6) \pm \sqrt{(-6)^2 - 4(1)(10)}}{2(1)}$
$= \dfrac{6 \pm \sqrt{36-40}}{2}$
$= \dfrac{6 \pm \sqrt{-4}}{2}$
No solution.

81. (a) $h = -16t^2 + 96t$
$t = \dfrac{-96}{2(-16)}$
$= 3$ seconds

$h = -16(3)^2 + 96(3)$
$= -16(9) + 288$
$= -144 + 288$
$= 144$ feet

## Quadratic Equations and Functions

(b) $h = -16(1)^2 + 96(1)$
$= 80$ feet

(c) $0 = -16t^2 + 96t$
$0 = -16t(t-6)$
$t = 0, t = 6$
6 seconds

83. $A = -24.25x^2 + 2428x - 23,100$

$x = \dfrac{-2428}{2(-24.25)}$
$= 50$

$A = -24.25(50)^2 + 2428(50) - 23,100$
$= 37,675$

85. $30,000 = -24.25x^2 + 2428x - 23,100$

$0 = -24.25x^2 + 2428x - 53,100$

$x = \dfrac{-2428 \pm \sqrt{(2428)^2 - 4(-24.25)(-53,100)}}{2(-24.25)}$

$x = 32, 68$

87. $x^2 = 2 - x$
$x^2 + x - 2 = 0$
$(x+2)(x-1) = 0$
$x = -2, x = 1$

$y = (-2)^2 = 4$ so $(-2, 4)$ is a solution.
$y = (1)^2 = 1$ so $(1, 1)$ is a solution.

89. $4 - x^2 = x^2 - 4$
$8 = 2x^2$
$4 = x^2$
$x^2 - 4 = 0$
$(x+2)(x-2) = 0$
$x = -2, x = 2$

$y = (-2)^2 - 4 = 0$ so $(-2, 0)$ is a solution.
$y = (2)^2 - 4 = 0$ so $(2, 0)$ is a solution.

91.

$\dfrac{\dfrac{-b + \sqrt{b^2 - 4ac}}{2a} + \dfrac{-b - \sqrt{b^2 - 4ac}}{2a}}{2}$

$= \dfrac{\dfrac{-b + \sqrt{b^2 - 4ac} - b - \sqrt{b^2 - 4ac}}{2a}}{2}$

$= \dfrac{\dfrac{-2b}{2a}}{2}$

$= \dfrac{\dfrac{-b}{a}}{2}$

$= \dfrac{-b}{2a}$

## Section 10.6 Functions

1. No, a function can have two pairs with the same second coordinate, but not the same first coordinate.

3. Function
   Domain = {3, 4, 5, 7}
   Range = {−3, 1, 2}

5. Not a function, since 3 has two outputs.

7. Function
   Domain = {1, 2 3 4}
   Range = {2, 3, 4, 5}

9. {(1, −2), (2, −1), (3, 0), (4, 1)}
   Function
   Domain = {1, 2, 3, 4,}
   Range = {−2, −1, 0, 1}

11. {(4, 0), (4, 1), (4, 2), (4, 3), (4, 4), (4, 5)}
    Not a function, since 4 has more than one output.

13. {(−1, 4), (0, 4), (1, 4)}
    Function
    Domain = {−1, 0, 1}
    Range = {4}

15. $y$ is not a function of $x$, since the input 1 has two different outputs, −1 and 1.

17. Function.

19. Function.

21. Not a function. The graph is a circle.

23. Function.

25. Not a function, since (−1, 1) and (−1, −1) both satisfy the equation.

27. Function.

29. For $y = x + 1$, each value of $x$ corresponds to exactly one value of $y$. For the inequality $y \leq x + 1$, for any value of $x$ there are many values of $y$.

31. Function, since passes Vertical Line Test.

33. Function, since passes Vertical Line Test.

35. Function, since passes Vertical Line Test.

37. Not a function, since fails Vertical Line Test.

39. Not a function, since fails Vertical Line Test.

41. Function, since passes Vertical Line Test.

43. The domain of $f$ is $\mathbb{R}, x \neq -4$ and the domain of $g$ is $\mathbb{R}$.

45. Domain: $\mathbb{R}$
    Range: $\mathbb{R}$

47. Domain: $\mathbb{R}$
    Range: $y \geq 0$

49. Domain: $x \geq 0$
    Range: $y \geq 0$

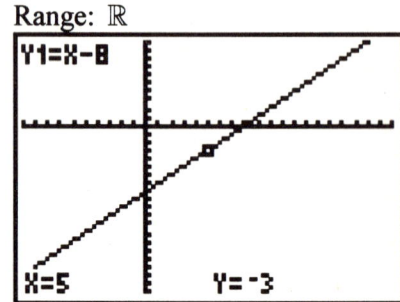

51. Domain: $\mathbb{R}$
    Range: $\mathbb{R}$

53. Domain: $\mathbb{R}$
    Range: $y \leq 9$

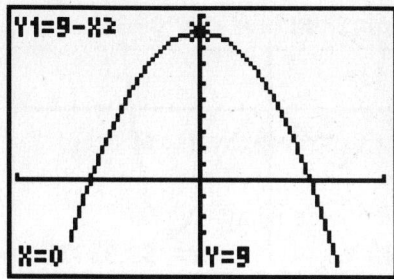

    [Graph: Y1=9-X², showing downward parabola with vertex at X=0, Y=9]

55. Domain: $x \leq 9$
    Range: $y \geq 0$

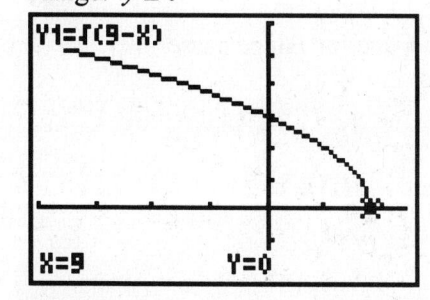

    [Graph: Y1=√(9-X), with point X=9, Y=0]

57. Domain: $\mathbb{R}$
    Range: $y \geq -1$

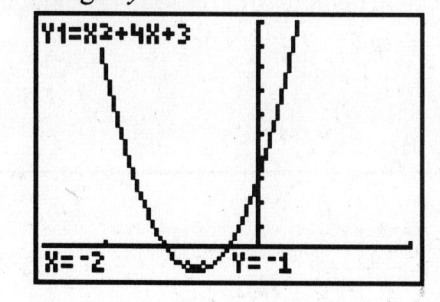

    [Graph: Y1=X²+4X+3, with vertex at X=-2, Y=-1]

59. $\mathbb{R}$

61. $\mathbb{R}, x \neq 1$

63. $x + 7 \geq 0$
    $x \geq -7$

65. $\mathbb{R}$

67. $3 - x \geq 0$
    $3 \geq x$
    $x \leq 3$

69. $\mathbb{R}, x \neq 0$

71. $\mathbb{R}$, since the denominator can never be 0.

73. $5 + x^2 \geq 0$
    $x^2 \geq -5$
    $\mathbb{R}$, since the inequality is true for all values of $x$.

75. $\mathbb{R}$, since we can calculate the cube root of positive and negative numbers.

77. $C(n) = \dfrac{420}{n}$

    (a) Domain: {1, 2, 3, 4, 5, 6}
        Range: {420, 210, 140, 105, 84, 70}

    (b) {(1, 420), (2, 210), (3, 140), (4, 105), (5, 84), (6, 70)}

79. (a) $f(x) = (10,000 - 500x)x$

    (b) The revenue per truck must be nonnegative, so
    $10,000 - 500x \geq 0$
    $10,000 \geq 500x$
    $20 \geq x$

    The domain is $0 \leq x \leq 20$.

    (c) $f(x) = 10,000x - 500x^2$
    $= -500x^2 + 10,000x$

    The vertex is at
    $x = \dfrac{-10,000}{2(-500)}$
    $= 10$ trucks

    (d) $f(10) = 10,000(10) - 500(10)^2$
    $= 100,000 - 500(100)$
    $= 100,000 - 50000$
    $= \$50,000$

    (e) $0 \leq y \leq 50,000$

81. (a) $S(x)$ is of the form $y = mx + b$, so the function is linear.

(b) The $y$-intercept represents the cancer survival rate in year 0 (1960). It was 39.6%.

(c) $S(30) = 39.6 + 0.56(30)$
$= 39.6 + 16.8$
$= 56.4$

The estimated 1990 survival rate is 56.4%.

83. Domain: $\mathbb{R}, x \neq 0$
Range: $\{-1, 1\}$

85. Domain: $\mathbb{R}$
Range: $y \geq -53.33$

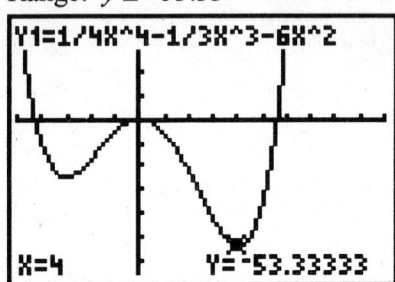

87. $f(x) = mx + b$
$3 = m(0) + 6$
$b = 3$

$f(x) = mx + 3$
$5 = m(1) + 3$
$2 = m$

$f(x) = 2x + 3$

## Chapter 10 Review Exercises

1. $-3, 8$

3. $x^2 = 28$
$\sqrt{x^2} = \sqrt{28}$
$|x| = \sqrt{4}\sqrt{7}$
$x = \pm 2\sqrt{7}$

5. $(6-x)^2 = 9$
$\sqrt{(6-x)^2} = \sqrt{9}$
$|6-x| = 3$
$6 - x = \pm 3$
$-x = -6 \pm 3$
$x = 6 \pm 3$
$x = 3, 9$

7. $x^2 - 2x + 1 = 4x + 8$
$x^2 - 6x - 7 = 0$
$(x-7)(x+1) = 0$
$x = 7, x = -1$

9. $x(x+2) = 2x+9$
$x^2 + 2x = 2x+9$
$x^2 = 9$
$\sqrt{x^2} = \sqrt{9}$
$|x| = 3$
$x = \pm 3$
The integer is to be positive, so $x = 3$ is the small integer. The large integer is 5.

11. The third term is the square of half the coefficient of $x$: $\left(\dfrac{1}{2} \cdot 10\right)^2 = 25$.

13. $x^2 - 8x + 16 = 4 + 16$
$(x-4)^2 = 20$
$\sqrt{(x-4)^2} = \sqrt{20}$
$|x-4| = 2\sqrt{5}$
$x - 4 = \pm 2\sqrt{5}$
$x = 4 \pm 2\sqrt{5}$

15. $x^2 - 10x = -34$
$x^2 - 10x + 25 = -34 + 25$
$(x-5)^2 = -9$

No real number solution, since $(x-5)^2 \geq 0$.

17. $x(90 - x) = 1625$
$90x - x^2 = 1625$
$x^2 - 90x + 1625 = 0$
$x = \dfrac{-(-90) \pm \sqrt{(-90)^2 - 4 \cdot 1 \cdot 1625}}{2 \cdot 1}$
$= \dfrac{90 \pm \sqrt{8100 - 6500}}{2}$
$= \dfrac{90 \pm \sqrt{1600}}{2}$
$= \dfrac{90 \pm 40}{2}$
$x = 25, 65$

The smaller angle is 25°.

19. $2x^2 - x - 5 = 0$
$a = 2$
$b = -1$
$c = -5$

21.

23.

24. $2 = 3x^2 - 6x$
$0 = 3x^2 - 6x - 2$
$x = \dfrac{-(-6) \pm \sqrt{(-6)^2 - 4(3)(-2)}}{2(3)}$
$= \dfrac{6 \pm \sqrt{36 + 24}}{6}$
$= \dfrac{6 \pm \sqrt{60}}{6}$
$= \dfrac{6 \pm 2\sqrt{15}}{6}$
$= \dfrac{3 \pm \sqrt{15}}{3}$

25. $x^2 - 4x + 3 = 6$
   $x^2 - 4x = 3$
   $x^2 - 4x + 4 = 3 + 4$
   $(x-2)^2 = 7$
   $\sqrt{(x-2)^2} = \sqrt{7}$
   $|x-2| = \sqrt{7}$
   $x - 2 = \pm\sqrt{7}$
   $x = 2 \pm \sqrt{7}$

27. $16x^2 - 24x + 9 = 0$
   $(4x-3)^2 = 0$
   $4x - 3 = 0$
   $4x = 3$
   $x = \dfrac{3}{4}$

29. $6x^2 - 6x + 1 = 0$
   $b^2 - 4ac = (-6)^2 - 4(6)(1)$
   $= 36 - 24$
   $= 12$

   Two real number solutions.

31. $i\sqrt{11}$

33. $-2i + 3i + 4$
   $= 4 + i$

35. $-21i - 28i^2$
   $= -21i - 28(-1)$
   $= 28 - 21i$

37. $-\dfrac{4}{2} + \dfrac{6i}{2}$
   $= -2 + 3i$

39. $x^2 - x + 3 = 0$
   $x = \dfrac{-(-1) \pm \sqrt{(-1)^2 - 4(1)(3)}}{2(1)}$
   $= \dfrac{1 \pm \sqrt{1 - 12}}{2}$
   $= \dfrac{1 \pm \sqrt{-11}}{2}$
   $= \dfrac{1 \pm i\sqrt{11}}{2}$
   $= \dfrac{1}{2} \pm \dfrac{\sqrt{11}}{2}i$

41. Downward, since $a < 0$.

43. $y = -x^2 + 4$
   $x = -\dfrac{0}{2(-1)}$
   $= 0$
   $y = -0^2 + 4$
   $= 4$
   vertex: $(0, 4)$

45. (a) The vertex is the highest point of the graph.

   (b) The $y$-coordinate of the vertex is the maximum value of $y$.

47. $y = x^2 - 8x + 7$
   $= (x-1)(x-7)$

   $x$-intercepts: $(1, 0)(7, 0)$
   $y$-intercept: $(0, 7)$

49. The parabola is upward, since the $y$-intercept is above the vertex. Since the vertex is below the $x$-axis, there are two $x$-intercepts.

51. No two ordered pairs of a function can have the same first coordinate. Thus (i) is a function and (ii) is not a function.

53. (i) Function

(ii) $y^2 = x+1$
$\sqrt{y^2} = \sqrt{x+1}$
$|y| = \sqrt{x+1}$
$y = \pm\sqrt{x+1}$
Not a function.

55. $4 - x \geq 0$
$4 \geq x$
Domain: $x \leq 4$

57. Range: $y \geq 0$

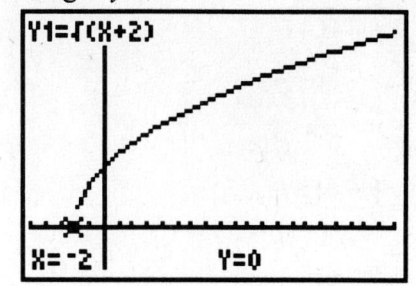

59. (i) Function

(ii) Not a function, since it fails the Vertical Line Test.

## Chapter 10 Test

1. $-4.45, 0.45$

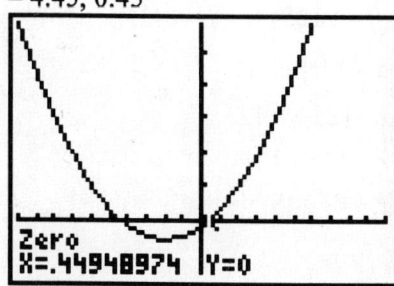

3. $(x-3)^2 = 16$
$\sqrt{(x-3)^2} = \sqrt{16}$
$(x-3) = 4$
$x - 3 = \pm 4$
$x = 3 \pm 4$
$x = 7, x = -1$

5. $-2x^2 - 3x + 1 = 0$
$x = \dfrac{-(-3) \pm \sqrt{(-3)^2 - 4(-2)(1)}}{2(-2)}$
$= \dfrac{3 \pm \sqrt{9+8}}{-4}$
$= \dfrac{3 \pm \sqrt{17}}{-4}$
$= \dfrac{-3 \pm \sqrt{17}}{4}$

7. $3x^2 + 5x + 4 = 0$
$b^2 - 4ac = 5^2 - 4(3)(4)$
$= 25 - 48$
$= -23$
No real number solutions.

9. $\dfrac{3-i}{3+i} \cdot \dfrac{3-i}{3-i}$

$= \dfrac{(3-i)(3-i)}{3^2 - i^2}$

$= \dfrac{9 - 6i + i^2}{9 - (-1)}$

$= \dfrac{9 - 6i - 1}{10}$

$= \dfrac{8 - 6i}{10}$

$= \dfrac{4}{5} - \dfrac{3}{5}i$

11. Parabola

13. $x = -\dfrac{4}{2(2)}$

$= -1$

$y = 2(-1)^2 + 4(-1) - 3$
$= 2 - 4 - 3$
$= -5$

vertex: $(-1, -5)$

15. (i) Domain: {1, 2, 3}
Range: {2, 3}

(ii) Domain: $x \geq -1$
Range: $y \geq 0$

(iii) Domain: $\mathbb{R}, x \neq -1$
Range: $\mathbb{R}, y \neq 0$

17. $(300 - 2x)(200 - 2x) = 56,064$

$60,000 - 600x - 400x + 4x^2 = 56064$

$4x^2 - 1000x + 60,000 = 56064$

$4x^2 - 1000x = -3936$

$x^2 - 250x = -984$

$x^2 - 250x + 15625 = -984 + 15625$

$(x - 125)^2 = 14641$

$\sqrt{(x-125)^2} = \sqrt{14641}$

$|x - 125| = 121$

$x - 125 = \pm 121$

$x = 125 \pm 121$

$x = 4, 246$

The sidewalk is 4 feet wide.

# Cumulative Test, Chapters 9 – 10

1. $\sqrt{(3x)^2}$
$= 3x$

3. $\dfrac{\sqrt{5x^2}}{\sqrt{20x^6}}$

$= \dfrac{\sqrt{5}\sqrt{x^2}}{\sqrt{4}\sqrt{5}\sqrt{x^6}}$

$= \dfrac{\sqrt{5}\,x}{2\sqrt{5}\,x^3}$

$= \dfrac{x}{2x^3}$

$= \dfrac{1}{2x^2}$

## Quadratic Equations and Functions

5. $\sqrt{4}\sqrt{3x} + \sqrt{25}\sqrt{3x}$
   $= 2\sqrt{3x} + 5\sqrt{3x}$
   $= 7\sqrt{3x}$

7. $x - \sqrt{5x-1} = -1$
   $x + 1 = \sqrt{5x-1}$
   $(x+1)^2 = (\sqrt{5x-1})^2$
   $x^2 + 2x + 1 = 5x - 1$
   $x^2 - 3x + 2 = 0$
   $(x-1)(x-2) = 0$
   $x = 1, \; x = 2$

9. $x =$ first leg
   $x + 5 =$ second leg

   $x^2 + (x+5)^2 = (25)^2$
   $x^2 + x^2 + 10x + 25 = 625$
   $2x^2 + 10x - 600 = 0$
   $x^2 + 5x - 300 = 0$
   $(x+20)(x-15) = 0$
   $x = 15$ inches

11. $x^2 - 5x = 16 - 5x$
    $x^2 = 16$
    $\sqrt{x^2} = \sqrt{16}$
    $|x| = 4$
    $x = \pm 4$

13. $x^2 - 6x = -4$
    $x^2 - 6x + 9 = -4 + 9$
    $(x-3)^2 = 5$
    $\sqrt{(x-3)^2} = \sqrt{5}$
    $|x-3| = \sqrt{5}$
    $x - 3 = \pm\sqrt{5}$
    $x = 3 \pm \sqrt{5}$

15. $(2i-3)(-2i-3)$
    $= -4i^2 + 9$
    $= -4(-1) + 9$
    $= 13$

17. No, if an equation has an imaginary solution, then the conjugate is also a solution.

19. Either $-2$ or $5$ results in two pairs with the same first coordinate.